DIGITAL BiCMOS INTEGRATED CIRCUIT DESIGN

THE KLUWER INTERNATIONAL SERIES IN ENGINEERING AND COMPUTER SCIENCE

VLSI, COMPUTER ARCHITECTURE AND DIGITAL SIGNAL PROCESSING

Consulting Editor

Jonathan Allen

Latest Titles

Neural Models and Algorithms for Digital Testing, S. T. Chakradhar, V. D. Agrawal, M. L. Bushnell,
ISBN: 0-7923-9165-9

Monte Carlo Device Simulation: Full Band and Beyond, Karl Hess, editor
ISBN: 0-7923-9172-1

The Design of Communicating Systems: A System Engineering Approach, C. J. Koomen
ISBN: 0-7923-9203-5

Parallel Algorithms and Architectures for DSP Applications, M. A. Bayoumi, editor
ISBN: 0-7923-9209-4

Digital Speech Processing: Speech Coding, Synthesis and Recognition A. Nejat Ince, editor
ISBN: 0-7923-9220-5

Sequential Logic Synthesis, P. Ashar, S. Devadas, A. R. Newton
ISBN: 0-7923-9187-X

Sequential Logic Testing and Verification, A. Ghosh, S. Devadas, A. R. Newton
ISBN: 0-7923-9188-8

Introduction to the Design of Transconductor-Capacitor Filters, J. E. Kardontchik
ISBN: 0-7923-9195-0

The Synthesis Approach to Digital System Design, P. Michel, U. Lauther, P. Duzy
ISBN: 0-7923-9199-3

Fault Covering Problems in Reconfigurable VLSI Systems, R.Libeskind-Hadas, N. Hassan, J. Cong, P. McKinley, C. L. Liu
ISBN: 0-7923-9231-0

High Level Synthesis of ASICs Under Timing and Synchronization Constraints D.C. Ku, G. De Micheli
ISBN: 0-7923-9244-2

The SECD Microprocessor, A Verification Case Study, B.T. Graham
ISBN: 0-7923-9245-0

Field-Programmable Gate Arrays, S.D. Brown, R. J. Francis, J. Rose, Z.G. Vranesic
ISBN: 0-7923-9248-5

Anatomy of A Silicon Compiler, R.W. Brodersen
ISBN: 0-7923-9249-3

Electronic CAD Frameworks, T.J. Barnes, D. Harrison, A.R. Newton, R.L. Spickelmier
ISBN: 0-7923-9252-3

VHDL for Simulation, Synthesis and Formal Proofs of Hardware, J. Mermet
ISBN: 0-7923-9253-1

Wavelet Theory and its Applications, R. K. Young
ISBN: 0-7923-9271-X

DIGITAL BiCMOS INTEGRATED CIRCUIT DESIGN

by

Sherif H.K. Embabi
Texas A & M University

Abdellatif Bellaouar
University of Waterloo

Mohamed I. Elmasry
University of Waterloo

Kluwer Academic Publishers
Boston/Dordrecht/London

Distributors for North America:
Kluwer Academic Publishers
101 Philip Drive
Assinippi Park
Norwell, Massachusetts 02061 USA

Distributors for all other countries:
Kluwer Academic Publishers Group
Distribution Centre
Post Office Box 322
3300 AH Dordrecht, THE NETHERLANDS

Library of Congress Cataloging-in-Publication Data

Embabi, S. H. K. (Sherif H. K.)
Digital BiCMOS integrated circuit design / by S.H.K. Embabi, A. Bellaour, M.I. Elmasry.
p. cm. -- (Kluwer international series in engineering and computer science)
Includes bibliographical references and index.
ISBN 0-7923-9276-0 (alk. paper)
1. Digital integrated circuits--Design and construction. 2. Metal oxide semiconductors, Complementary. 3. Bipolar integrated circuits. I. Bellaouar, A. II. Elmasry, Mohamed I., 1943-
III. Title. IV. Series.
TK7874.E52 1993
621.3815--dc20 92-29636
CIP

Printed on acid-free paper.

Printed in the United States of America

To

Afaf, Hassan, Dina, Ahmed, and Aya Embabi

Mohamed and Massaouda Bellaouar

Elizabeth, Carmen, Samir, Nadia, and Hassan Elmasry

Table of Contents

Preface

BiCMOS has faced, as any emerging technology, fierce opposition, but nevertheless has been widely accepted. The need for BiCMOS has been and still is a controversial issue. The cost of high-performance BiCMOS process speaks against it. However, there are other arguments supporting the need for such a technology. The comparison between the cost of CMOS and BiCMOS is not trivial especially if we account for the increasing expenses of scaling CMOS. In other words, adding bipolar to an existing CMOS process might be cheaper than scaling CMOS to obtain the same performance improvement. On the other hand, the high cost of BiCMOS may be justified by the high performance it can offer. In general, the application itself will dictate the technology that is to be used. If CMOS alone can satisfy the requirements, then it is more cost effective to use CMOS. Yet there are applications where the requirements can be only satisfied by using BiCMOS. There are currently commercial BiCMOS digital products such as high-speed SRAMs, fast microprocessors, gate arrays, in addition to digital, analog and mixed signal BiCMOS ASICs.

The technology for BiCMOS fabrication has matured over the last decade, yet the development of circuits and systems for such a technology is still lagging. There is a great interest among digital/analog circuit and system designers to develop techniques to exploit the flexibility that BiCMOS offers.

A large number of reports and technical papers have been published on the development of BiCMOS processes, circuits and systems. This book is an attempt to introduce many of the published ideas and concepts in order to assist students, researchers and design engineers to review the progress achieved so far. For example, in Chapter 2, we introduce the main BiCMOS processes and their performance/cost tradeoffs. In Chapter 7, we present the basic digital BiCMOS structures. The advantages of such circuits over CMOS are explained as well as their limitations. The application of BiCMOS for digital circuits and systems is illustrated in Chapter 8. Moreover, we have added two overview chapters; Chapter 5 on MOS IC's and Chapter 6 on bipolar current mode logic circuits. These two chapters were added because it is expected that future BiCMOS systems will not be limited to the use of BiCMOS circuit structures. The synergy of different structures and logic families will allow for better management of the overall speed and power. Therefore, we believe that those who work in this field should be familiar with the fundamentals of MOS and bipolar CML as well.

This book is also intended for senior undergraduate and graduate Electrical and Computer Engineering students who are interested in digital circuit design. Exposing the students at an early stage to the different technologies and design styles is of great benefit to them. To those students we have also included Chapters 3 and 4, which introduce important basic information regarding the structures, operation and modeling of MOS and bipolar transistors. Such basic knowledge is essential for circuit designers because of the close interaction between device characteristics and circuit performance.

This book is an early contribution to the field of digital BiCMOS circuit design. It contains a collection of the contributions of all those who have devoted their time and effort to help us learn more about this field. We do not claim that this text is complete. In many instances, we have tried to summarize the work of others. So those who have special interest in one of the topics have to refer to the given references.

There is still the need for more research, investigation and experimentation to find answers to many of the unanswered questions regarding the optimum use of BiCMOS circuit structures. We are sure that the relentless search for better performance and quality will continue and will result in interesting and innovative ideas.

We hope that the readers of this book will find it useful and that it will add to their knowledge and assist them in contributing to the state-of-the-art of BiCMOS digital circuit design.

S.H.K. Embabi
A. Bellaouar
M.I. Elmasry

Acknowledgements

We would like, first of all, to acknowledge the countless blessings of God Almighty throughout our lifes.

We would also like to thank our families whose support and endurance helped us to complete writing this book.

Writing this book would have been impossible without the knowledge contributed by the researchers and workers in this area. Their effort is deeply appreciated.

We also extend our thanks to Mr. Carl Harris from Kluwer Academic Publishers for encouraging us to work on this project.

We would also like to thank the reviewers for their careful and thorough review as well as their valuable comments and suggestions which resulted in some improvements. Special thanks to Drs. S. Rofail and Ali Salama for their critical review of the manuscript and fruitful discussions.

Our gratitude goes to all our colleagues in the Department of Electrical and Computer Engineering at the University of Waterloo for their support and encouragement, in particular, M. Elrabaa, W. Al-Mulla and P. Harvey.

The research work of the authors reported in this book has been supported in part by NSERC, BNR, NTE, ITRC and MICRONET. This support is greatly appreciated.

One of the authors, S.H.K. Embabi, would like to thank his colleagues in the Department of Electrical Engineering at Texas A&M University for their support, in particular, Professor E. Sanchez-Sinencio and Professor J.W. Howze.

We appreciate the effort of those who assisted us in various ways to prepare the manuscript and the figures, in particular, J. Pache, D. Pu, A. Dajani, P. Regier and T. Creasy.

List of Symbols

A_E	emitter area
A_C	collector area
β_f (or β_o)	maximum forward current gain.
β	forward current gain as function of the collector current.
BV_{DS}	drain-source breakdown voltage.
BV_{CBO}	collector-base breakdown voltage.
BV_{CEO}	collector-emitter breakdown voltage.
C_c	base-collector capacitance.
C_e	base-emitter capacitance.
C_s	collector-substrate capacitance.
C_{ox}	Gate capacitance.
C_d	source (drain) depletion capacitance.
C_{ov}	gate to source (drain) overlap capacitance.
C_L	external load capacitance.
C_D	diffusion capacitance.
C_o	total output capacitance.
D_{nB}	diffusion constant for the electrons in the base.
D_{pE}	diffusion constant for the holes in the emitter.
D_I	implantation dose to adjust the threshold voltage.
ε_o	permittivity of oxide.
ε_s	permittivity of silicon.
f_T	cutoff frequency.
γ	body effect coefficient.
γ_E	emitter injection efficiency.
I_{DS}	drain-source current.
I_{DSsat}	drain-source saturation current.
I_B	base current.
I_C	collector current.

I_E	emitter current.
I_K	forward knee current.
I_{HLI}	high level injection current.
I_s	transport saturation current.
I_{ss}	CML current
ITF	τ_{fo} high-current parameter.
J_K	forward knee current density.
k	Boltzman constant.
K_N	NMOS saturation current factor
K_P	PMOS saturation current factor
L	drawn channel length.
L_{eff}	effective channel length.
λ	channel length modulation factor.
N_s	channel doping concentration at the surface.
N_{bs}	channel doping concentration below the surface.
N_B	base doping concentration.
N_C	collector doping concentration.
N_E	emitter doping concentration.
N_{epi}	epitaxial layer doping concentration.
N_a	acceptor doping concentration.
N_d	donor doping concentration.
M_j	junction exponent (grading factor).
ϕ_j	junction built-in potential.
ϕ_s	surface potential of silicon.
ϕ_{ms}	gate electrode and silicon work function difference.
R_B	base resistance.
R_{Be}	extrinsic base resistance.
R_{Bi}	intrinsic base resistance.
R_C	collector resistance.
R_E	emitter resistance.

T	temperature.
τ_{fo}	minimum forward transit time.
τ_f	forward transit time.
t_{ox}	gate oxide thickness.
S	subthreshold swing.
V_{AF}	forward early voltage
V_{AR}	reverse early voltage
V_B	MOS bulk (substrate) voltage.
V_{BC}	base-collector voltage.
V_{BE}	base-emitter voltage.
V_{BEon}	turn ON base-emitter voltage.
V_{CE}	collector-emitter voltage.
V_{CEsat}	saturation collector-emitter voltage.
V_{FB}	flat band voltage.
V_{GS}	gate-source voltage.
V_{DS}	drain-source voltage.
V_{DSsat}	saturation drain-source voltage.
V_{dd}	power supply voltage.
V_{in}	input voltage.
V_o	output voltage.
V_{pt}	punchthrough voltage.
V_{rt}	reachthrough voltage.
V_s	Voltage swing.
V_t	thermal voltage.
VTF	τ_f V_{BC} dependence coefficient.
V_T	MOS threshold voltage.
V_{TO}	MOS threshold voltage when V_B=0.
V_{TN}	NMOS threshold voltage.
V_{TP}	PMOS threshold voltage.
v_{sat}	carrier saturation velocity.

W_{eff}	effective channel width.
W_D	depletion layer width in the bulk of the MOSFET.
W_N	NMOSFET device width.
W_P	PMOSFET device width.
W_E	emitter width.
W_B	base width.
W_C	collector width.
XTF	τ_f dependence coefficient.
x_j	source/drain junction depth.
x_{jE}	emitter junction depth.

DIGITAL BiCMOS INTEGRATED CIRCUIT DESIGN

CHAPTER 1

INTRODUCTION

The evolution of digital intergated circuit technology went through several phases starting with bipolar in the 1950s and 60s to NMOS in the 70s until the 1980s when CMOS became the mainstream technology. The demand for superior performance, which has motivated the relentless search and development of new technologies, was behind the emergence of BiCMOS. It is the technology which merges old rivals, namely, CMOS and bipolar, to complement each other so that the strength of one covers the weakness of the other. This concept was first proposed by Lin et al. in 1969 [1]. The progress achieved in the silicon processing technology in the 70s and the 80s motivated the revival of the idea of merging CMOS and bipolar. Since the mid 80s enormous resources have been devoted towards the development of several versions of BiCMOS processes, and real BiCMOS products and systems have been shipped out.

1.1. Why BiCMOS ?

CMOS and bipolar technologies have their weak and strong points. CMOS has qualified to be the most appropriate choice for VLSI applications because of its low DC power dissipation and its high packing density, yet, its speed is limited by the capacitive loading. On the other hand, bipolar digital circuits, CML in particular, outperform CMOS in terms of speed, but are power consuming. There is a performance gap between CMOS and ECL as shown in Fig. 1.1. The existence of this gap implies that neither CMOS nor ECL have the flexibility required to cover the full delay-power space. This can only be achieved by a technology such as BiCMOS. The objective of the synergy of bipolar and CMOS is to exploit the advantages of both at the circuit and system levels.

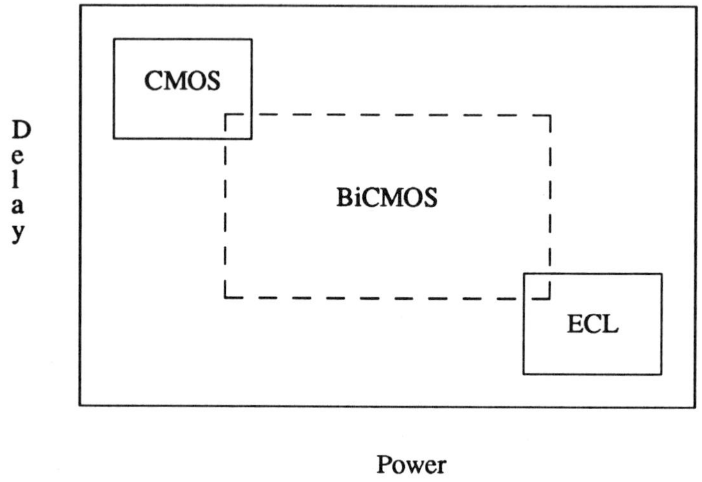

Figure 1.1 A comparison between CMOS, ECL and BiCMOS in terms of delay and power.

A variety of digital BiCMOS circuit structures have been developed. An example of such circuits is the BiCMOS totem-pole gate which was originally proposed by Lin et al[1]. This structure is currently the most commonly used BiCMOS gate [see Fig. 1.2]. The advantages of this circuit are:

(i) Low DC power consumption.

(ii) Superior driving capability of on- and off- chip loads.

(iii) Lower delay sensitivity to loading, which is important especially for semi-custom systems such as gate arrays.

Other MOS/bipolar merged current mode structures have also been proposed. Also BiCMOS sensing circuits have demonstrated superior performance over their MOS counterparts because of the high transconductance of bipolar transistors.

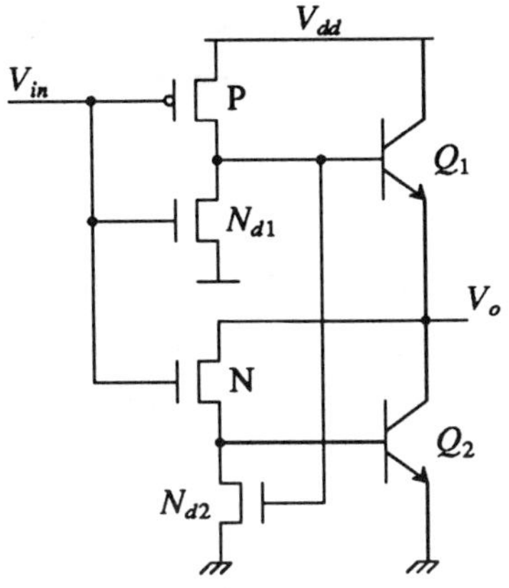

Figure 1.2 BiCMOS inverter.

BiCMOS has been exploited at the system level as well. It has been used for fast ECL SRAMs, microprocessors, etc. The availability of CMOS and bipolar devices on the same chip can also be used for mixing slow/low-power and fast/high-power logic families. This offers a great deal of flexibility to optimize the overall system performance. The low-power circuits should be used to realize the non-critical paths, while, the fast circuits should be limited to the critical paths only. This approach allows for efficient mangement of speed and power.

In addition to the advantages that BiCMOS has to offer, there are practical considerations which make this technology seem more attractive than ever.

a- The cost of scaling CMOS is increasing, while, the performance improvement is incremental. In other words the ratio of the performance improvement to the investment is decreasing. During the 1980s the scaling of CMOS was rewarding because the supply voltage remained constant which led to an appreciable speed improvement from one generation to the other. This trend is expected to change dramatically in the 1990s when the CMOS transistors are to be scaled below $0.5\mu m$. This can be referred to both, the reduction of supply voltage, which is necessary to avoid high electric fields, and the velocity saturation of the carriers in the channel. At the same time, the gate oxide capacitances will increase because of the scaling of the gate oxide thickness.

Also the junction capacitances are expected to increase as the doping levels increase. Finally, the interconnect capacitances will not improve significantly with scaling[2]. All these factors will contribute to the lower rate of speed increase as the devices are scaled below $0.5\mu m$.

b- Although the synergy of CMOS and bipolar was proposed in the late 60s, yet, it was not applied in practice for technological reasons. By the mid 80s the development of BiCMOS process was greatly encouraged by the fact that the state-of-the-art bipolar and CMOS structures have been converging. For instance, the buried- and epitaxial- layers, which were exclusive for bipolar transistors, were adopted for CMOS to mute the latchup susceptibility. Polysilicon contacts are now equally used for the gate of a MOS transistor and the emitter and base of a bipolar transistor. The similarity between the process steps required for both devices, CMOS and bipolar, means that these steps can be shared for both of them. This implies that the extra masks required for BiCMOS compared to CMOS can be reduced. Adding a bipolar NPN transistor, with a $7GHz$ cut-off frequency, to a CMOS baseline process requires from 3 to 7 extra masks[3].

1.2. This Book

More than half a decade has been devoted to develop BiCMOS process. The interest is now shifting from process to circuits and systems development. Researchers have to come up with new ideas to exploit the BiCMOS technology. Circuit designers have to be educated to design in BiCMOS. A whole set of CAD tools for BiCMOS have to be developed.

This book targets two types of audiences; the senior undergraduate and postgraduate university students and the practicing circuit designers in industry. The students, who are the future designers, deserve to learn how to design CMOS, bipolar and BiCMOS circuits at an early stage. Our previous experience shows that the students who were taught to analyse and design in one technology have a hard time switching to another technology. For this audience we have tried to cover the basics of CMOS, bipolar and BiCMOS process technologies. The fundamentals of MOS and bipolar transistors are presented to allow the students to appreciate the significance of the accuracy of the models they typically use for circuit design. They are then exposed to the analysis and design of the three mainstream logic families, namely CMOS, CML and BiCMOS. The emphasis on CMOS and CML in this text stems from our belief that future BiCMOS systems will include a mixture of CML, CMOS and BiCMOS circuits. The book includes typical design rules and model parameters for a 0.8μm BiCMOS process which are useful for exercises. Finally, we present a brief introduction to existing BiCMOS systems so they can appreciate how BiCMOS affects the system performance. While the majority of practicing designers are currently CMOS-oriented there are those who are involved in designing in bipolar and a small group is already designing in BiCMOS. This book should be useful for practicing CMOS and BJT circuit designers as well. The book includes an extensive list of references at the end of each chapter which allows the readers to study in depth any of the topics they find interesting.

This book is organized into eight chapters. The first chapter is an introduction to the topics of Chapters 2 to 8 and will be presented in the following sections.

1.3. Process Technology

Chapter 2 deals with CMOS, bipolar and BiCMOS process technologies. In the first two sections we review the process flow of advanced CMOS (N-well and twin-tub) and bipolar transistors. Some of the features of the state-of-the-art device structures, such as LDD MOS transistors and trench-isolated self-aligned bipolar transistors, are presented. A comparison between conventional and advanced trench-isolated self-aligned structures shows the significant improvement achieved in the bipolar technology in terms of area and reduction of parasitic capacitors. In section three, the reader is introduced to the isolation techniques used in CMOS and bipolar technologies. The issue of isolation is given special attention because of its importance to the BiCMOS technology especially for VLSI applications where the packing density is an important factor. The discussion on the convergence of CMOS and bipolar technologies, in section four, shows that a number of process steps can be shared by both types, which makes the integration of CMOS and bipolar feasible. The fifth section introduces BiCMOS technology. Three classes of BiCMOS processes are discussed to illustrate the tradeoff between performance and cost. The least expensive process requires only one extra mask (for the base) to add a bipolar transistor to the CMOS baseline. The high-performance BiCMOS process requires four extra masks. Finally, we present the flow of a complementary BiCMOS (CBiCMOS) process, in which a vertical isolated PNP transistor is added to the NPN and CMOS transistors. Chapter 2 includes a set of design rules for a 0.8μ BiCMOS process.

1.4. Device Design Considerations

Chapter 3 deals with device design of advanced MOS and bipolar transistors. The first two sections are devoted to study the relation between key process and device parameters for MOS and bipolar transistors. Among the parameters of MOSFETs, which have been considered, are the threshold voltage, the body effect, the breakdown voltage, etc. In addition, we have included a discussion on the short-channel effects, the hot carrier effects and latchup in CMOS. The discussion on the bipolar transistor covers the current gain, the cutoff frequency, the breakdown voltage, the reach- and punch- through phenomena and the junction capacitances. The high current effects are also addressed because of their relevance to BiCMOS circuits. The third section deals with the tradeoffs involved in the design of BiCMOS devices with special emphasis on the design of the n-well and its impact on the performance of the CMOS and the bipolar transistors. The last section in Chapter 3 is devoted to MOS and bipolar device scaling. The analytical expressions which relate the device parameters to the scaling factors are given to the reader to use them for a first order prediction of the device performance under scaling. These scaling rules take into account a number of the phenomena and features of advanced structures such as the velocity saturation of short-channel devices and the bandgap narrowing and polysilicon emitter contact of bipolar transistors.

1.5. Device Modeling

The topic of device modeling is of interest to those readers who need to analyse, design and/or simulate circuits. Chapter 4 introduces simple device (bipolar and MOS) models which can be used for first order (hand calculation) analysis or design. It also introduces commonly used device models in SPICE and discusses their limitations. This should help the reader, who intends to use SPICE (the most commonly used circuit simulator) for circuit simulation and design, to understand the capabilities and limitations of each of the models. This is important to make an educated decision when a model is to be chosen.

1.6. MOS Digital Integrated Circuits

The first part of Chapter 5 focuses on NMOS logic circuits. The importance of NMOS stems from the role they played historically. Basic concepts that apply for both NMOS and CMOS are introduced in this part. The NMOS inverter is treated first. The DC analysis of different configurations of the NMOS inverter is discussed. This is followed by a transient analysis of the NMOS inverter to identify the circuit and device parameters which determine the circuit's speed and the importance of each of these parameters. The reader is exposed to a variety of NMOS configurations such as the NMOS push-pull driver, NMOS transmission gates and NMOS current mode logic (CML). Although some of these configurations might seem to be obsolete they are, however, presented here for the readers who have the creativity and might find some use for such circuits or modify them for certain applications. For example, the NMOS current mode logic might be used to implement mixed CMOS/NMOS CML systems in a CMOS technology which is analogous to the mixed CMOS/bipolar CML systems proposed for BiCMOS technologies. In the second part the emphasis is on the fundamentals of the design of CMOS digital circuits. The important design equations of a CMOS inverter are derived from the DC and transient analyses. This is followed by a review of a variety of static and dynamic CMOS circuit structures.

1.7. Bipolar CML Integrated Circuits

Chapter 6 deals with bipolar Current Mode Logic (CML) with special emphasis on Emitter Coupled Logic (ECL) and Emitter Function Logic (EFL). Current mode logic, in general, finds widespread applications for mainframe computers. It is also expected to be used for mixed ECL/CMOS systems in BiCMOS technologies. The first section of Chapter 6 is an introduction to the design of the different blocks of the ECL gate. It also includes a transient analysis of the ECL gate to explain its transient behavior and to derive an analytical expression for its delay. Such an expression can be useful for understanding the importance of the device and circuit parameters and their effect on the speed. The second part of Chapter 6 introduces the Emitter Function Logic (EFL), which is based on an unconventional but rather interesting concept. Several circuits blocks are shown to demonstrate its high gate functionality.

1.8. BiCMOS Digital Integrated Circuits

Chapter 7 is devoted to demonstrate the synergy of bipolar and CMOS at the circuit level. This chapter starts with a comparison of bipolar and MOS transistors with special emphasis on the comparison of their transconductances. This comparative analysis leads to the evolution of the totem-pole BiCMOS gate. The conventional totem-pole BiCMOS inverter is treated in detail for its importance as the most commonly used BiCMOS gate at the present time. The DC and transient analysis of the totem-pole BiCMOS are developed. An analytical delay expression is derived. It relates the device parameters and transistor sizing to the delay so that it can be applied for first order circuit design. The impact of the variation of the device parameters on the speed of the totem-pole are discussed. It is shown that some of the parameters such as the forward transit time have a significant impact on the circuits' delay, while, others, such as the knee current, become important under heavy loading conditions. The implications of sizing the MOS and bipolar transistors are also explained. This is followed by a comparison of CMOS and BiCMOS inverters. The second part of Chapter 7 discusses the performance degradation of the conventional totem-pole BiCMOS gate due to power supply voltage scaling. A number of techniques, which have been developed to improve the voltage swing of BiCMOS gates, are presented and compared with each other. Chapter 7 then introduces another class of BiCMOS circuits, namely, the current mode BiCMOS structures. This class of circuits has not been widely used in practice, yet, it seems to be promising for VLSI applications where the critical paths can be replaced, fully or partly, by such circuits in order to reduce their delay.

1.9. BiCMOS Digital Circuit Applications

The last chapter, Chapter 8, deals with the application of BiCMOS for different digital subsystems. The chapter starts with arithmetic logic units such as adders and multipliers. Different types of adders and carry lookahead adders are considered. It is shown that bipolar sense amplifiers can be used to improve the speed of CMOS carry propagation adders. This is followed by a discussion on BiCMOS memories including RAM, ROM and DRAM. The decoding, cell, read/write and sensing circuits are discussed to show the blocks which should be implemented in either bipolar, CMOS or BiCMOS. This chapter also addresses the application of BiCMOS for semi-custom systems with emphasis on programmable logic arrays and gate arrays.

The reader can notice the large collection of digital circuit configurations in this text. This stems from our belief that future circuit designers have to be familiar with the different concepts and circuit techniques in MOS, bipolar and BiCMOS to design in any environment.

References

1. H.C. Lin, J.C. Ho, R.R. Iyer, and K. Kwong, "CMOS-Bipolar Transistor Structure," *IEEE Trans. Electron Devices*, **Vol. ED-6, No. 11** pp. 945-951 (November 1969).

2. A.P. Chandrakasam et al., "Low-Power CMOS Digital Design," *IEEE J. Solid-State Circuits*, pp. 473-484 (April 1992).

3. R.H. Eklund, "BiCMOS Process Integration," *IEDM Short Course*, (Dec. 1991).

CHAPTER 2

PROCESS TECHNOLOGY

This chapter serves as an introduction to IC fabrication of CMOS, bipolar and BiCMOS devices. Section 2.1. is a review of CMOS process technologies. Examples for an N-well CMOS process and a twin-tub CMOS process are considered. Section 2.2. deals with bipolar technology with emphasis on advanced bipolar structures. The topic of the isolation techniques used for both bipolar and CMOS is addressed in Section 2.3. In Section 2.4. we discuss the similarities between advanced CMOS and advanced bipolar transistor structures to demonstrate how both technologies are indeed converging. The BiCMOS technologies are introduced in Section 2.5. with emphasis on CMOS-based processes. Three BiCMOS technologies, with different performance/cost, are presented. Section 2.6. introduces a complementary BiCMOS structure, where a vertical isolated PNP transistor is merged with an NPN transistor in a CMOS process. Finally, in Section 2.7, a table with the design rules of a $0.8\mu m$ BiCMOS process is supplied.

2.1. CMOS Process Technology

The idea of CMOS was first proposed by Wanlass and Sah [1]. In the 1980's, it was widely acknowledged that CMOS is *the* technology for VLSI because of its unique advantages, such as low power, high noise margin, wider temperature and voltage operation range, overall circuit simplification and layout ease. The development of VLSI in the 80's has driven the integration density to millions of transistors on a single chip.

In this section we review two CMOS technologies: N-well and twin-tub processes. Other processes such as retrograde-well and SOI (Silicon On Insulator) technologies are not discussed. The N-well process is suitable for circuits rich in NMOS transistors such as dynamic CMOS, DRAM, and BiCMOS applications.

2.1.1. N-well CMOS Process

In the N-well CMOS process, the P-channel transistor is formed in the N-well itself and the N-channel in the P-substrate. Fig. 2.1 illustrates cross-sectional views and process steps of a typical N-well process.

The process starts by growing an oxide on the wafer. The oxide is then patterned to open N-well windows. Phosphorus atoms are implanted into the silicon followed by a high-temperature annealing to diffuse the well [Fig. 2.1(a)]. The LOCOS (LOCal Oxidation of Silicon)[1] technique is used to isolate the different active areas. After removing the nitride used in the LOCOS process, a photoresist layer is deposited and is then patterned by a P-well mask (new mask). This is followed by low energy ion implantation of boron (B I/I) to adjust the threshold voltage of the N-channel transistor [Fig. 2.1(b)]. A second ion implantation can be applied to eliminate punchthrough in the short channel device. Similarly, the threshold voltage of the P-channel transistor is adjusted [Fig. 2.1(c)]. A thin gate oxide is then grown and a layer of polysilicon is deposited and doped with phosphorus. The polysilicon is patterned to form the gates of all the transistors and interconnection layer [Fig. 2.1(d)]. The source and drain regions are then implanted by using a photoresist mask. Boron is used for the P^+ regions of the P-channel transistors and arsenic for N-channel transistors [Fig. 2.1(e)]. The N^+ and P^+ regions are also used N- and P-wells contacts, respectively. The photoresist is removed and a thick oxide is deposited by Chemical Vapor Deposition (CVD) as an isolation layer between the polysilicon layer and the subsequent metal layer. Contact holes are opened in the oxide layer and metal (usually aluminum) is deposited on the whole wafer. At this stage, the metal is patterned and annealed at relatively low-temperature (450 C) [Fig. 2.1(f)]. One or two other metal layers are usually added. At the end, the wafer is passivated and windows are patterned over the metal bonding pads to provide electrical contacts with pins.

---

1 For more details on the LOCOS isolation see Section 2.3.1.

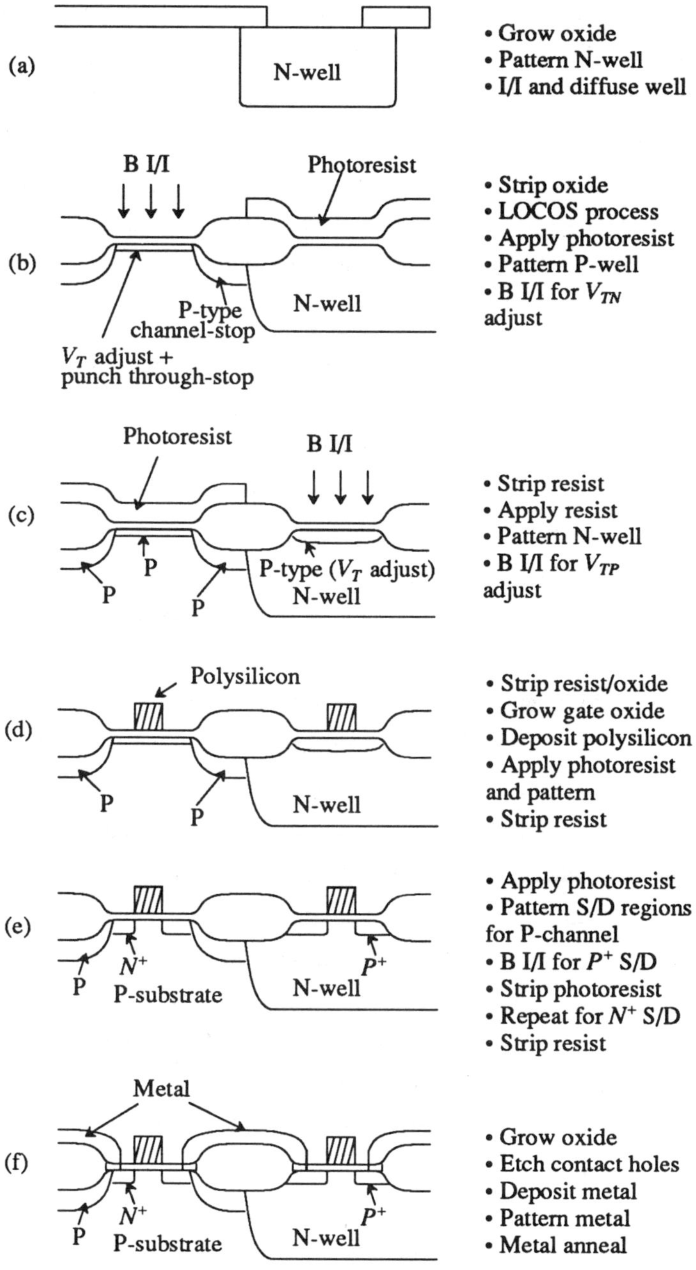

Figure 2.1 Cross-sectional views of an N-well process.

2.1.2. Twin-Tub CMOS Process

An alternative approach for CMOS devices fabrication is to use two separate wells (tubs) for N- and P-channel transistors in a lightly doped N- or P-type substrate. This "twin-tub" CMOS technology uses a single mask that allows it to form two independently doped and self-aligned tubs [2]; hence both CMOS devices types are optimized independently. This flexibility in selecting the substrate type with no change in the process flow is the major advantage of twin-tub CMOS. This technology is also more attractive when the devices are scaled down to submicron dimensions.

Fig. 2.2 shows the major steps involved in a typical twin-tub process. The starting material is a lightly doped P-epitaxial material over a P^+ substrate to reduce latch-up. In addition to the conventional N-tub process, another N-type (arsenic) shallow implant is used to increase the surface doping of the N-tub to prevent punchthrough (for short channel devices). It is also used to form the channel-stoppers[2] for the P-channel transistors [Fig. 2.2(a)]. The photoresist is stripped and a selective oxidation of the N-tub is performed. The nitride/pad oxide layers are removed to implant boron, which is driven in to form the P-tub. This is followed by a second boron ion implantation for the channel-stoppers for the N-channel device [Fig. 2.2(b)]. The N-tub oxide is then stripped. So far only one mask (N-tub mask, MASK#1) is required for self-aligned wells and channel-stopper processes. Both tubs are driven in. LOCOS isolation is developed to isolate between the devices using MASK#2, which defines the active areas. After the LOCOS process, boron is implanted through the pad oxide (used in the LOCOS) to reduce the threshold voltage of the P-channel transistor using MASK#3. This process results in a *buried-channel* PMOS transistor. The pad oxide is then removed. The remaining steps are similar to those used in the N-well process where MASK#4 is needed to pattern the polysilicon [Fig. 2.2(c)]. MASK#5 and MASK#6 are required to form the N^+ and P^+ sources/drains (S/D), respectively. MASK#7 for contact openings, and MASK#8 for patterning the metal [Fig. 2.2(d)].

---

[2] For more details on the channel-stoppers refer to Section 2.3.

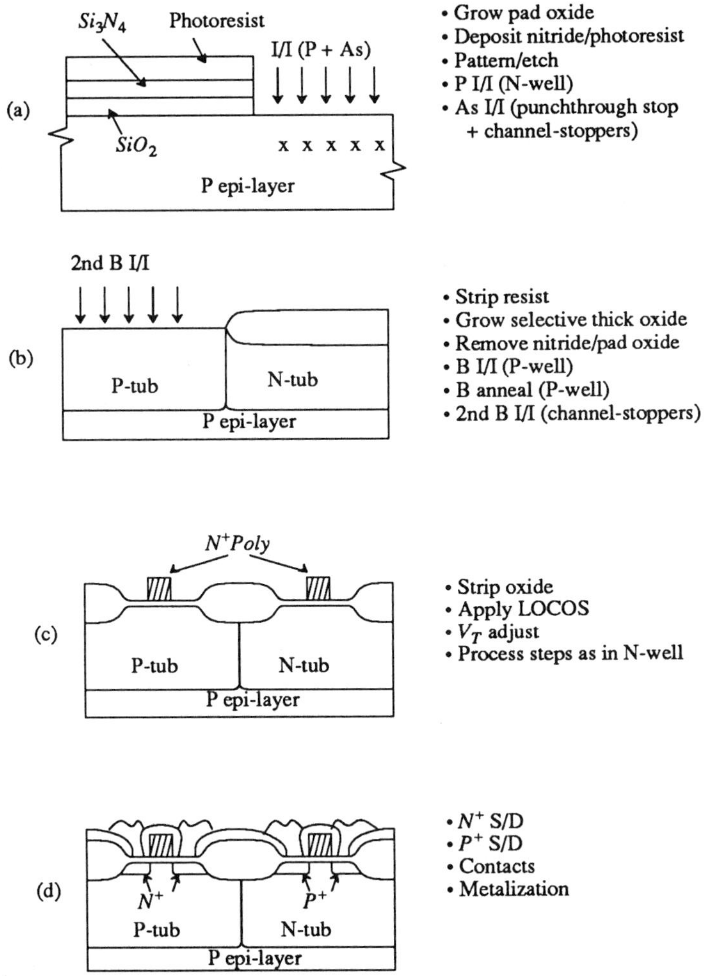

Figure 2.2 Twin-tub process sequence.

The fabrication of submicron MOS transistors requires additional process steps to avoid hot carrier effects. Fig. 2.3 illustrates a CMOS twin-tub structure with Lightly Doped Drain (LDD). Both NMOS and PMOS devices have *lightly doped* extensions to the source and drain regions. The electric field near the drain is reduced due to its light doping. This prevents the generation of hot carriers. This phenomena will be discussed in Section 3.1.5. The major process steps to fabricate the LDD structure are shown in Fig. 2.4.

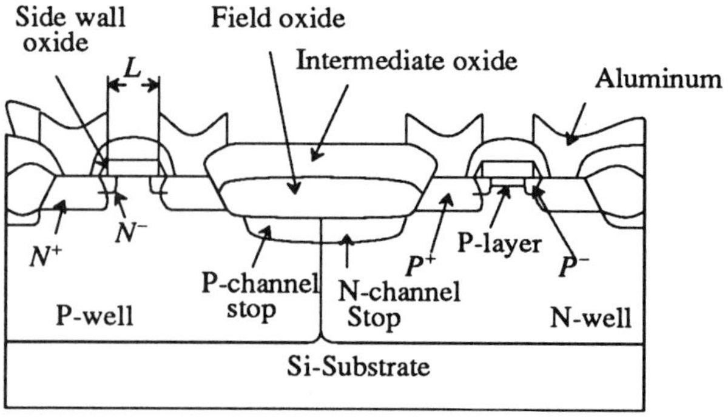

Figure 2.3 CMOS twin-tub structure with LDD, (L_{eff}≈1μm).

2.2. Bipolar Process Technology

The technology of epitaxial growth gave rise to the economical manufacturing of monolithic bipolar IC's as it allows a high-quality thin film of semiconductor to be grown on the top of a substrate. Junction-isolation and epitaxy techniques triggered the progress of bipolar technology. Although, most of the focus has been on the development of CMOS for the last ten years, yet, we find that bipolar technology has achieved significant progress as well. Impressive high-speed results were demonstrated at the 1985 ISSCC (International Solid-State Circuits Conference) and thereafter. CML (Current Mode Logic) gate delays close to 20 ps have been reported[3]. It was shown that advanced silicon bipolar technologies, although quite complex, could be integrated at the LSI level and operate at frequencies above those of CMOS circuits. Since then, the interest in advanced

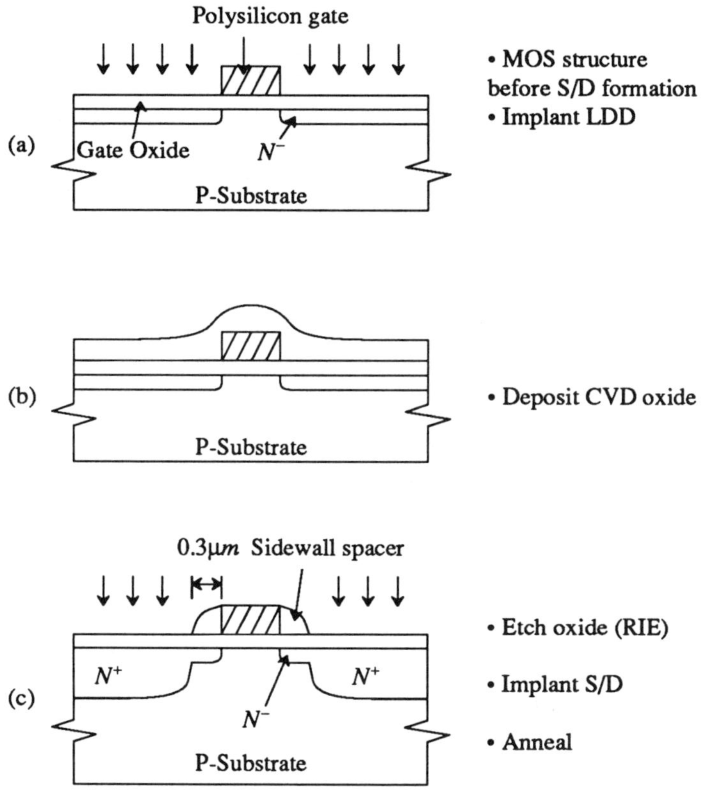

Figure 2.4 Process steps of lightly doped S/D.

bipolar processes has increased. The key features for such technologies are: i) self-aligned base, ii) advanced isolation techniques such as deep-trench, and iii) polysilicon emitter contact.

Fig. 2.5 depicts the differences between the structures of a conventional bipolar transistor[3] and an advanced one[4]. The parasitical capacitances and resistances have improved by reducing the areas of the emitter, base and collector. This has been

[3] In this context, the term "conventional bipolar transistors" refers to those bipolar transistors which have a metal emitter contact with oxide-isolation.

achieved by using self-alignment and trench isolations.

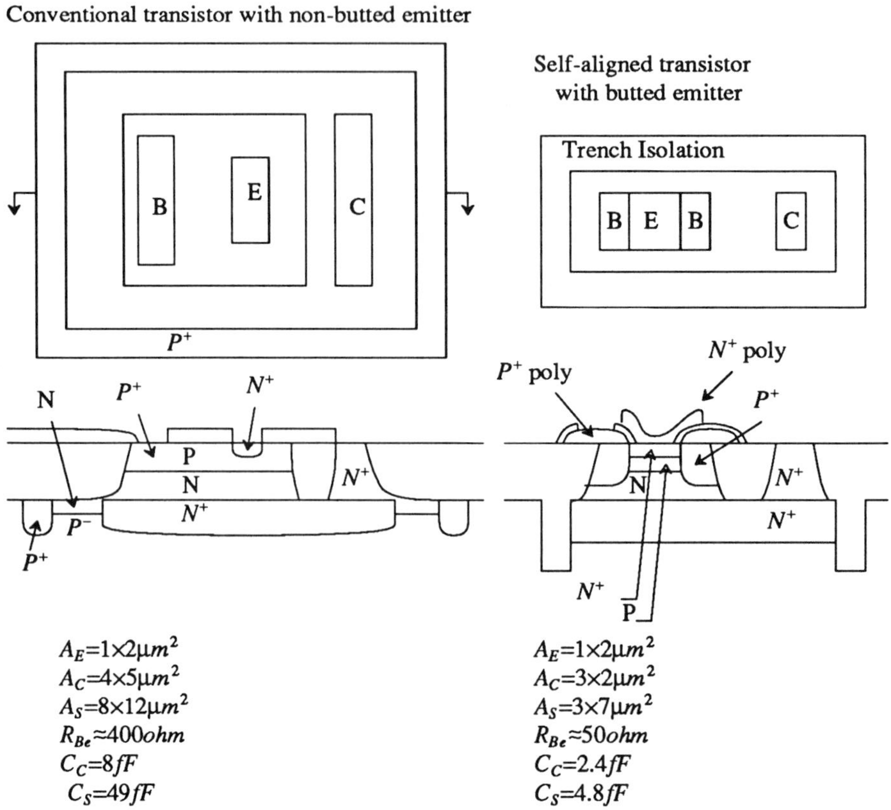

Figure 2.5 Comparison of a conventional and trench-isolated self-aligned bipolar devices with $1\mu m$ feature size[4].

Further reduction in parasitics can be achieved by using the so-called *SICOS* (*SIde wall base electrode COntact Structure*) process [Fig. 2.6][5]. The extrinsic base regions required for base contacts in the conventional bipolar structures have been replaced by the side wall base electrodes. This allows the base area to be almost as large as the emitter. The SICOS structure is suitable for VLSI applications because of its density and low parasitics.

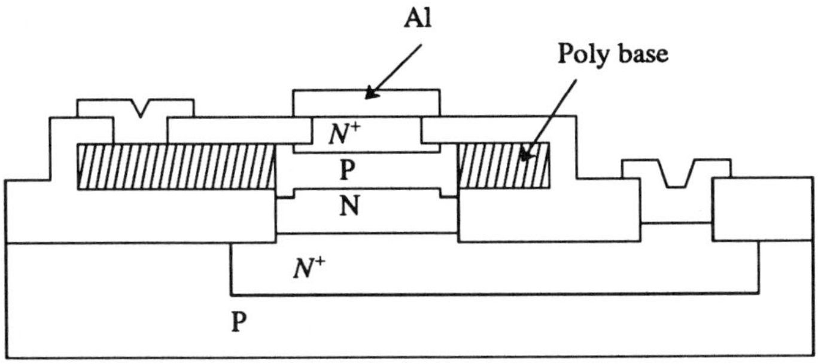

Figure 2.6 Cross-sectional view of the SICOS bipolar device structure[5].

One of the features of advanced bipolar transistors is the replacement of aluminum by polysilicon for the contact of the emitter. This step has led to noticeable improvement in the current gain of bipolar transistors. For further reading on polysilicon emitter BJTs refer to [4,6,7].

In this section, we introduce a typical Double-Polysilicon Self-Aligned (DPSA) process technology as an example of the advanced bipolar technologies[4].

Any bipolar process typically starts with creating the buried layers and the epitaxial layer. Fig. 2.7 illustrates the major steps of the epitaxial growth with an N^+ buried layer (BL). This buried layer is introduced to reduce the collector resistance of a bipolar transistor. While the epitaxial layer offers the high-quality silicon host for the bipolar transistor. The steps involved in Fig. 2.7 are the following. First, an oxide layer is grown on the substrate and is then patterned using the buried layer mask. The photoresist on the oxide serves as a mask against etching and ion implantation. After etching the oxide, the exposed regions of the silicon surface are implanted by arsenic or antimony to form the N^+ buried layers. The photoresist is then removed and an annealing step is carried out. All oxide is then stripped. An N-epitaxial layer is grown on the substrate as shown in Fig. 2.7(b). The thickness of

---

[4] A review of conventional bipolar technology using the junction isolation techniques can be found in [8].

this epitaxial layer can be as low as 0.8μm for advanced digital bipolar technology. The problems limiting the scaling down of the thickness of epitaxial layer are the autodoping and out-diffusion of the buried layer.

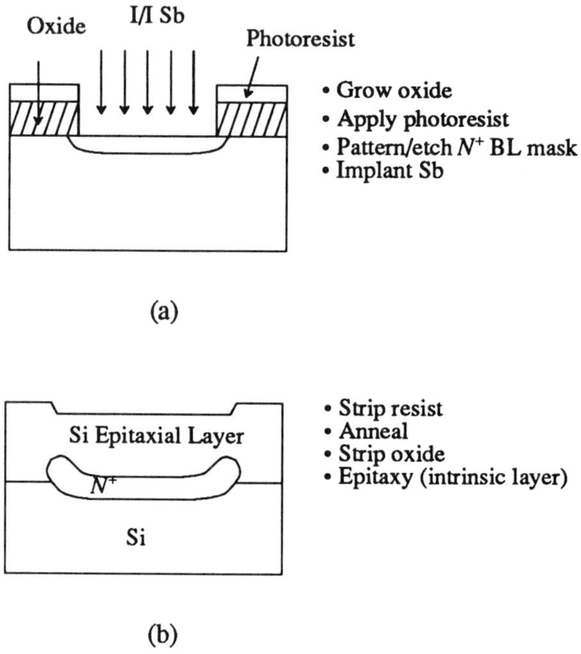

Figure 2.7 Buried and epitaxial layers growth.

Fig. 2.8 illustrates the sequence of a DPSA process assuming a starting structure with N^+ buried layer, N-epitaxial layer and isolation oxide as shown in Fig. 2.8(a). First, photoresist is deposited and patterned to define the collector contact region (deep N^+ collector sink). This region is then implanted with phosphorus to increase its doping level. The photoresist is stripped and a P-type base is implanted through a pre-implantation oxide as shown in Fig. 2.8(b). The resist and the oxide are then removed. A combination of P^+ polysilicon and oxide layers are deposited over the wafer. These layers are then etched as shown in Fig. 2.8(c). A CVD oxide is deposited over the wafer. The oxide is then dry etched using reactive ion etching

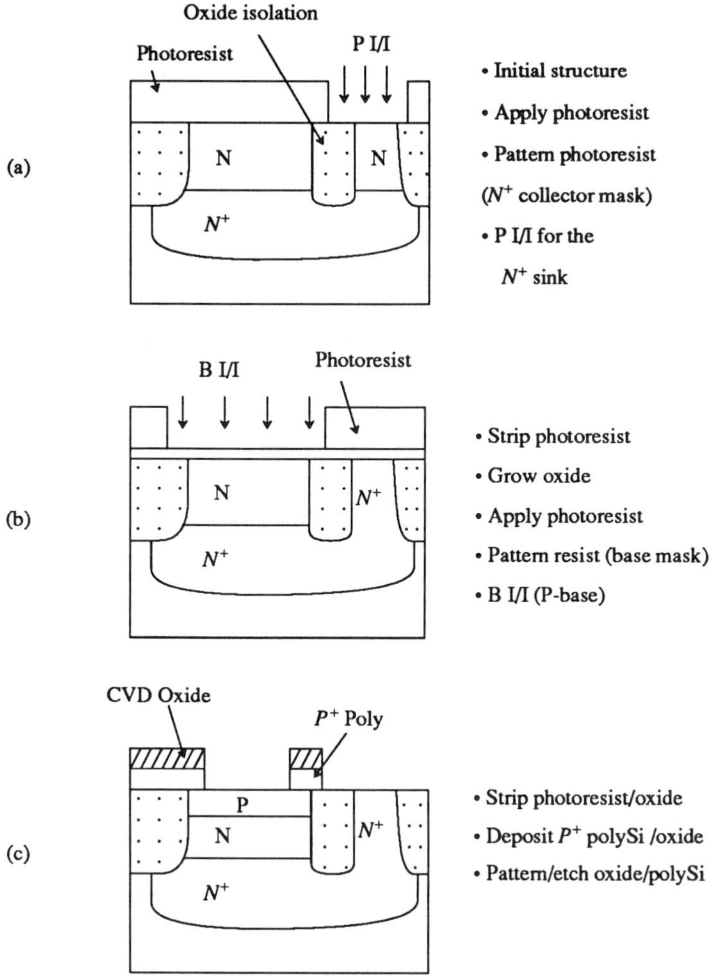

Figure 2.8 Fabrication steps of a self-aligned bipolar transistor.

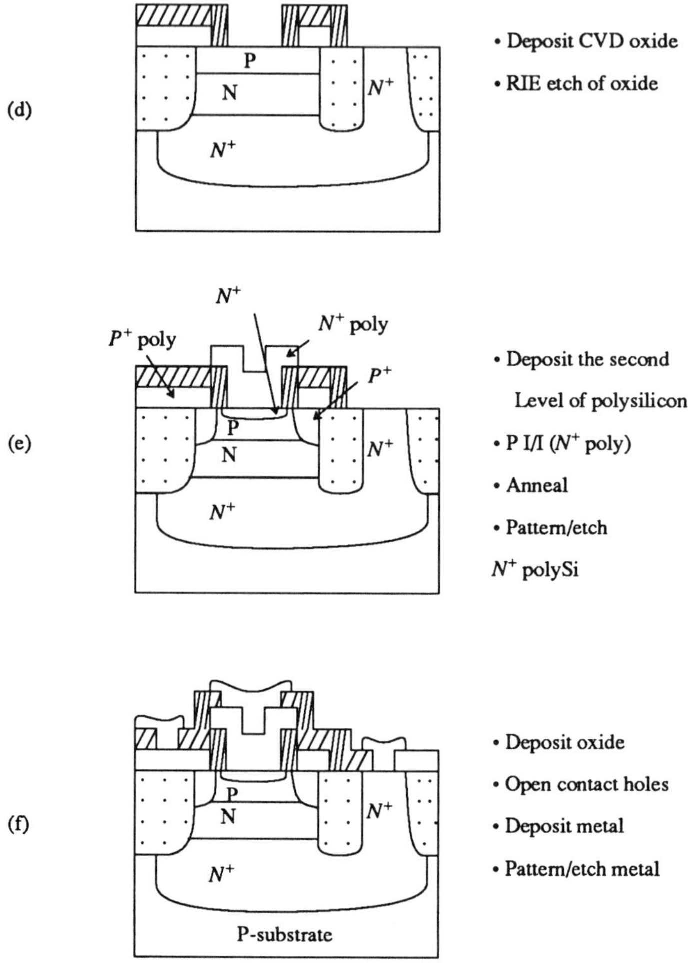

Figure 2.8 (Continued)

(RIE). The P^+ polysilicon is walled with the oxide (called sidewall spacer) [Fig. 2.8(d)]. The second level of polysilicon is deposited and implanted with phosphorus that will ultimately form the diffused emitter junction. At this stage, the wafer is annealed to drive the dopants from the P^+ and N^+ polysilicon layers. Fig. 2.8(e) illustrates the structure after patterning the N^+ polysilicon. The P^+ diffusion under the polysilicon forms the extrinsic base. The contact openings to the P^+ and N^+

polysilicon, and collector are etched. This is followed by the metallization step. At the end, the metal is patterned as shown in Fig. 2.8(f).

2.3. Isolation in CMOS and Bipolar Technologies

2.3.1. CMOS Device Isolation Techniques

Isolation in an integrated circuit means to electrically isolate similar or different transistors. In a CMOS chip, where more than one million transistors can be integrated, $1\mu A$/transistor of leakage current due to a bad isolation can lead to a few watts of DC power consumption. Moreover this leakage current provokes susceptibility to the latch-up as discussed in Section 3.1.6.

Isolation in CMOS is required to separate the devices electrically by eliminating the inversion layers, which might be induced by the interconnection layer between the transistors. The principle of isolation in CMOS is based on a field oxide formation between two active areas [Fig. 2.9]. The width of the isolation region should be minimized to attain dense layout and particularly for VLSI circuits.

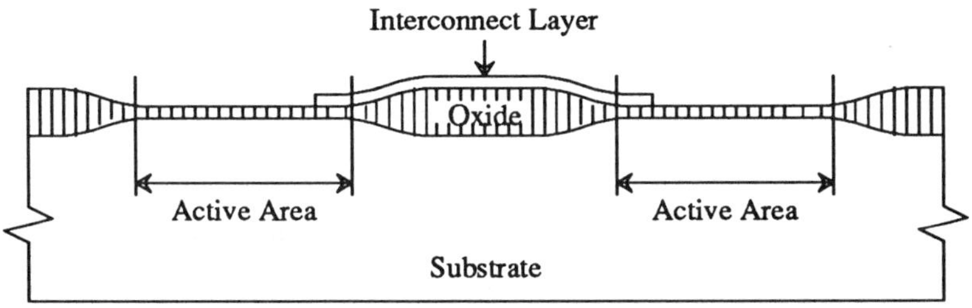

Figure 2.9 Field oxide isolation in MOS integrated circuits.

Several isolation techniques have been proposed and used. The most popular are LOCOS (LOCal Oxidation of Silicon) [9], trench isolation [10-13], and selective epitaxy [14]. Selective epitaxy is not studied in this chapter.

(A) Local Oxidation of Silicon (LOCOS)

LOCOS is a relatively simple process for the isolation of active devices in CMOS technology. It is realized by forming a thick field oxide (FOX) between the active areas. FOX is very thick (0.4–0.6μm), hence the corresponding field threshold voltage is high. The condition for preventing an inversion layer under FOX and between two active regions is that this field threshold voltage should be higher than the highest power supply voltage used on chip. The field threshold voltage can be further increased by raising the doping level under the FOX. This can be achieved by selectively implanting the regions over which the FOX is subsequently grown. These regions are commonly known as *channel-stoppers.*

The steps of the LOCOS process are illustrated in Fig. 2.10. A pad oxide of 40nm is grown and is followed by chemical vapor deposition of a 100nm thick nitride layer, which masks the active region. The pad oxide is called stress-relief-oxide (SRO) because it protects the silicon from stress caused by the nitride during subsequent high temperature processes. Silicon nitride is used as a mask to protect the active region from oxidation. A layer of photoresist is applied to the wafer and then patterned using the mask of the active areas. The nitride/oxide layers are etched [Fig. 2.10(a)]. A P-type dopant is implanted to form the channel-stoppers [Fig. 2.10(b)]. The photoresist, which is used for protection against ion implantation, is stripped and a thick thermal oxide is grown; i.e. FOX. Only local oxidation is realized because the nitride masks the regions beneath it. At the end, the nitride/oxide are removed [Fig. 2.10(c)]. During this LOCOS process, 56% of the FOX thickness is under the silicon surface because the oxidation consumes some of the silicon. This process is called semi-recessed LOCOS isolation. One problem associated with this process is the lateral extension of the field oxide under the nitride during the oxidation, forming what is called bird's beak encroachment [Fig. 2.10(c)]. A typical value of this encroachment is 0.5μm/side. This encroachment limits the scaling of the active areas and the channel width of the MOS device. Moreover, this bird's beak introduces imprecise channel widths.

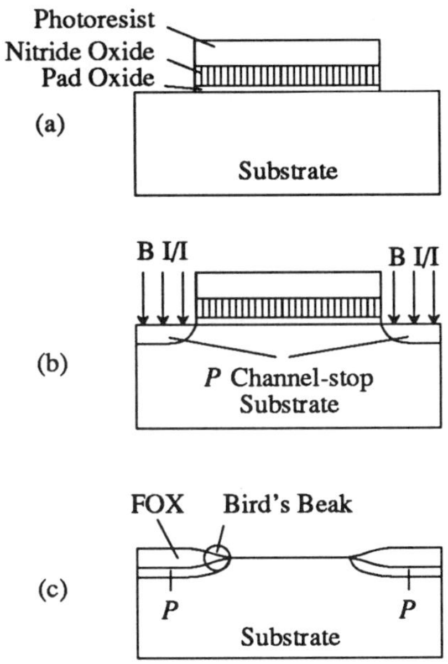

Figure 2.10 LOCOS process flow with channel-stopper's ion implantation and bird's beak encroachment.

The *Poly Buffer LOCOS* process was developed to reduce the bird's beak encroachment[15]. In this modified LOCOS process, the nitride mask thickness has been increased to 240*nm* and a polysilicon stress relief buffer layer of 50*nm* has been added between the nitride and a 10*nm* pad oxide [Fig. 2.11(a)]. This arrangement prevents deep lateral extension of the field oxide under the nitride layer [Fig. 2.11(b)]. A 0.8μm field oxide thickness results in 0.15μm/side of encroachment and 2.2μm minimum isolation pitch. Other techniques to solve the problem of the bird's beak encroachment can be found in [16-18].

(B) Trench Isolation

Trench Isolation is another alternative to LOCOS isolation process. This technology has been accepted relatively quickly by the industry [19]. It addresses the isolation problem between opposite type devices (like N-channel and P-channel

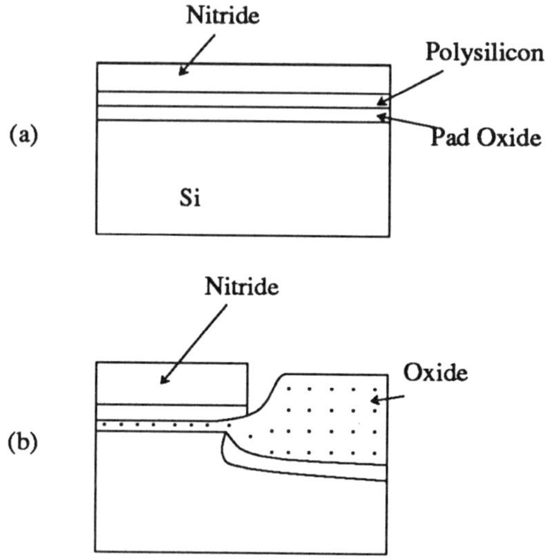

Figure 2.11 Poly buffered LOCOS process.

MOSFET's in CMOS technology). The advantages of the trench isolation are: i) no bird's beak encroachment, ii) latch-up free structure, and iii) planar surface.

Fig 2.12 illustrates the steps of the trench isolation process. First, the pad oxide, the nitride and the thick oxide layers are patterned using the mask of the active areas. The thick oxide serves as a mask in the trench processing [Fig. 2.12(a)]. A deep trench is formed by dry etching (RIE). This is followed by a boron implant to create the P^+ channel-stoppers at the bottom of the trench. The top thick oxide is removed, and the trench sidewalls are oxidized [Fig. 2.12(b)]. The polysilicon is deposited over the whole wafer, filling the trenches. The polysilicon is used as the trench dielectric because it uniformly fills the trenches better than other dielectrics. The surface polysilicon is then etched to yield the structure shown in Fig. 2.12(c). The wafer is oxidized using the nitride as a mask. The nitride is finally removed as illustrated in Fig. 2.12(d). At this stage, conventional processing can be used to integrate the CMOS devices.

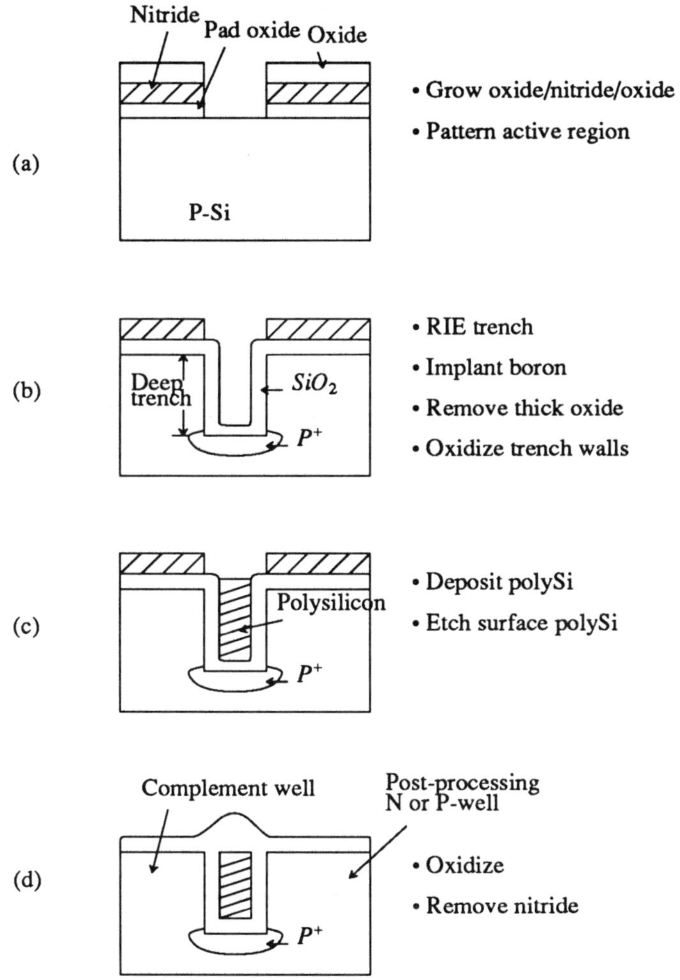

Figure 2.12 The major process steps of the trench isolation.

Although trench isolation permits reduction of the separation between the active regions; it has several drawbacks: i) it is a costly process because of the large number of processing steps, and ii) it can not be used as an isolation region for the inactive parts of the chip. In this case, LOCOS is usually used. The description of other trench isolation processes can be found in[20].

2.3.2. Bipolar Device Isolation Techniques

The first technique used for bipolar isolation was based on collector/substrate junction isolation [Fig. 2.13]. The N-wells (N collectors) of the adjacent transistors were separated by P^+ isles, which are deeply diffused to reach the P-type substrate. By tying these isles and the substrate to the most negative voltage, the junctions between them and the N-type collectors are reverse biased. Thus, all the components in different N-wells (N collectors) are isolated. The area consumed by the isolation isles is large relative to the transistor area.

The packing density of the bipolar technology can be improved by replacing the junction isolation with LOCOS isolation. An additional advantage of LOCOS isolation is the reduction of the parasitic collector-substrate capacitance. Fig 2.14 illustrates the cross-sectional view of an NPN bipolar transistor with LOCOS isolation. The area occupied by the oxide isolation is proportional to the epitaxial layer thickness. As the epitaxial thickness is being reduced for higher device performance the oxide isolation area becomes smaller, which means that LOCOS may become a practical isolation technique for advanced bipolar and BiCMOS technologies.

Fig. 2.15 illustrates the process steps for oxide isolation in a bipolar process. After epitaxy growth, a thin layer of SiO_2 is grown and a layer of Si_3N_4 is deposited. A photoresist layer is applied and patterned with an isolation mask [Fig. 2.15(a)]. Then the nitride/pad oxide layers and approximately half of the epitaxial layer are dry etched. Boron implant is performed to form the channel-stopper [Fig. 2.15(b)]. The photoresist is then removed and the wafer is oxidized to grow the thick isolation oxide. This oxide is called *recessed oxide*. The Si_3N_4 and the pad oxide are stripped at this stage. The resulting structure is almost planar. In this structure the bird's beak is formed as in the MOS case [Fig. 2.15(c)].

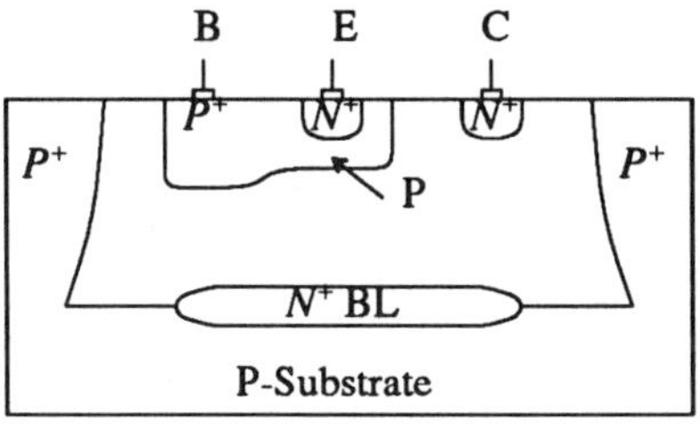

Figure 2.13 Cross-section of a junction isolated NPN bipolar transistor (with a buried layer).

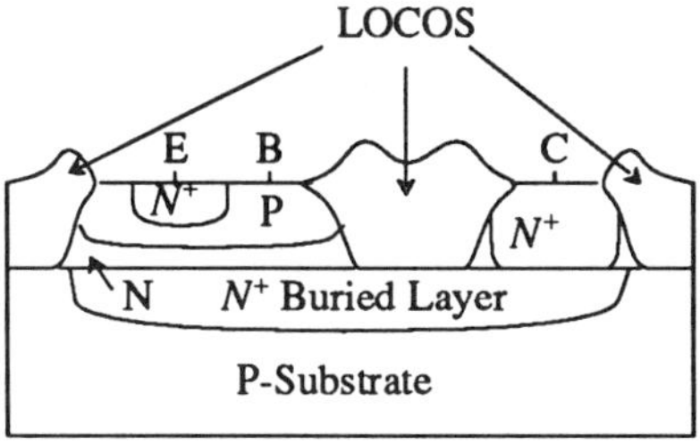

Figure 2.14 Cross-sectional view of an NPN bipolar transistor with LOCOS isolation.

In the early 1980's, new isolation techniques such as grooves and trenches [21-23] were demonstrated. These techniques reduced the collector-substrate capacitance and increased the packing density. Hence they improve circuit speeds. The fabrication process is the same as the one described in CMOS trench isolation.

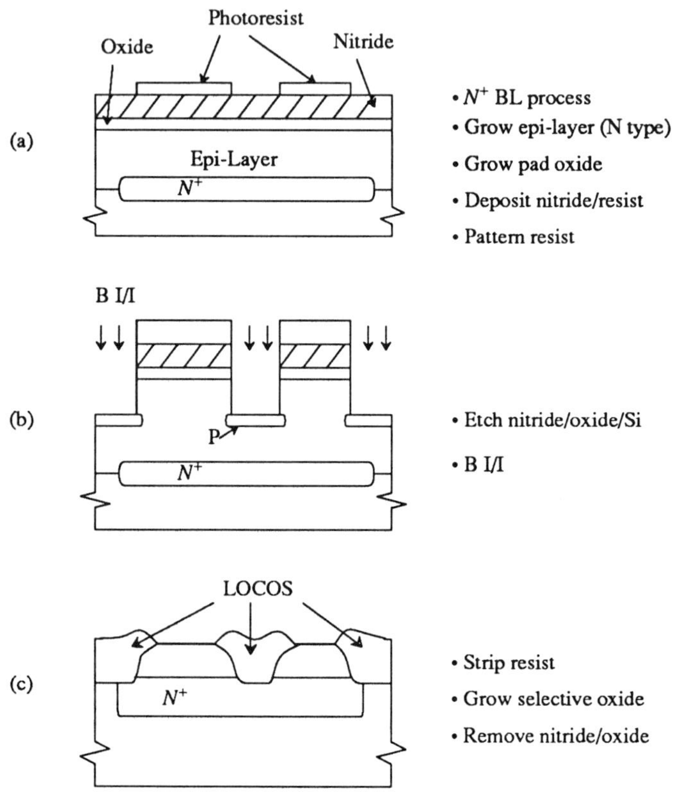

Figure 2.15 The major process steps involved in the fabrication of oxide-isolation isles in a bipolar process.

2.4. CMOS and Bipolar Processes Convergence

An interesting exchange of process technology know-how between the CMOS and the bipolar domains has taken place over the years. We have seen that epitaxial and buried layers have been used for CMOS to mute the latch-up. At the same time LOCOS, which was originally developed for CMOS, has been used for isolating bipolar transistors. The use of polysilicon for creating self-aligned MOS transistors was later adapted for self-aligned poly emitter bipolar transistors. Another example of the convergence between bipolar and CMOS is the use of oxide spacers in CMOS for formation of LDD regions, while, it has been used in bipolar to reduce the separation between the base contact and the emitter. The convergence of both technologies made the attractive idea of merging bipolar and CMOS seem more rational and feasible than ever.

Many of the steps of the advanced CMOS and bipolar processes are similar, hence, they can be shared for the fabrication of MOS and bipolar transistors when they are integrated in a BiCMOS process. Some examples of these steps are:

1) The N-well, which can be used as the body of the PMOS transistor and as the N-collector of the NPN transistor;

2) The N^+ buried layer of the NPN can be used to form a retrograde well for the PMOS to reduce the latch-up susceptibility;

3) The polysilicon can be used for the CMOS gates and for the emitter contacts;

4) The shallow P-type implantation can be shared by the PMOS S/D and the self-aligned extrinsic base of the NPN transistor;

5) The shallow N-type implantation can be shared by the NMOS S/D and the emitter of the NPN transistor;

6) The final annealing steps match.

However, as more steps are being shared by the different device, the device characteristics have to be compromised. There is a tradeoff between the process complexity and device quality.

2.5. BiCMOS Technology

Although the idea of merging bipolar and CMOS on the same chip originated 20 years ago [24], it was not feasible from a practical point of view because of the lack of adequate process technology. With the technological progress achieved in recent years, this idea has been revived. There are many techniques to merge bipolar and CMOS devices as reported in the literature [25-30]. There are two ways of classifying BiCMOS processes. One way is to classify them according to the baseline process. A CMOS-based BiCMOS process is a CMOS baseline process, to which a bipolar transistor is added. Similarly, a bipolar-based BiCMOS process is a bipolar baseline process, to which CMOS transistors are added. In both cases, the added device would have to be compromised, which means that its characteristics can not be optimized. Alternatively, BiCMOS processes can be classified according to their cost/performance. In this regard, three categories can be identified:

(i) Low-cost;

(ii) Medium-performance;

(iii) High-performance (high-speed).

In this section, we present three examples of BiCMOS processes. The first one represents a low-cost process. It needs only one mask to incorporate the bipolar device in a CMOS-based process. The second example shows a medium-performance BiCMOS process, which requires 3 extra masks to a CMOS process. The third example illustrates a high-performance process in which polysilicon emitter and self-aligned structures are used.

2.5.1. Example 1: Low-Cost BiCMOS Process

In a low-cost BiCMOS process, a bipolar transistor is added to a CMOS process with minimum additional process steps. A typical N-well CMOS/bipolar process sequence is listed in Fig. 2.16(a). The N-well of the PMOS is used for the collector of the vertical NPN. The base is implanted in a separate step using an additional mask. The P^+ S/D and the extrinsic base share the same implantation step. The emitter and the N^+ S/D of the NMOS are also implanted in the same step. Fig. 2.16(b) illustrates the cross-section of an N-well BiCMOS structure. The process complexity is comparable to that of the CMOS. However, there are many trade offs in designing the emitter, base, and collector of the NPN. If the CMOS process is optimized, some of the bipolar device parameters, such as the breakdown voltage and the gain, may be satisfactory, but many others are degraded. For example, due to the absence of the buried layer and the deep N^+ collector in the NPN, the collector resistance is high. Hence, the cut-off frequency is low, the current drive is poor, and the collector-emitter saturation voltage is high.

2.5.2. Example 2: Medium-Performance BiCMOS Process

Fig. 2.17 shows a cross-sectional view of a BiCMOS structure, which can be realized by adding an NPN to a baseline twin-tub CMOS process. This structure has an N^+ buried layer and a deep N^+ collector sink which enhance the collector conductivity. The N^+ buried layer, under the PMOS, with the uniform N-well form a desired retrograde N-well. Similarly, the P^+ buried layer creates a retrograde P-well for the NMOS transistor. It also acts as an isolation region between the N^+ buried layers. A thin epitaxial layer ($1\mu m - 2\mu m$) is used to increase the cutoff frequency of the NPN transistor and to reduce the required width of the isolation isles between the bipolar transistors. The N collector is formed at the same time with N-well of the PMOS transistor. After the formation of LOCOS a deep N^+ sink is implanted and driven in. The P^+ extrinsic base is implanted at the same time with P^+ S/D regions of the PMOS transistor. The N^+ emitter and the N^+ S/D share the same implantation step. In this process an aluminum emitter contact is used. Therefore, the size of the emitter is larger compared to the case where a self-aligned polysilicon emitter contact is used. This process uses only 3 extra masks to form the bipolar transistor. The first mask is needed for N^+ buried layer. The second mask is used to implant the N^+ deep collector, and the third one for the base implantation.

The BiCMOS process described above can be optimized to be used for high performance circuits. The collector resistance is low in comparison to the low-cost process (example 1). For a $0.8\mu m$ process, the cut-off frequency (f_T) of a bipolar can be as high as $5GHz$.

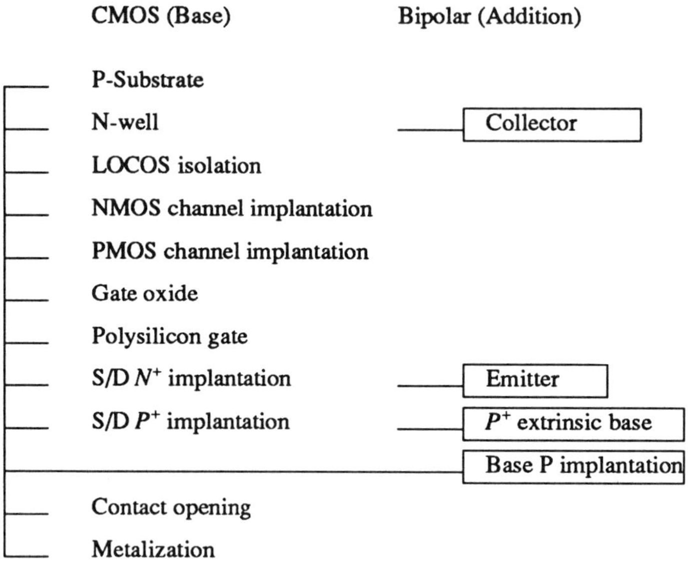

(a)

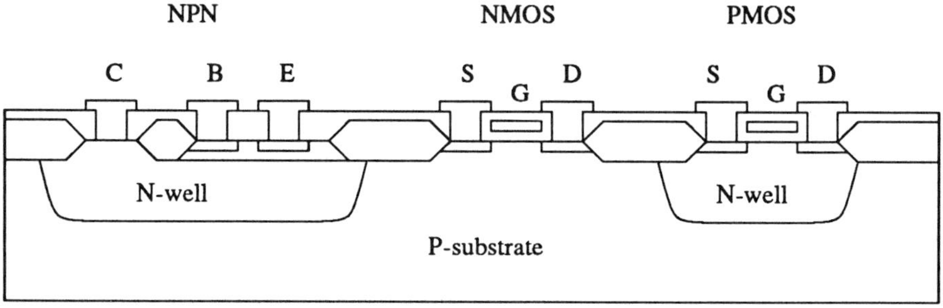

(b)

Figure 2.16 CMOS based BiCMOS process

(a) Process flow; (b) Device cross-sectional view.

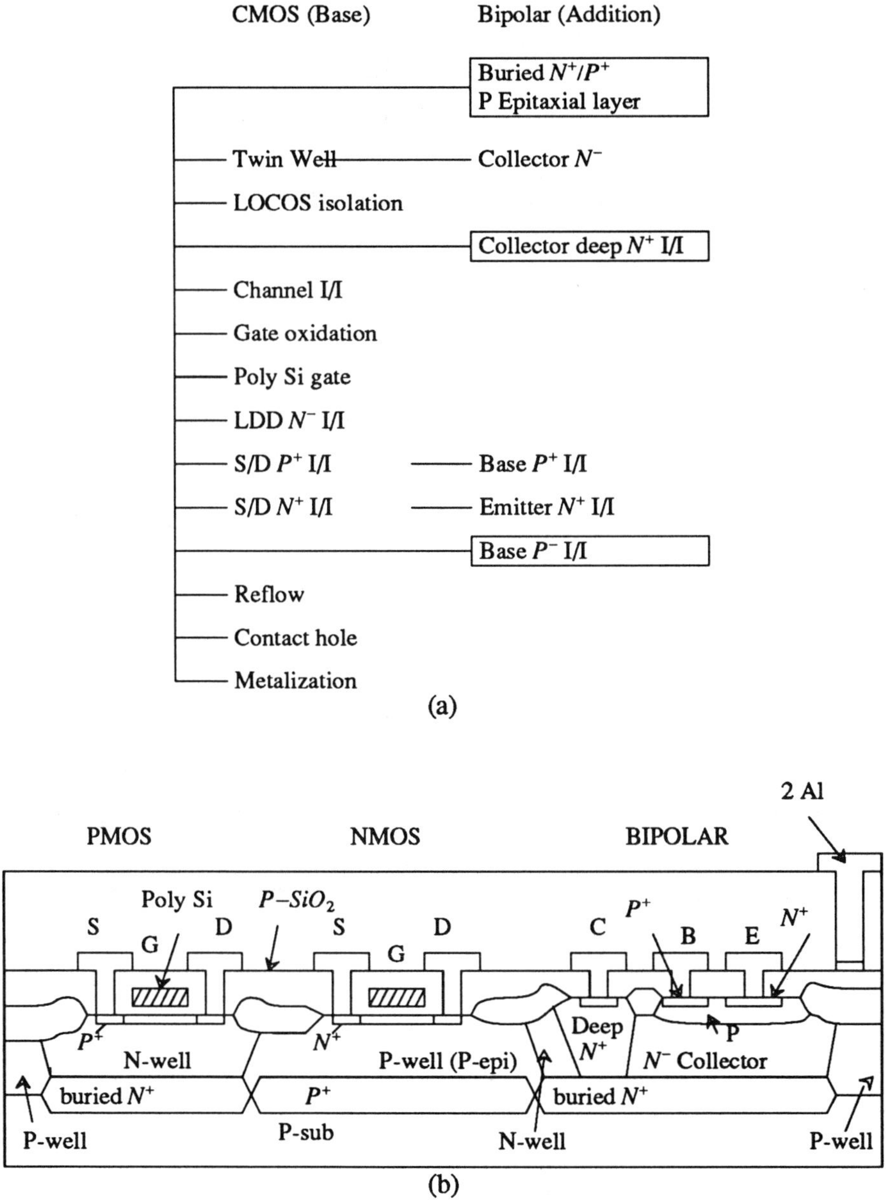

Figure 2.17 Medium-performance BiCMOS process
(a) Process flow; (b) Device cross-sectional view.

2.5.3. Example 3: High-Performance BiCMOS Process

A high-performance BiCMOS process can be achieved by replacing the N^+ S/D implant, used to form the emitter in example (2), by a doped polysilicon emitter. One extra mask is required to open the emitter window of the bipolar transistor. The ion implantation of the poly emitter and MOS gates is developed simultaneously. As shown in Fig. 2.18, four additional mask levels (N^+ buried layer, N^+ deep collector, P-base, and emitter window) are required to obtain an advanced BiCMOS.

After the formation of the N^+/P^+ buried layers, the conventional twin-tub process is carried out. LOCOS is developed to isolate the devices. The deep collector N^+ is implanted and driven in, and the P-base is then patterned and implanted. The threshold voltages of the MOS transistors are adjusted by additional ion implantations. After the gate oxide growth, a thin polysilicon is deposited as shown in Fig. 2.19(a). The emitter window is then patterned and a second polysilicon layer is deposited [Fig. 2.19(b)]. The polysilicon is then doped by implantation and patterned to define the CMOS gates and polysilicon emitter [Fig. 2.19(c)]. Next, implants are selectively carried out to form the LDD regions for CMOS. Before implanting the N^+/P^+ S/D regions, a sidewall oxide is formed near the emitter and gate edges. Fig. 2.18(b) shows the final cross-section of this BiCMOS process.

The BJTs realized in the presented high-performance BiCMOS process have low collector resistance (because of the buried layer and deep sink), high current gain (because of the poly emitter contact) and low parasitic capacitances (because of the self-alignment). With this BiCMOS process f_T's greater than 5 GHz can be achieved.

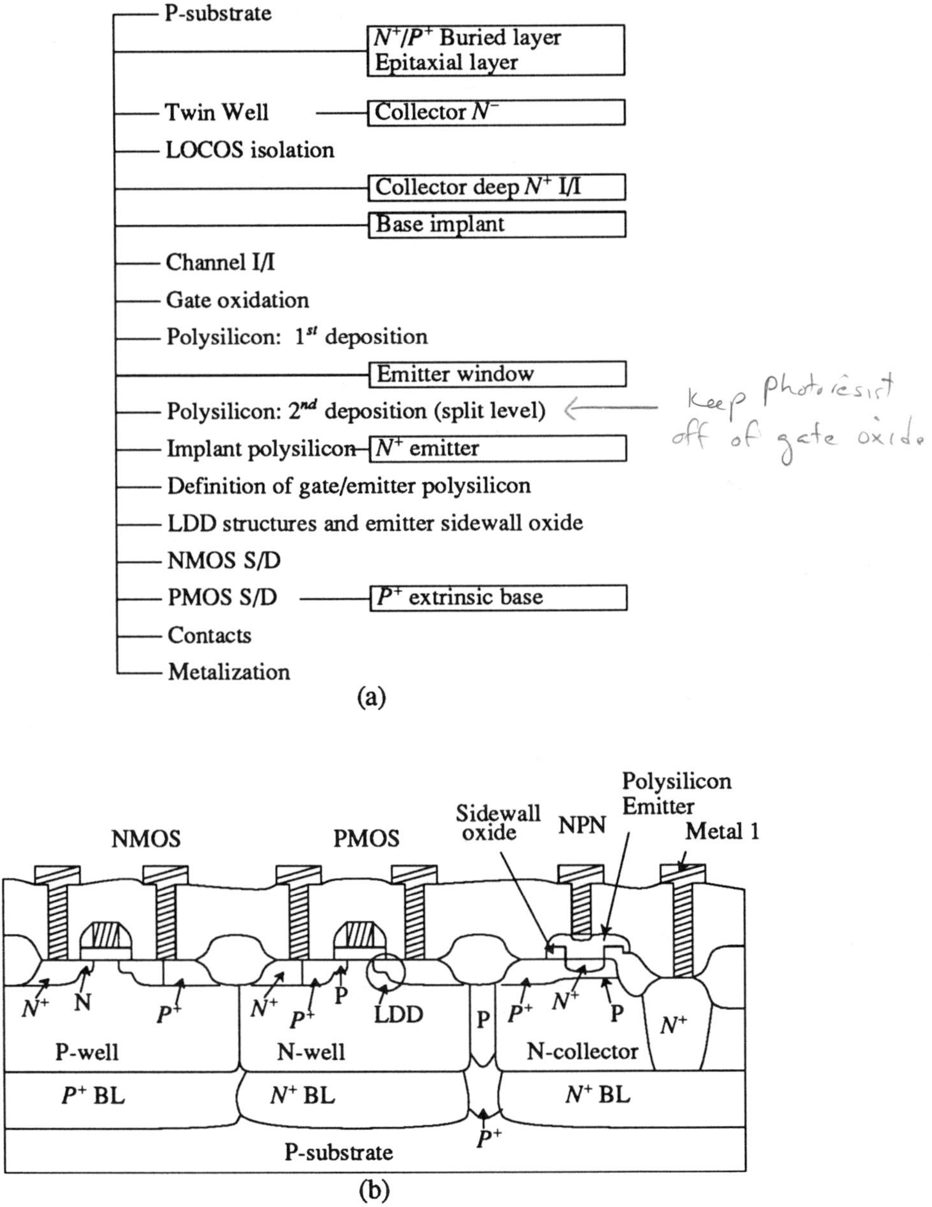

Figure 2.18 High-performance BiCMOS process

(a) Process flow; (b) Device cross-sectional view.

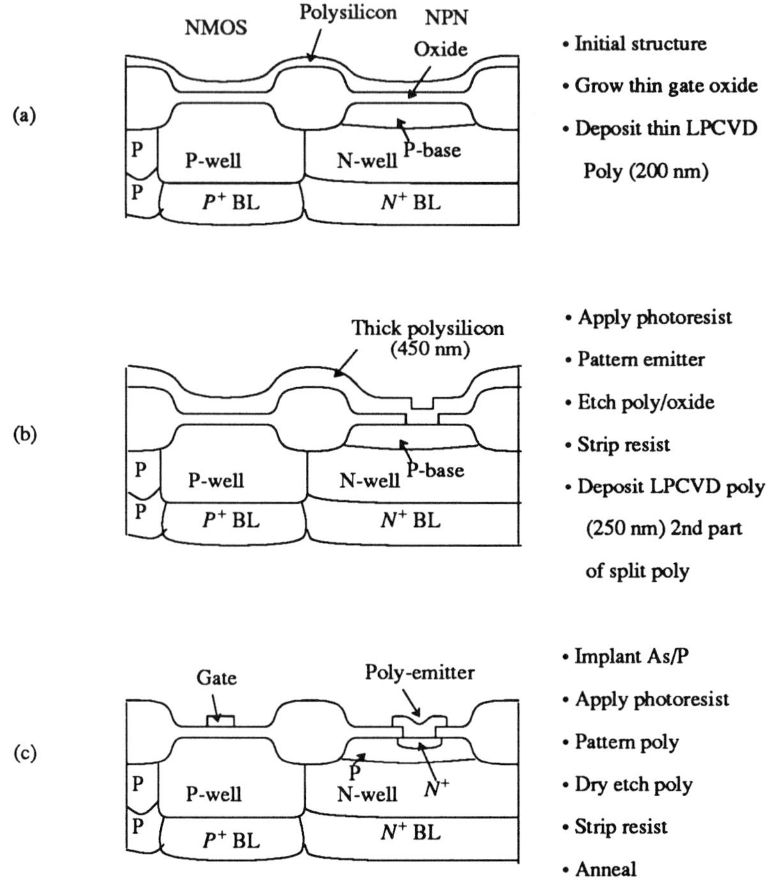

Figure 2.19 Cross-sectional view of the split-level polysilicon technique.

2.6. Complementary BiCMOS Technology

In a Complementary BiCMOS (CBiCMOS) process both vertical NPN and PNP transistors are merged with CMOS on the same chip. Recent investigations indicate that CBiCMOS allows for improving the performance of BiCMOS gates at low supply voltages[31,32]. Further discussion on CBiCMOS circuits can be found in Section 7.3.5. Although, to date, the NPN has shown superior performance to that of PNP, future trend indicates that PNP is approaching NPN. Some of the problems associated with the PNP transistor are its high collector resistance, low current gain, and high base transit time.

It has been recently reported that CBiCMOS processes can offer NPNs with f_T's of 8–17GHz and PNPs with 2–7GHz f_T[33-36]. Fig. 2.20 shows a cross-sectional view and process flow of a CBiCMOS [34]. The N^+ buried layer of the NPN transistor creates a retrograde well for the PMOS transistor. The P^+ buried layer is only used for isolation isles between NPN transistors. After the epitaxial layer growth, twin-well and LOCOS processes are performed. The P-well of the NMOS device is used as the collector of PNP transistor. A second high energy (600 keV) boron ion implantation is carried out to form the retrograde well (2nd P-well) for the NMOS and the P^+ buried layer for PNP device. The S/D implants of MOS transistors are used simultaneously for the extrinsic bases of the NPN and the PNP transistors. The emitters of the NPN and the PNP are formed by the self-aligned contact doping technique to simplify the process flow. Finally, the metal is deposited and patterned.

Complementary BiCMOS offers a technology with versatile devices. It adds flexibility for mixed bipolar/MOS circuit design. The CBiCMOS technology promises further improvements to BiCMOS circuits performance.

- P-substrate
- N^+/P^+ buried layer
- N-type epitaxial layer
- N/P twin well (1st P-well for PNP)
- Field isolation
- Collector deep N^+
- Deep P^+ I/I for NMOS retrograde well and 2nd P-well for PNP (P^+ buried layer)
- Gate (CMOS)
- NMOS S/D (N^+ extrinnic base for PNP)
- PMOS S/D (P^+ extrinnic base for NPN)
- NPN Base
- PNP Base
- Contact hole
- N^+ and P^+ emitters implant
- Metalization

(a)

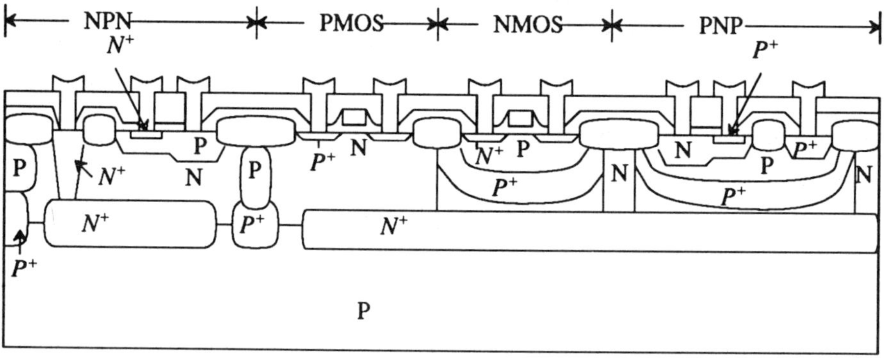

(b)

Figure 2.20 (a) Fabrication process flow and (b) Cross-sectional view of CBiCMOS[34].

2.7. BiCMOS Design Rules

In this section, a set of lambda-based design rules of a typical BiCMOS process is presented[5]. The minimum length of the MOS gate is 2λ and the minimum length and width of the bipolar emitter contact is 2λ and 4λ respectively. Table 2.1 describes the basic masks used in the layout design of BiCMOS devices. The rest of the masks are generated automatically.

Table 2.2 lists the design rules for the (design) masks only of a typical BiCMOS technology in terms of the parameter λ. The corresponding graphical representation of design rules is illustrated in Plate I. Plate II shows the layouts of minimum size PMOS, NMOS and bipolar transistors in a $0.8\mu m$ BiCMOS technology.

2.8. Chapter Summary

In this chapter, we have studied the fabrication steps of CMOS and bipolar transistors in different technologies. We have shown that the advanced CMOS and bipolar processes are converging, and many process techniques can be shared for the fabrication of both devices. The different options for merging bipolar and CMOS devices are then discussed. Three examples for BiCMOS processes with different complexities are presented. The complementary BiCMOS process is also considered. A table of design rules for a state- of-the-art BiCMOS technology is given for layout exercises.

[5] The given design rules are typical of a generic $0.8\mu m$ high-performance BiCMOS process (for $0.8\mu m$, $\lambda = 0.4\mu m$). The corresponding device parameters are presented in Chapter 4.

Table 2.1 Basic BiCMOS Design Masks

N-well (NW)	The NW mask is used to define the N substrate (bulk) of the PMOS and the N-collector of the NPN transistor.
N^+ deep collector (CN)	The CN mask defines the area which is exposed for the N^+ sink implantation.
P base (CP)	The CP mask defines the region which is to receive an P-implant to create the base diffusion.
Polysilicon (PO)	The PO mask defines the gate and the emitter electrodes, and the polysilicon interconnect layer.
Emitter window (EW)	The EW mask defines the opening for the emitter window.
N^+ (DN)	The DN mask defines the N^+ source and drain regions of the N-channel device within the P-well, and the body contact regions in the N-well.
P^+ (DP)	The DP mask defines the P^+ source and drain regions of the P-channel device within the N-well, and the body contact regions in the P-well.
Contact (CO)	The CO mask defines the contact openings.
Metal 1 (M1)	The M1 mask defines the metal 1 interconnects.
Via (VIA)	The VIA mask defines the openings of the via that connects metal 1 to metal 2.
Metal 2 (M2)	The M2 mask defines the metal 2 interconnects.

Table 2.2 Design rules for a typical BiCMOS Process

		Dimensions
1.	N-well (NW)	
	1.1 minimum width	12λ
	1.2 minimum spacing	12λ
2.	N^+ -diffusion (DN)	
	2.1 minimum width	3λ
	2.2 minimum spacing	3λ
	2.3 minimum NW overlap of DN	0λ
	2.4 minimum NW to external DN spacing	6λ
3.	P^+ -diffusion (DP)	
	3.1 minimum width	3λ
	3.2 minimum spacing	3λ
	3.3 minimum NW overlap of DP	4λ
	3.4 minimum NW to external DP spacing	4λ
	3.5 minimum space to DN (same potential)	0λ
	3.6 minimum space to DN (different potential)	3λ
4.	N-collector plug (CN)	
	4.1 minimum width	4λ
	4.2 minimum spacing	12λ
	4.3 minimum space to NW	10λ
	4.4 minimum NW overlap of CN	3λ
	4.5 minimum space to DN	6λ
	4.6 minimum space to DP	5λ
5.	P-base diffusion (CP)	
	5.1 minimum width	4λ
	5.2 minimum spacing	4λ
	5.3 minimum NW overlap of CP	3λ

	5.4 minimum space to CN	5λ
	5.5 minimum space to DN	3λ
	5.6 minimum space to DP	3λ
6.	Polysilicon (PO)	
	6.1 minimum width	2λ
	6.2 minimum spacing	3λ
	6.3 minimum space to DP or DN	2λ
	6.4 gate overhang of DP or DN	2λ
	6.5 minimum space to CN or CP	1λ
7.	Emitter window (EW)	
	7.1 minimum width	2λ
	7.2 minimum length	4λ
	7.3 minimum spacing	3λ
	7.4 minimum CP overlap of EW	2λ
	7.5 minimum poly overlap of EW	2λ
8.	Contact (CO)	
	8.1 minimum size (single)	2λ×2λ
	8.2 minimum size (double)	2λ×4λ
	8.3 minimum spacing	3λ
	8.4 minimum DN or DP overlap of CO	1λ
	8.5 minimum space to gate	2λ
	8.6 minimum PO overlap of CO	1λ
	8.7 minimum CN or CP overlap of CO	1λ
	8.8 minimum PO to CO spacing in P base	2λ
	8.9 minimum poly emitter CO to CP spacing	2λ
9.	Metal 1 (M1)	
	9.1 minimum width	2λ
	9.2 minimum spacing	3λ
	9.3 minimum M1 overlap of CO	1λ

	9.4 maximum current density	1 $mA/\mu m$
10.	Metal 2 (M2)	
	10.1 minimum width	2λ
	10.2 minimum spacing	3λ
	10.3 maximum current density	1 $mA/\mu m$
11.	Via (VIA)	
	11.1 minimum size	$3\lambda \times 3\lambda$
	11.2 minimum spacing	3λ
	11.3 minimum M1 or M2 overlap of VIA	1λ
	11.4 minimum VIA to CO spacing	2λ
	11.5 minimum PO to VIA spacing	2λ
	11.6 minimum PO overlap of VIA	2λ

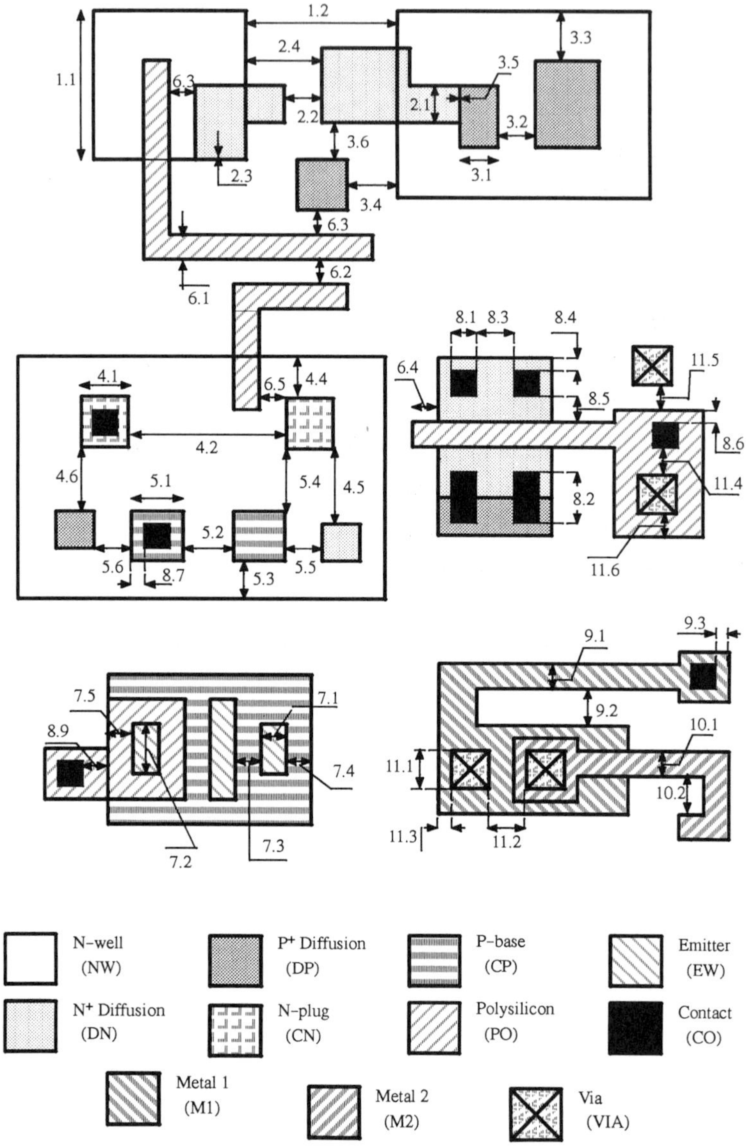

Plate I: Design Rules of Table 2.2.

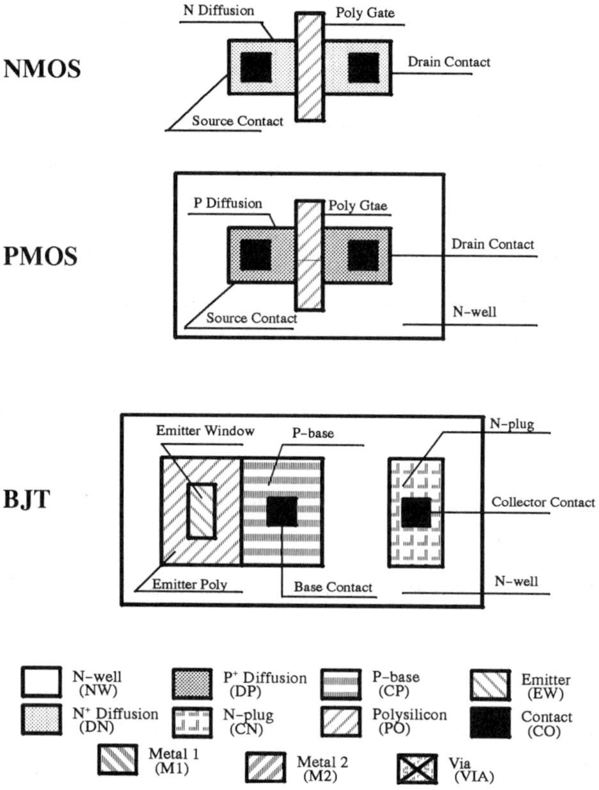

Plate II: Layouts of minimum size PMOS, NMOS and bipolar transistors.

References

1. F.M. Wanlass, and C.T. Sah, "Nanowatt Logic using Filed-Effect MOS Triodes," *International Solid-State Circuits Conference Tech. Dig.*, p. 32 (1963).

2. L.C. Parrillo, R.S. Payne, R.E. Davis, G.W. Reutlinger, and R.L. Field, "Twin-Tub CMOS: A Technology for VLSI Circuits," *International Electron Devices Meeting Tech. Dig.*, pp. 752-755 (1980).

3. K. Ehinger et al., "Narrow BF2 Implanted Bases For 35 GHz/24ps High-Speed Si Bipolar Technology," *International Electron Devices Meeting Tech. Dig.*, pp. 459-462 (1991).

4. T.H. Ning, and D.D. Tang, "Bipolar Trends," *Proc. IEEE*, **Vol. 74, No. 12** pp. 1669-1677 (December 1986).

5. T. Nakamura, T. Miyazaki, S. Takahashi, T. Kure, T. Okabe, and M. Nagata, "Self-Aligned Bipolar Transistor with Polysilicon Sidewall Base Electrode for High Packing Density and High Speed," *IEEE Journal of Solid-State Circuits*, **Vol. SC-17, No. 2** pp. 226-230 (April 1982).

6. T.H. Ning, and R. D. Isaac, "Effect of Emitter Contact on Current Gain of Silicon Bipolar Devices," *IEEE Electron Device Letters*, **ED-27** pp. 2051-2055 (Nov. 1980).

7. A.K. Kapoor and D.J. Roulston, *Polysilicon Emitter Bipolar Transistors*, IEEE Press (1989).

8. M.I. Elmasry, *Digital Bipolar Integrated Circuits*, John Wiley & Sons, New York (1983).

9. E. Kooi, J.G. Van Lierop, and J.A. Appels, "Formation of Silicon Nitride at a Si-SiO_2 Interface during Local Oxidation of Silicon and During Heat Treatment of Oxidized Silicon in NH_3 Gaz," *J. Electrochem. Soc.*, **Vol. 123** p. 1117 (1976).

10. R.D. Rung, H. Momose, and Y. Nagakubo, "Deep-Trench Isolated CMOS Devices," *International Electron Devices Meeting Tech. Dig.*, p. 6 (1982).

11. T. Yamaguchi, S. Morimoto, G. Kawamoto, H.K. Park, and G.C. Eiden, "High-Speed Latch-up Free 0.5 μm-Channel CMOS using Self-Aligned Ti-Si and Deep-Trench Isolation Technologies," *International Electron Devices Meeting Tech. Dig.*, p. 522 (1983).

12. R.D. Rung, "Trench Isolation Prospects for Application in CMOS VLSI," *International Electron Devices Meeting Tech. Dig.*, pp. 574-577 (1984).

13. H. Mikoshiba, T. Homma, and K. Hamano, "A New Trench Isolation Technology as a Replacement for LOCOS," *International Electron Devices Meeting Tech. Dig.*, pp. 578-581 (1984).

14. P. Singer, "Selective Epitaxial Growth Finds New Applications," *Semiconductor International*, p. 15 (January 1988).

15. R.A. Chapman, et al., "An 0.8 μm CMOS Technology for High-Performance Logic Applications," *International Electron Devices Meeting Tech. Dig.*, p. 362 (1987).

16. K.Y. Chiu, R. Fang, J. Lin, and J.L. Moll, "The SWAMI- A Defect Free and Near-Zero Bird's Beak Local Oxidation Technology for VLSI," *Symp. on VLSI Technology Tech. Dig.*, pp. 28-29 (1982).

17. K.Y. Chiu, J.L. Moll, and J. Manoliu, "A Bird's Beak Free Local Oxidation Technology Feasible for VLSI Circuits Fabrication," *IEEE Trans. on Electron Devices*, **Vol. ED-29** pp. 536-540 (1982).

18. J. Hui, P. Vande Voorde and J. Moll, "Scaling Limitations of Submicron Local Oxidation Technology," *International Electron Devices Meeting Tech. Dig.*, pp. 392- (1985).

19. H.B. Pogge, "Trench Isolation Technology," *Bipolar Circuits and Technology Meeting Tech. Dig.*, pp. 18-25 (September 1990).

20. Y. Niitsu, "Latch-up Free CMOS Structure using Shallow Trench Isolation," *International Electron Devices Meeting Tech. Dig.*, p. 509 (December 1985).

21. H. Yamamoto, O. Mizuno, T. Kubota, M. Nakamae, H. Shiraki, and Y. Ikushima, "High-Speed Performance of a Basic ECL Gate with 1.25 Micron Design Rule," *Symp. on VLSI Technology Tech. Dig.*, pp. 38-39 (1981).

22. Y. Tamaki, T. Shiba, N. Honma, S. Mizuo, and A. Hayazaka, "New U-Groove Isolation Technology for High-Speed Bipolar Memory," *Symp. VLSI Technology Tech. Dig.*, pp. 24-25 (1983).

23. D.D. Tang, P.M. Solomon, T.H. Ning, R.D. Isaac, and R.E. Burger, "1.25 μm Deep-Groove-Isolated Self-Aligned Bipolar Circuits," *IEEE Journal of Solid-State Circuits*, **Vol. SC-17** pp. 925-931 (1982).

24. H.C. Lin, J.C. Ho, R.R. Iyer, and K. Kwong, "CMOS-Bipolar Transistor Structure," *IEEE Trans. on Electron Devices*, **Vol. ED-6, No. 11** pp. 945-951 (November 1969).

25. H. Momose, "High Performance 1.0 Micron N-Well CMOS/Bipolar Technology," *Symp. on VLSI Technology Tech. Dig.*, pp. 40-41 (1983).

26. T. Ikeda, A. Watanabe, Y. Nishio, I. Masuda, N. Tamba, M. Okada, and K. Ogiue, "High-Speed BiCMOS Technology with a Buried Twin Well Structure," *IEEE Trans. on Electron Devices*, **Vol. ED-34, No. 6** pp. 1304-1309 (June 1987).

27. H. Momose, K.M. Cham, C.I. Drowley, H.R. Grinolds, and H.S. Fu, "0.5 Micron BiCMOS Technology," *International Electron Devices Meeting Tech. Dig.*, pp. 838-840 (Dec. 1987).

28. A.R. Alvarez, J. Teplik, D.W. Schucker, T. Hulseweh, H.B. Liang, M. Dydyk,and I. Rahim, "Second Generation BiCMOS Gate Array Technology," *Bipolar Circuits and Technology Meeting Tech. Dig.*, pp. 113-117 (1987).

29. B. Bastani, C. Lage, L. Wong, J. Small, R. Lahri, L. Bouknight, T. Bowman, J. Manoliu, and T. Tuntasood, "Advanced 1 Micron BiCMOS Technology for High Speed 256k SRAM's," *Symp. on VLSI Technology Tech. Dig.*, pp. 41-42 (1987).

30. T. Yamaguchi and T.H. Yuzuriha, "Process Integration and Device Performance of a Submicron BiCMOS with 16-GHz f_t Double Poly-Bipolar Devices," *IEEE Trans. on Electron Devices*, **36, No. 5** pp. 890-896 (May 1989).

31. Shin, "Performance Comparison of Driver Configurations and Full-Swing Techniques for BiCMOS Logic Circuits," *IEEE Journal of Solid-State Circuit*, **Vol. 25, No.3** pp. 863-865 (June 1990).

32. S.H.K. Embabi, A. Bellaouar, M.I. Elmasry, and R.A. Hadaway, "New Full-Voltage-Swing BiCMOS Buffers," *IEEE Journal of Solid-State Circuits*, **Vol. SC-26** pp. 150-153 (Feb. 1991).

33. Y. Kobayashi, C. Yamaguchi, Y. Amemiya, and T. Sakai, "High Performance LSI Process Technology: SST CBiCMOS," *International Electron Devices Meeting Tech. Dig.*, pp. 760-763 (December 1988).

34. K. Higashitani, H. Honda, K. Ueda, M. Hatanaka, and S. Nagao, "A Novel CBi-CMOS Technology by DIIP Process," *Symp. on VLSI Technology*, pp. 77-78 (1990).

35. T. Maeda, K. Ishimaru, and H. Momose, "Lower Submicron FCBiMOS (Fully Complementary BiMOS) Process with RTP and MeV Implanted 5GHz Vertical PNP Transistor," *Symp. on VLSI Technology*, pp. 79-80 (1990).

36. W.R. Burger, C. Lage, B. Landau, M. DeLong, and J. Small, "An Advanced 0.8 Micron Complementary BiCMOS Technology for Ultra-High Speed Circuit Performance," *Bipolar Circuits and Technology Meeting Tech. Dig.*, pp. 78-81 (1990).

CHAPTER 3

DEVICE DESIGN CONSIDERATIONS

In this chapter, we discuss device design considerations of MOS and bipolar transistors and the effect of process parameters on device parameters [Sections 3.1. and 3.2.]. The trade offs involved in designing BiCMOS are presented in Section 3.3. The study of Sections 3.1. to 3.3. is based on classical first order analyses, whose details can be found in [1-3]. The topic of device scaling has been addressed in this chapter because of the significant impact of scaling on the device design. A first order scaling theory for MOS and bipolar transistors is derived in Section 3.4.

3.1. Design Considerations for MOSFETs

This section presents some of the device limits and design considerations of an MOS transistor and discusses the significance of some of the parasitic effects. The relationship between process and device parameters are presented. The process parameters include the channel doping concentration at the surface (N_s), the channel doping concentration below the surface (N_{bs}), the effective channel length (L_{eff}), the source/drain junction depth, and the oxide thickness (t_{ox}). The device parameters include the threshold voltage (V_T), the body effect, the breakdown voltage, the short-channel effects, and the parasitic resistances and capacitances. The hot carriers effects and latchup are also discussed.

3.1.1. Threshold Voltage

The threshold voltage of a long-channel MOS transistor is given by[2] [1] :

$$V_{T0} = V_{FB} + 2|\phi_f| + \frac{1}{C_{ox}}\sqrt{2q\varepsilon_s N_s (2|\phi_f|)} + q\frac{D_I}{C_{ox}} \tag{3.1}$$

For a given gate oxide thickness, the threshold voltage is chosen to be low enough to provide high current drive, but high enough to prevent leakage current when the device is in the subthreshold region. The surface channel doping concentration N_s, established by a shallow boron channel implant, controls V_T (threshold adjust).

The threshold voltage of a PMOS transistor (V_{TP}) is in general higher in magnitude ($\approx$ 1.5V) than is desirable (< 1V), because N-polysilicon is used as a gate for both types of MOS devices. To adjust (or lower in magnitude) V_{TP} ($\approx$ 0.8V), a boron implant is performed in the channel region with enough high dose to overcompensate the N-well surface. If the channel region is inverted to P-type due to the boron I/I, the device becomes a "buried channel PMOS".

When the effective channel length is very small, the threshold voltage is reduced and Equation (3.1) breaks down. In a short-channel MOS transistor, a fraction of the bulk depletion charge is created by the source and drain junctions. Hence, a smaller gate voltage is needed to initiate the surface inversion. Fig. 3.1 illustrates the reduction of the threshold voltage as the effective channel length is scaled down[4]. For short-channel, V_T is also sensitive to x_j.

3.1.2. Body Effect

If V_B, the voltage across the reverse biased source-bulk junction increases, the width of the channel-substrate depletion layer also increases, leading to an increase in the threshold voltage by[2]:

---

[1] The derivation of the Equation (3.1) can be found in Section 4.1.1.

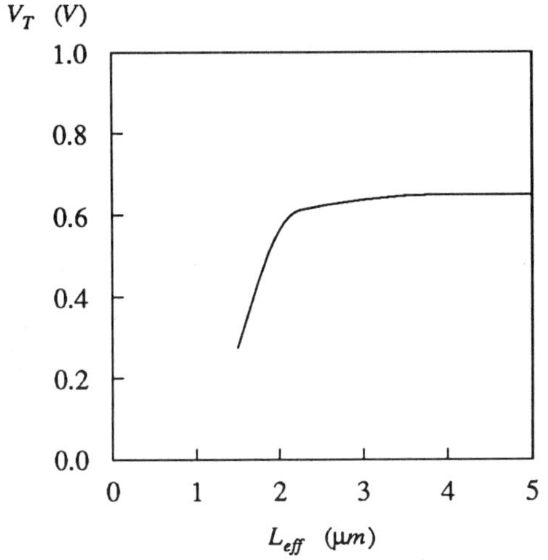

Figure 3.1 Threshold voltage vs. effective channel length for NMOS transistor[4].

$$\Delta V_T(V_B) = \gamma[\sqrt{|V_B|+2|\phi_f|} - \sqrt{2|\phi_f|}] \tag{3.2}$$

where the body effect coefficient γ is given by:

$$\gamma = \frac{\sqrt{2q\varepsilon_s N_{bs}}}{C_{ox}} \tag{3.3}$$

It is clear that the doping concentration of the bulk close to the surface (N_{bs}) and the oxide thickness (t_{ox}) must be low to reduce γ.

3.1.3. Breakdown Voltage

The drain-substrate junction is normally reverse biased. The breakdown of this junction is due to avalanche multiplication. This breakdown voltage is given by[2]:

$$BV_{DS} = \frac{\varepsilon s E_{br}^2}{2qN_{bs}} \tag{3.4}$$

Low N_{bs} is desired to have a reasonable breakdown voltage.

Although the breakdown voltage is an important constraint, yet, the maximum voltage in CMOS circuits is limited by the gate oxide breakdown, which, in turn, is limited by the oxide breakdown voltage ($5\times10^{6} V/cm$).

3.1.4. Short-Channel Effects

As the channel length is scaled down and approaches the junction depth, several short-channel effects arise due to the resulting 2-D potential distribution in the channel. Such effects cause the threshold and the punch-through voltages to decrease, and degrade the subthreshold characteristics of an MOS transistor. If the power supply voltage is kept at 5V while the device dimensions are scaled down, high electric fields are generated in the device causing hot carrier generation and impact ionization. Proper design of the device can prevent it from behaving as a short-channel device, even if its channel seems short. The following subsections discuss briefly some of the important short-channel effects.

(a) Subthreshold

When the MOS device is operating in the weak inversion mode (subthreshold region), the drain-source current (I_{DS}) is dominated by the diffusion current of the minority carrier in the channel. The subthreshold behavior of an MOS transistor is defined by the following two characteristics:

1- The drain current is proportional to W/L.

2- The drain current varies exponentially with V_{GS}, and becomes independent on V_{DS} when $V_{DS} > 3\frac{KT}{q}$.

The subthreshold behavior is also characterized by what is called the subthreshold swing (S), which is the gate voltage swing required to reduce the drain current by one decade. The subthreshold swing is given by[2]:

$$S = 2.3\left(k\frac{T}{q}\right)\left[1+\frac{C_d}{C_{ox}}\right] \quad V/decade \tag{3.5}$$

where C_d is the depletion-layer capacitance of the source/drain junctions. The subthreshold swing determines the gate swing needed to reduce the subthreshold current to an acceptable level. For example, if $S = 90mV/decade$ (which is a typical value for the subthreshold swing) then the gate voltage must be reduced by 0.55V to bring the drain current from $1\mu A$ down to $1pA$, which yields very low off-state (DC) power dissipation. Therefore, small values of S are desired.

The subthreshold swing is increased (i.e. subthreshold current is increased) when the channel length is scaled down. Whereas, reducing the oxide thickness and/or the substrate doping concentration causes S to decrease. An example of subthreshold characteristics for an MOS transistor with $0.8\mu m$ gate length is shown in Fig. 3.2.

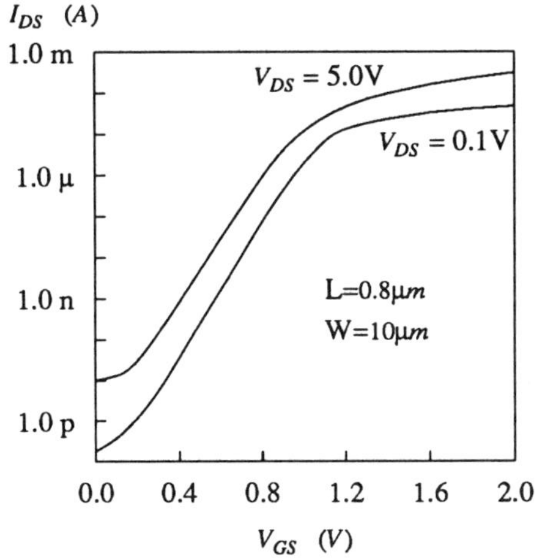

Figure 3.2 I_{DS} vs. V_{GS} **characteristics in the subthreshold region.**

(b) Drain Induced Barrier Lowering and Punchthrough

In the subthreshold mode, the current is a surface channel current. When V_{DS} is high and $V_{GS} < V_T$, an undesirable leakage (punchthrough) current, which flows below the surface of the channel can be observed [Fig. 3.3(a)]. This additional leakage current is due to what is called *Drain Induced Barrier Lowering* (DIBL). Reducing channel length and/or increasing drain-source voltage causes the drain and source depletion regions to move closer. This results in a field penetration from drain to source, which causes the potential barrier at the source junction to be lowered. The effect of the DIBL becomes more pronounced as the junction depth and the oxide thickness increase. Increasing the surface and/or the bulk doping mutes the DIBL effect.

The subthreshold swing, S, can be used as a monitor of the onset of the DIBL phenomenon. Fig. 3.3(b) illustrates the subthreshold swing as a function of electrical channel length for an NMOS transistor. For $L < 0.8\mu m$, S increases rapidly due to punchthrough current. To prevent this effect, an additional boron implant can be placed in the subsurface region as illustrated in Fig. 3.3(a).

3.1.5. Hot Carrier Effects

As the power supply voltage is kept constant (5V) and the dimensions of an MOS device are reduced, the electric field in the channel is increased. The maximum electric field (near the drain) can be given by [5]:

$$E_m = \frac{(V_{DS} - V_{DSsat})}{0.22 t_{ox}^{1/3} x_j^{1/2}} \tag{3.6}$$

The increase of E_m stems from the reduction of t_{ox}, x_j, and V_{DSsat}[2]. When E_m approaches $5 MV/cm$ the carriers near the drain gain high energies and are therefore described to be hot. For an NMOS device, hot electrons can be injected into the gate oxide and accelerated by the gate voltage to generate a gate current [Fig. 3.4]. Some of these injected carriers can be trapped in the oxide. The degradation of the MOS device, due to this problem, is manifested by an increase in the threshold voltage and a decrease in transconductance. The hot electrons can also cause impact ionization near the drain of the NMOS device. The generated holes flow towards the substrate and a substrate current is manifested as shown in Fig. 3.4. Hot carriers problem is more serious for N-channel than for P-channel transistor because; i) the impact ionization rate of electrons is higher than that of holes, and ii) the P^+ drain of the PMOS transistor is deeper [see Equation (3.6)].

---

[2] The value of V_{DSsat} decreases as the channel length is scaled down [see Section 4.1.3.].

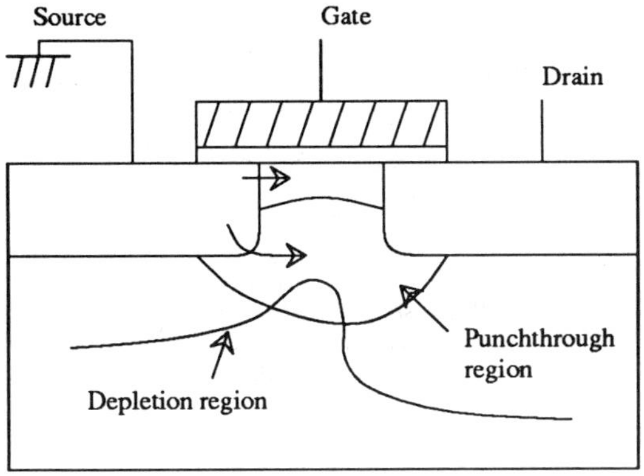

(a)

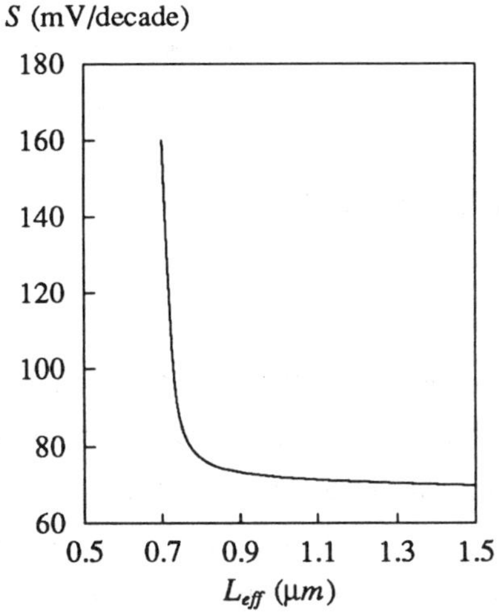

(b)

Figure 3.3 (a) Cross-sectional view of a MOSFET showing the punchthrough region;(b) Subthreshold swing vs. channel length for an NMOS device without a subthreshold punchthrough implant.

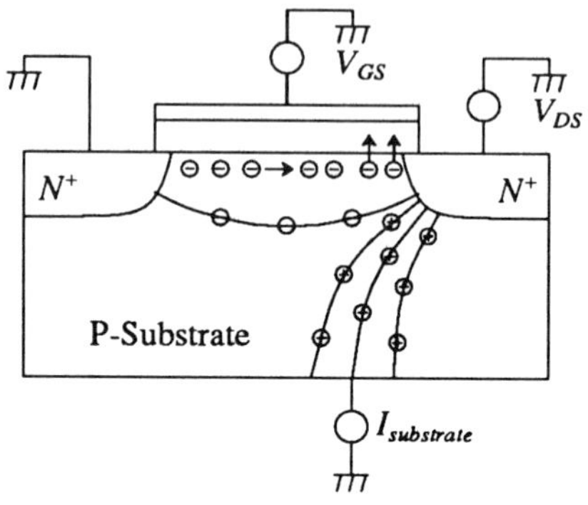

Figure 3.4 Hot electron phenomenon in an N-channel MOS device.

Lightly Doped Drain (LDD) structure described in Section 2.1.2. provides a good protection against the phenomenon of hot carriers. The "lightly doped" extension of the S/D region allows the drain voltage to be dropped across a wider region so that the maximum electric field is reduced.

3.1.6. Latchup in CMOS

A bulk CMOS technology has inherent parasitic bipolar transistors and resistors. These parasitic devices are different from the ones in a BiCMOS technology. For N-well CMOS technology shown in Fig. 3.5(a), the P^+ source, N-well, P-substrate and the N^+ source form a PNPN parasitic SCR (*Semiconductor-Controlled Rectifier*). This SCR may be triggered under certain conditions such as leakage currents, ionization radiation, terminal overvoltage, etc.

The lumped model of the parasitic devices is shown in Fig. 3.5(b). When the well current I_w and the substrate current I_s are sufficiently large such that the I_wR_w drop in the well and I_sR_s drop in the substrate approach $\approx 0.7V$, the base-emitter junctions for NPN and PNP transistors are forward-biased and the bipolar devices are active. The PNP and NPN devices form a loop with a gain equal to the product $\beta_{PNP}\beta_{NPN}$, where β_{PNP} and β_{NPN} are the current gains for the PNP and the NPN transistors. If the loop

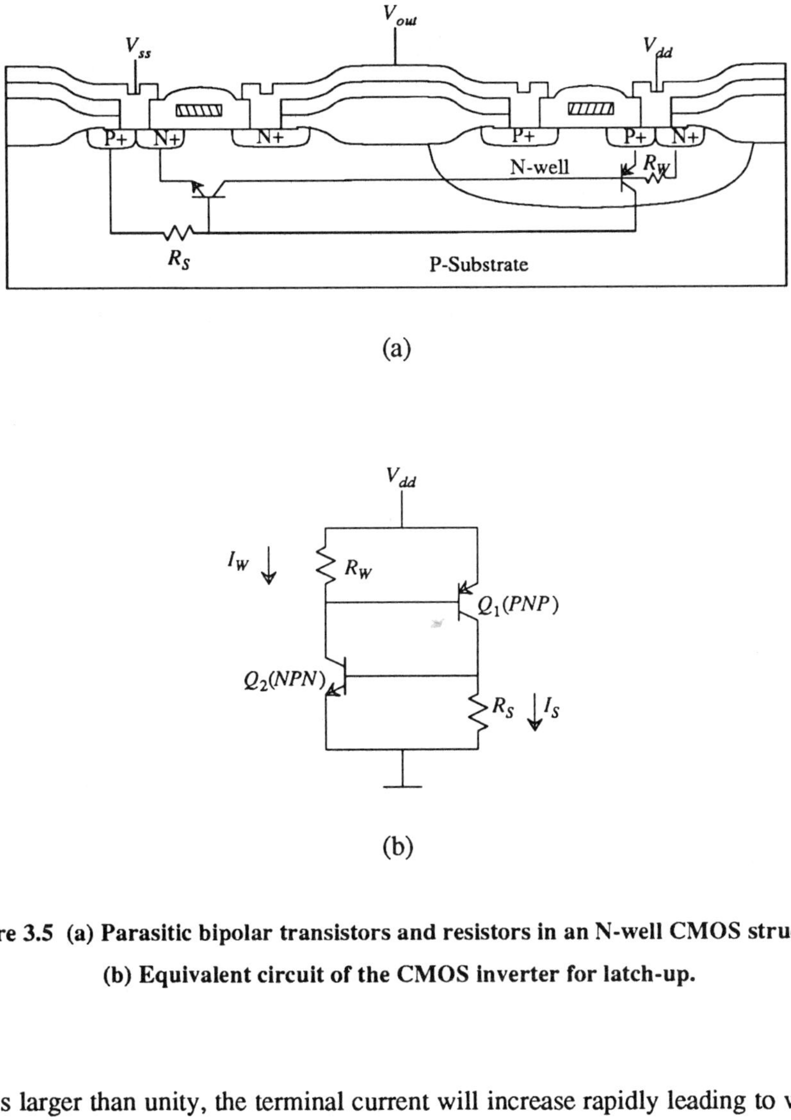

Figure 3.5 (a) Parasitic bipolar transistors and resistors in an N-well CMOS structure, (b) Equivalent circuit of the CMOS inverter for latch-up.

gain is larger than unity, the terminal current will increase rapidly leading to what is commonly known as *latchup*. This phenomenon is destructive and can cause permanent damage to the circuit. For further details about latchup, the reader can consult Troutman's book [6].

Several methods have been suggested to avoid or to prevent latchup. Their objective is to reduce the well and substrate resistances, and/or reduce the current gain of the bipolar transistors. Reducing the resistances may be achieved by using a retrograde-well where the resistivity of the well decreases with the depth. To reduce the current gain, gold doping or neutron irradiation can be used to shorten the minority carrier lifetime. Latchup can be controlled by horizontal design rules and guard rings. Heavily doped guard rings are placed around the MOS transistors, as shown in Fig. 3.6. The guard rings reduce the current gain of parasitic bipolar devices and the resistances. However, this solution would be at the expense of lower packing density.

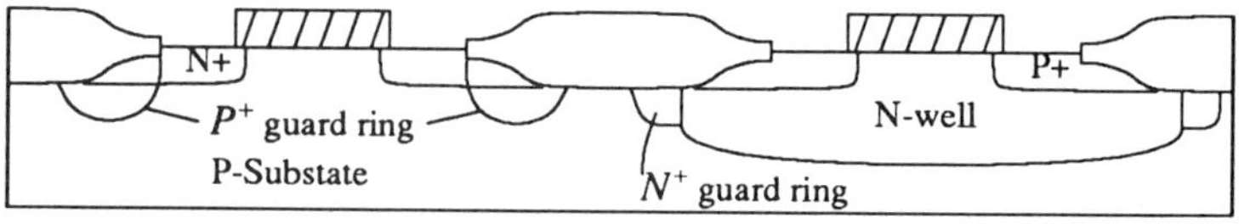

Figure 3.6 Guard rings for latchup prevention.

Finally, the CMOS devices could be isolated by using the trench isolation technique which prevents the latchup by blocking the current flow path.

3.2. Design Considerations for Bipolar Transistors

This section introduces some key relations between process and device parameters, which are to be considered when designing bipolar devices. The process parameters include base Gummel number (Q_B), peak base doping (N_B), base width (W_B), emitter Gummel number (Q_E), emitter junction depth (x_{jE}), collector doping concentration (N_C), and collector width (W_C). The device parameters of interest are, the DC current gain (β_f), cutoff frequency (f_T), collector saturation current (I_{cs}), breakdown voltages (BV_{CB0} and BV_{CE0}), reachthrough voltage (V_{rt}), punchthrough between the emitter and the base (V_{pt}), emitter resistance (R_E), base resistance (R_B), collector resistance (R_C), and parasitic capacitances of the base-emitter and base-collector junctions (C_E and C_C). Fig. 3.7 illustrates the parasitic resistances and capacitances in polysilicon emitter self-aligned bipolar transistors. Some effects such as, high level injection (HLI) and hot carrier effect (HCE) are also studied in this section.

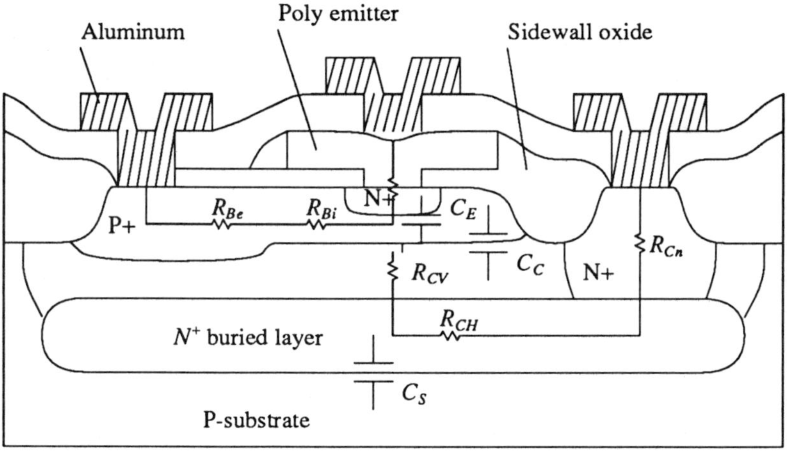

Figure 3.7 Parasitical components in a bipolar transistor.

3.2.1. Current Gain

The current gain is given by[3] :

$$\beta_f = \frac{D_{nB} N_E W_E}{D_{pE} Q_B} \tag{3.7}$$

This equation indicates that the trends for high gain are:

- High emitter doping concentration

[3] This equation is simplified for transparent emitter (i.e, $x_{jE} \ll L_{pE}$). The polysilicon emitter and band gap narrowing are not taken into account. Their effect is shown in Equation (3.31) [Section 3.4.2.].

- Wide emitter
- Narrow base
- Low base Gummel number

Modern advanced bipolar transistors are near the physical limits in the doping level of the emitter; i.e. solid solubility limit. Hence the only parameters that control β_f are, the base doping, the base width, and the use of polysilicon emitter. Techniques such as rapid thermal annealing (RTA) permits the reduction of W_B. Current gains over 50 are obtained for advanced bipolar devices.

Equation (3.7) indicates that the collector current has no effect on the current gain; unfortunately, this is not true. For low currents, the base current is dominated by the recombination in the emitter-base depletion layer which implies that the ratio of the base to emitter current is increased or, in other words, the current gain is reduced. For high currents, the gain falls off under high level injection (HLI) in the base, or, due to base pushout (Kirk effect)[4]. Fig. 3.8 shows the effect of the collector current on the current gain.

The HLI in the base is triggered when the emitter-base junction of the bipolar transistor is highly forward-biased such that the density of the injected minority carriers in the base approaches the majority concentration. The injected carriers effectively increase the base charge (i.e. Q_B), I which in turn causes the gain to roll off. The collector current that marks the onset of high level injection is given by[3]:

$$I_{HLI} = \frac{qD_{nB}N_B}{W_B}A_E \tag{3.8}$$

For $I_C > I_{HLI}$, the gain may be estimated by[3]:

$$\beta_f = \beta_{f0}\left(\frac{I_{HLI}}{I_C}\right) \tag{3.9}$$

[4] Further discussion on the high current effects on the characteristics of the BJT can be found in Section 4.2.

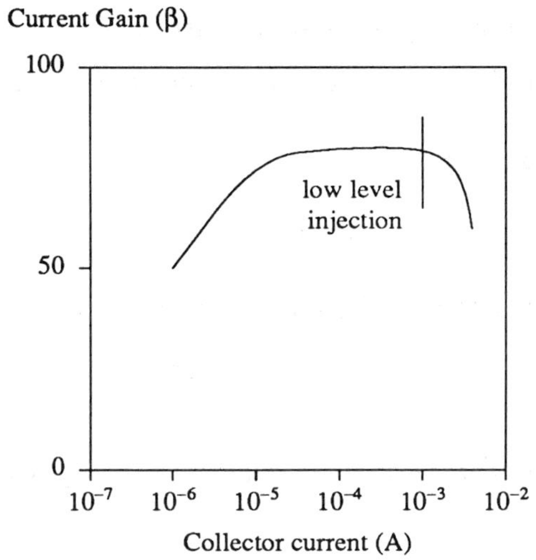

Figure 3.8 Current gain versus the collector current.

When the collector doping concentration is low, the Kirk effect (also referred to as base-pushout) becomes the main reason for the gain degradation. The base-pushout is triggered when the collector current increases to the extent that the concentration of the carriers, which originate from the emitter and reach the base-collector space charge layer, exceeds the collector doping concentration. Under such conditions, the space charge layer moves from the base-collector junction to the NN^+ interface (around the edge of the N^+ buried layer). As a result the effective base width increases. The collector current at which the base-pushout is triggered is given by[3]:

$$I_K = qv_{\max}N_C\left(1+\frac{2\varepsilon_s V_{BC}}{qN_C W_C^2}\right)A_E \tag{3.10}$$

In modern bipolar transistors, with a lightly doped collector, the Kirk effect dominates over the HLI mechanism. To increase the current I_K the collector doping concentration must be increased and/or its width, W_C, reduced.

3.2.2. Cutoff Frequency

The cutoff frequency f_T, is defined as the frequency at which the common emitter current gain of the transistor equals unity. For a bipolar transistor, $f_T = \frac{1}{2\pi\tau_t}$, where the total delay τ_t is given by:

$$\tau_t = \tau_E + \tau_f + \tau_d + \tau_C \tag{3.11}$$

where the emitter-base depletion layer charging time is given by:

$$\tau_E = \frac{kT}{qI_C}(C_C + C_E) \tag{3.12}$$

The base forward transit time is expressed by the following simple equation:

$$\tau_f = \frac{W_B^2}{2D_{nB}} \tag{3.13}$$

The base-collector depletion layer transit time τ_d is:

$$\tau_d = \frac{W_{scl}}{2v_{max}} \tag{3.14}$$

The final component τ_C is given by:

$$\tau_C = R_C C_C \tag{3.15}$$

Obviously all the components of τ_t must be minimized in order to maximize f_T. To increase f_T: (i) the base width should be reduced, (ii) N_C should be kept low enough to reduce C_C and (iii) the collector width (i.e. the epitaxial layer) should be decreased to reduce R_C [see Fig. 3.9]. Equation (3.12) shows that f_T is current-dependent. It increases with the increase of the current, but when the current approaches I_K, f_T begins to fall off (due to the Kirk effect). Fig. 3.10 shows the effect of collector current on f_T.

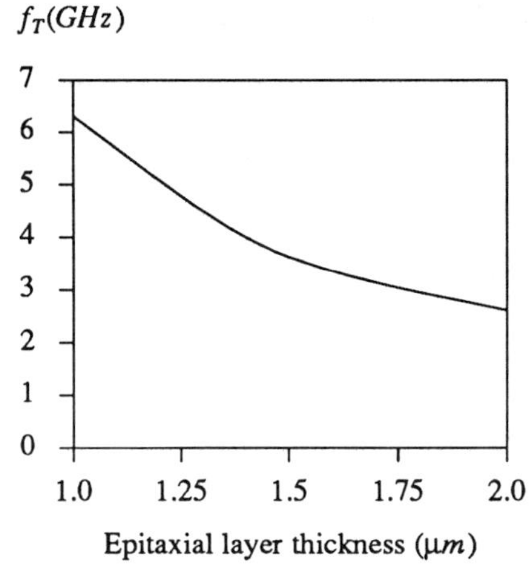

Figure 3.9 The dependence of f_T on the epitaxial layer thickness.

3.2.3. Breakdown Voltages

The breakdown voltage of the base-collector junction BV_{CB0} can be approximated by the following expression[3]:

$$BV_{CB0}\,(plane) = \frac{\varepsilon_s}{2q}\frac{E_{br}^2}{N_C} \qquad (3.16)$$

where the critical field E_{br} is approximately $2\text{x}10^5\ V/cm$. While increasing the collector doping and reducing the collector width have a positive effect on some parameters, they degrade BV_{BC0}.

The emitter-collector breakdown voltage (base is open) is less than BV_{CB0} and may be approximated by[3]:

$$BV_{CE0} = \frac{BV_{CB0}}{(\beta)^{\frac{1}{n}}} \qquad (3.17)$$

typically, n is in the range between 3 and 6. The breakdown voltage BV_{CE0} sets the minimum epitaxial layer thickness. The lower limit of BV_{CE0} is determined by the

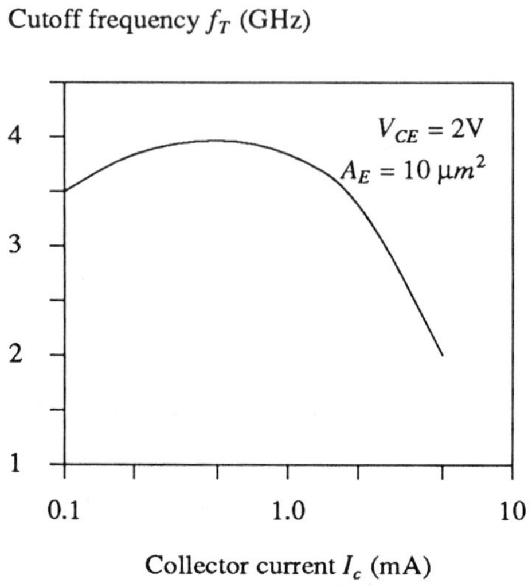

Figure 3.10 Cutoff frequency versus collector current.

supply voltage. For $5V$ supply voltage BV_{CE0} should be around $9V$.

3.2.4. Reachthrough Voltage

The reachthrough phenomenon occurs when the edge of the depletion layer in the collector side reaches the buried layer edge. In the event of reachthrough, avalanche breakdown can be easily triggered. The collector-base voltage required to cause the edge of the depletion layer to reach through to the buried layer is called the reachthrough voltage and is given by[3]:

$V_{rT} = 9.12$, $N_c = 3e15$, $W_c = 2\mu$
2.2, $N_c = 3e15$, $W_c = 1\mu$

$$V_{rt} = \frac{qN_C}{2\varepsilon_s} W_C^2 \tag{3.18}$$

Equation (3.18) states that the reachthrough voltage is limited by the width and the doping level of the collector. Therefore, the thickness and the doping level of the epitaxial layer have to be chosen such that the reachthrough voltage is greater than the supply voltage.

3.2.5. Base-Emitter Punchthrough

Punchthrough occurs when the edges of the emitter-base and collector-base depletion layers in the base touch each other. The base-collector voltage which causes the collector-base depletion layer to extend to the emitter-base depletion layer is called the punchthrough voltage and is given by[2] [5] :

$N_B = 5e17,\ W_B = .3\mu\ \ V_{pt} = 34V$

$$V_{pt} = \frac{q}{2\varepsilon_s} N_B W_B^2 \tag{3.19}$$

The width and doping level of the base have to be large enough to guarantee that the punchthrough voltage is greater than the supply voltage. Typically, the peak of the base doping concentration is increased while the base is kept as narrow as possible to satisfy the gain and cutoff frequency specifications.

Fig. 3.11 shows the effect of the base doping concentration and width on the punchthrough voltage with avalanche limits[7].

3.2.6. Parasitical Resistances

(a) Emitter Resistance

The emitter resistance R_E of conventional bipolar transistors (metal emitter contact) does not appear to cause significant voltage drop. However, when the area of the emitter is decreased, R_E increases by the same factor and it may become a limiting factor to the performance of ECL and BiCMOS circuits. The resistance of a polysilicon emitter contact could also limit the device performance. The use of advanced techniques such as RTA processing for polysilicon emitter reduces the emitter resistance significantly [8].

(b) Base Resistance

The base resistance is a key parameter for some circuits especially for current mode structures. The base resistance is composed of an intrinsic resistance (R_{Bi}) and an extrinsic resistance (R_{Be}) as shown in Fig. 3.7 and can hence be expressed as:

---

[5] This definition is based on the assumption that the bipolar device operates in the forward active region.

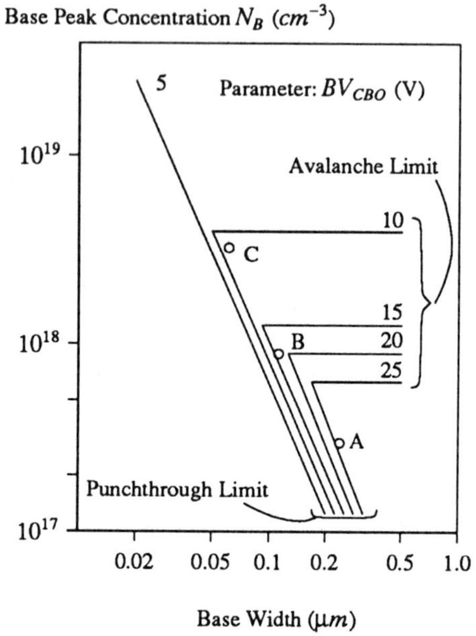

Figure 3.11 Punchthrough voltage as a function of the base profile with avalanche limit[7].

$$R_B = R_{Bi} + R_{Be} \tag{3.20}$$

The intrinsic base resistance is the resistance of the active base under the emitter. Its sheet resistance is given by:

$$R_{sB} = \frac{1}{\int_0^{W_B} q\mu_{pB} N_B(x)dx} \approx \frac{1}{q\mu_{pB} N_B W_B} \tag{3.21}$$

The approximate expression in Equation (3.21) is based on the assumption that the base doping is constant and equal to N_B. The intrinsic base resistance can hence be approximated by[3]:

$$R_{Bi} = \frac{R_{Bis}}{G} \tag{3.22}$$

where the G is the base resistance reduction factor, which accounts for the distributed nature of the base current and is dependent on the layout topology and base contact locations. R_{Bis} is given by:

$$R_{Bis} = R_{sB}\frac{L}{B} \approx \frac{L}{q\mu_{pB}N_B W_B B} \tag{3.23}$$

L and B are the emitter length and width [6] respectively (i.e. $A_E = L \times B$). The base resistance can be improved by widening the base or by increasing its doping. Unfortunately, the enhancement of the base resistance comes at the expense of the current gain[9] and the breakdown voltage. The extrinsic base resistance depends strongly on the type of bipolar technology. For example, using self-alignment helps to improve the extrinsic base resistance [Fig. 2.5].

(c) Collector Resistance

The collector resistance has a significant impact on the performance of some types of BiCMOS logic circuits [Section 7.3.2.3.]. It is therefore important to identify the process parameters that determine the value of the collector resistance. Practically all modern bipolar transistors use a buried layer structure. The collector resistance is hence dominated by the vertical resistance due to the N-collector under the base which is given by[3]:

$$R_C \approx \frac{W_C}{q\mu_n N_C A_E} \tag{3.24}$$

This equation shows that the increase of the collector doping concentration and/or the decrease of W_C improve R_C.

In Section 2.5, we have seen that the low-cost BiCMOS process offers bipolar transistors without a buried layer. In this case, the collector resistance would be too large. Equation (3.24) has to be modified to account for the horizontal resistance in

---

[6] B is to be distinguished from W_E. W_E is actually the emitter thickness but is commonly known as the emitter width. B is the emitter width from the layout point of view.

the low-doped collector region.

3.2.7. Junction Capacitances

Maintaining small junction capacitances is important for high performance circuits. The junction capacitances of a BJT can be lumped into: (i) an emitter-base capacitance (C_E), (ii) a collector-base capacitance (C_C), and (iii) a collector-substrate capacitance (C_s) [see Fig. 3.7]. Assuming zero-bias conditions, these capacitances can be approximated by:

$$C_E = A_E \left(q\varepsilon_s \frac{N_B}{2\phi_j} \right)^{1/2} \tag{3.25}$$

$$C_C = A_C \left(q\varepsilon_s \frac{N_C}{2\phi_j} \right)^{1/2} \tag{3.26}$$

$$C_S = A_S \left(q\varepsilon_s \frac{N_{substrate}}{2\phi_j} \right)^{1/2} \tag{3.27}$$

A_E, A_C and A_S represent the emitter-base, collector-base and collector-substrate areas respectively. These capacitances are reduced by using advanced isolation and self-alignment techniques. C_C can be reduced by lowering the collector doping concentration which is traded off against Kirk current (I_K) and BV_{CE0}.

3.3. BiCMOS Device Design Considerations

In a BiCMOS technology, it is important to design high performance CMOS and bipolar transistors for minimum cost. The device design complexity for BiCMOS depends on which bipolar structure is integrated with CMOS, or more precisely, the common process steps which can be shared by the CMOS and bipolar devices. In the following discussion it is assumed that the N-well of the PMOS acts as the N-collector of the NPN, and that an epitaxial layer is used.

Table 3.1 shows the effect of the N-well doping concentration and epitaxial layer thickness on the important BiCMOS device parameters. The arrows of the first column (called "Impact") indicate the desired trend for each parameter. For example, the fact that the body effect should be minimized is indicated by the downward arrow. The arrows in the last two columns show the way the epitaxial layer thickness and doping should change to improve each parameter. It is clear that many critical trade offs are involved in choosing the profile of the N-well and the thickness of the epitaxial layer. For example, thicker, lightly doped epi layer is preferred for CMOS while a thinner and more doped epitaxial layer is desired for the bipolar transistor.

Table 3.1 Effects of N-well doping and epitaxial layer thickness on BiCMOS device parameters.

Parameter	Impact	Epi-layer thickness	N-well doping Concentration
MOS			
Body effect	↓	↑	↓
Subthreshold swing	↓	↑	↓
DIBL	↓	↓	↑
C_j	↓	↑	↓
HCE	↓	↑	↓
BV_{DS}	↑	↑	↓
Latchup	↓	↓	↑
Bipolar			
f_T	↑	↓	↑
β_f	↑	↓	↑
I_K	↑	↓	↑
R_C	↓	↓	↑
C_C	↓	↑	↓
BV_{CBO}	↑	↑	↓

Table 3.2 illustrates the limitations imposed by different process parameters on the parameters of the MOS and the bipolar transistors.

Table 3.2 Limits of process parameters vs. device parameters constraints

Parameter	Lower Bound	Upper Bound
Gate oxide thickness	- HCE and oxide BV	- BE, S, SCE, current drive
Surface channel doping	- surface SCE	- S, mobility
S/D junction depth	- HCE, sheet resistance	- SCE, overlap capacitance
Base doping	* punchthrough, R_B	* current gain
Base width	* punchthrough	* f_T, current gain, R_B
N-well (N collector) doping	- SCE, punchthrough * Kirk effect, R_C	- BE, C_j, BV * BV, C_C
Epi thickness	- BE, C_j * BV, C_C	* R_C, Kirk effect

- CMOS constraints * Bipolar constraints

BE: body effect, BV: Breakdown voltage, SCE: short-channel effects, HCE: hot carriers effects.

Using these limitations, the permissible ranges for the different profiles can be defined. For example, the breakdown voltage and the junction capacitance of the PMOS set an upper bound for the N-well doping. While the Kirk current density and the collector resistance add a lower bound. The epitaxial layer is also bounded by the breakdown voltage and the parasitic capacitances from one side and by the Kirk effect and the collector resistance form the other side. The device parameters are functions of the circuit requirements for optimal performance. These device considerations set the process profiles. The design of the NMOS device should be relatively easy because it has nothing in common with the NPN and PMOS except the epitaxial layer.

One of the important design issues for BiCMOS devices is the latchup. The parasitic PNP and NPN transistors are inherent in the CMOS component of the BiCMOS structure [Fig. 3.12]. However, BiCMOS structures are less susceptible to latchup for the following reasons. The presence of buried layers under the MOS devices reduce the parasitic resistances significantly. Moreover, the parasitic bipolar transistors have reduced current gains. As an example, the triggering current in BiCMOS technology with N^+/P^+ buried layers is 5-10 times larger than that of conventional CMOS [4]. However, in BiCMOS technology there is an extra current source established by the active bipolar transistor. When the active vertical NPN is switched into saturation, the base-collector junction becomes forward-biased and current is injected into the substrate [10]. This injected current is decreased by the presence of a buried layer in the collector. Consequently, BiCMOS technology with inherent buried layer offers better latchup hardness over the conventional CMOS technology.

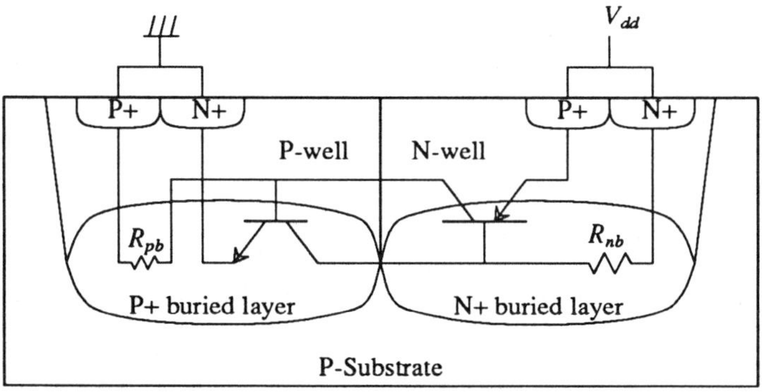

Figure 3.12 Latch-up circuit for the twin-tub CMOS part of a BiCMOS process.

3.4. BiCMOS Device Scaling

Scaling device feature size has been used to increase packing density and speed. Because CMOS was the main VLSI technology, scaling of MOS transistors has received more attention than scaling of bipolar transistors. With the emergence of BiCMOS as a strong candidate for many VLSI applications the interest in understanding the effects of scaling on bipolar and MOS devices is growing. In this section we present a first-order scaling theory for bipolar and MOS transistors.

3.4.1. MOS Device Scaling

MOSFET scaling can follow three theories:

1- Constant Electric Field (CE) scaling [11].

2- Constant Voltage (CV) scaling [12].

3- Quasi-Constant Voltage (QCV) scaling [12].

In the CE scheme all horizontal and vertical dimensions and voltages scale linearly with the same factor. In the QV scheme, the dimensions are scaled, while the voltages are kept constant. While the constant electric field scaling is natural from the device physics point of view, the constant voltage scaling is more practical from the systems standpoint. Changing the supply voltage every technology generation (when the feature sizes are scaled) is too expensive because multiple power supply generators will be required for each PC board. Therefore, the $5V$ remained to be the standard supply voltage. However, as the channel length scales below about $0.6\mu m$ the $5V$ supply voltage must be reduced for reliability reasons (e.g. Hot carrier effects, breakdown, etc). The quasi-constant voltage scaling is an intermediary scheme between the CE and CV views. The scaling factors of the horizontal dimensions, the vertical dimensions and the voltage are denoted by k_h, k_v and k_u respectively. Table 3.3 summarizes the scaling of the important device parameters according to the three theories as a function of the horizontal scaling factor (k_h). Note that in the QCV scheme, the dimensions scale more aggressively than the voltage ($k_u = k_h^{0.5}$).

Table 3.3 Scaling laws of the MOS device[12].

Parameter	Constant Field	Constant Voltage	Quasi-Constant Voltage
Dimensions	k_h	k_h	k_h
Gate oxide thickness	k_h	$k_h^{0.5}$	k_h
Doping	k_h	k_h	k_h
Voltage	k_h	1	$k_h^{0.5}$

A more general set of scaling rules, in which the dimensions scale independently of voltage, was presented by Baccarani et al[13]. [see Table 3.4]. The electric field scales by $k_v k_h^{-1}$. From Poisson's equation, and in order to maintain the shape of the electric field in the channel, the doping concentration must increase by $k_v k_h^{-2}$. For the drain current, two cases are considered. In the first case (column one), it is assumed that the horizontal electric field is smaller than the vertical field, which is the case for long-channel transistors. The saturation drain current can be obtained from the classical (quadratic law) Shockley model. In the second case (second column), the drain current is modeled by the linear expression [see Equation (4.98)[7]], which is valid for short-channel devices. Note that the performance of the device improves in both cases. The last entry in Table 3.4 shows how the gate delay scales assuming that the gate output capacitance is proportional to k_h. The delay of the short-channel transistors (second column) does not scale as well as it does in the case of long-channel transistors if a CV or QCV scaling scheme is adopted. This leads to the important conclusion that the scaling of MOSFETs in the submicronic range will not improve as it used to in the 1980's when they were scaled down to 1μm.

[7] The derivation of Equation (4.98) can be found in Section 4.1.3.

Table 3.4 Generalized Scaling laws of the MOS device[13].

Parameter	Expression	Scaling Factor
Dimensions	W, L, t_{ox}, x_j	k_h
Voltages	V_{dd}, Φ_j	k_u
Electric Field	E	$k_u k_h^{-1}$
Doping Concentration	N	$k_u k_h^{-2}$
Capacitances	AC_{ox}, AC_j	k_h
Current (long-channel)	$\mu C_{ox}\frac{W}{L}(V_{GS}-V_T)V_{DS}$	$k_h^{-1}k_u^2$
Current (short-channel)	$Kv_{\max}C_{ox}W(V_{GS}-V_T)$	k_u
Power	$I_{DS}V_{dd}$	$k_h^{-1}k_u^3$ k_u^2
Gate Delay	$C_g\frac{V_{dd}}{I_{DS}}$	$k_h^2 k_u^{-1}$ k_h

In Table 3.4, we have assumed, for simplicity, that the factor K of the saturation current for short-channel transistors is not affected by scaling. This assumption is not accurate. Equation (4.97) describes the relation between the factor K and the supply voltage and some device parameters. Because of the complexity of this relation, it is difficult to obtain the scaling law for the saturation current for short-channel MOSFETs analytically. However, one can develop an empirical scaling formula from the experimental data and simulation using the BSIM model[14]. For $V_{GS}=V_{dd}$, the saturation current was found to scale as $k_h^{-0.4}k_v^{1.3}$. For $V_{GS}=V_{dd}-V_{BEon}$ (important for BiCMOS totem pole gate), the saturation current is proportional to $k_h^{-0.4}k_u^{2.2}$. These empirical relations are valid for L_{eff} between $0.3\mu m$ and $1.0\mu m$, t_{ox} between 50 and $250A^o$, x_j less than $0.3\mu m$ and $V_{GS}-V_T$ greater than $V_{dd}/2$. Using these new scaling formulas of the saturation current lead to the same conclusions regarding the comparison of the delays of the short- and long- channel MOSFETs.

3.4.2. Bipolar Device Scaling

Scaling of bipolar devices have received less attention than that of MOSFETs. The scaling properties and limitations of bipolar circuits have been studied in[7, 15-17]. Two different schemes have been considered[18]; the constant collector-current density (CJC) scaling, and the constant collector current (CIC) scaling. It was shown that the CIC scaling, where the leakage current resulting from the stress of emitter-base junction has less impact on the current gain, is preferable[18].

We will consider both scaling schemes; the constant-collector-current (CIC) and the constant-collector-current density (CJC).

(i) **CIC Scaling**

The knee current must be kept constant. Because the gain roll-off is most probably due to base pushout (Kirk effect), the knee current can be given by :

$$I_K = qv_{\max}A_E N_C \tag{3.28}$$

The collector (epitaxial layer) doping is hence proportional to k_h^{-2}. From Equation (3.19), the base doping is related to the punchthrough voltage by the following expression:

$$N_B = \frac{2\varepsilon_s}{q}\frac{V_{pt}}{W_B^2} \tag{3.29}$$

scales by $k_v^{-2}k_u$ if it is assumed that the base punchthrough voltage scales as the supply voltage. The same assumption would also be natural for the collector reachthrough voltage (V_{rt}). Therefore, the collector depth[8] , which can be expressed as[3]:

[8] The collector thickness is strongly related to the epitaxial layer thickness. Hence, Equation (3.30) can be used to estimate the epitaxial thickness after adding the base and emitter thicknesses to W_C.

$$W_C = \left(\frac{2\varepsilon_s}{q}\frac{V_{rt}}{N_C}\right)^{1/2} \tag{3.30}$$

scales by $k_h k_u^{0.5}$.

As a first order approximation, the current gain, for transparent polysilicon emitter (i.e, $W_{jE} << L_p$) can be expressed as[19, 20]:

$$\beta_{f0} \approx \frac{D_{nB} N_E C_N(N_E)}{D_{pE} N_B C_N(N_B)} \frac{1}{W_B} f(\mu_1, W_1) \tag{3.31}$$

$f(\mu_1, W_1)$, which is called the polysilicon term, is given by:

$$f(\mu_1, W_1) = \frac{D_{pE} L_{p1}}{D_{p1}} \tanh(\frac{W_1}{L_{p1}}) \tag{3.32}$$

where W_1 is the thickness of the polysilicon layer and L_{p1} is the hole diffusion length in the polysilicon. $C_N(N_E)$ and $C_N(N_B)$ [Equation (3.31)] are correction factors of the emitter and base dopings respectively, which account for the effect of the bandgap narrowing. $C_N(N_E)$ and D_{pE} are not subject to scaling. The correction factor due bandgap narrowing in the base is given by[21]:

$$C_N(N_B) = e^{-\Delta E_g(N_B))/kT} \tag{3.33}$$

where

$$\Delta E_g(N_B) = \alpha ln((\frac{N_B}{N_0})^5 + 1.0) \tag{3.34}$$

where

$$\alpha = 3.74 \times 10^{-3} \ (e.V) \tag{3.35}$$

$$N_0 = 1 \times 10^{17} \ cm^{-3} \tag{3.36}$$

Assuming that $\frac{N_B}{N_0} \gg 1$ and using Equations (3.29), (3.33) and (3.34) it can be shown that $C_N(N_B)$ scales as $k_v^{1.4} k_u^{-0.7}$.

The electron mobility in the base is related to the base doping by the expression derived by Arora et al[22]. However, this expression is not useful for our scaling study due to its complexity. Instead, we used the following simple relation :

$$\mu_n \approx \mu_{0n}\left(\frac{N_B}{N_{ref}}\right)^{-0.45} \tag{3.37}$$

Where N_{ref} is a constant and is equal to $3\times10^{16} cm^{-3}$. Equation (3.37) agrees well with Arora's expression for $10^{17} < N_B < 10^{19}\ cm^{-3}$, which is the typical range for the base doping concentration. Using Equation (3.37) we find that the mobility μ_n scales as $k_v^{0.9} k_u^{-0.45}$.

The polysilicon term, $f(\mu_1, W_1)$, is mainly a function of the properties of the polysilicon emitter contact[23], such as Auger recombination lifetime, the mobility of the minority carriers and the thickness of the polysilicon layer. It is not affected by any of the scaling factors k_u, k_v, k_h, etc. We define a new scaling factor (k_f) which accounts for any possible variation of $f(\mu_1, W_1)$ due to the change of the properties of the polysilicon. Like the rest of the scaling factors, k_f is assumed to be less than 1. It is defined such that $f(\mu_1, W_1)$ increases when k_f decreases, hence we can say that $f(\mu_1, W_1)$ is proportional to k_f^{-1}.

Accounting for the scaling of N_B, W_B, $C_N(N_B)$, D_{nB} and $f(\mu_1, W_1)$ we find that the current gain at low-level injection scales as $k_v^{0.5} k_f^{-1} k_m^{-0.75}$. In Table 3.5 we take the factor $k_\beta = k_v^{-0.5} k_f$.

The forward base transit time, which is given by:

$$\tau_f = \frac{W_B^2}{2D_{nB}} \tag{3.38}$$

scales to a first order by $k_v^{1.1} k_u^{0.45}$.

The base resistance is typically dominated by the intrinsic component. Therefore, we can assume that the base resistance is given by Equation (3.22), which is rewritten here for convenience.

$$R_B = \frac{1}{G}\left(\frac{L}{q\mu_p N_B W_B B}\right) \tag{3.39}$$

The base resistance reduction factor, G, is not scalable. We can, therefore, conclude that the base resistance is proportional to $kv^{0.3}ku^{-0.65}$.

The collector resistance is dominated by the vertical resistance in the bulk of the collector ($W_C/q\mu_n N_C A_E$) [Equation (3.24)]. This is only true for bipolar transistors with buried layers and deep collector contacts, which is the case for the medium to high performance BiCMOS technologies. The collector resistance is hence expected to scale by $k_h k_v^{-0.9} k_u^{0.95}$.

The emitter and collector depletion capacitances can be approximated by Equations (3.25) and (3.26). Using the scaling factors of N_C and N_B, which were derived from Equations (3.28) and (3.29), we can show that the collector-base and the emitter-base capacitances scale by k_h and $k_h^2 k_v^{-1} k_u^{0.5}$ respectively.

(ii) **CJC Scaling**

In the case of constant-collector-current density, the current is scaled by k_h^2. If we start by the knee current, which is also scaled by k_h^2, it can be shown that N_{epi} is to remain constant. The epi-layer thickness has to be scaled by $k_u^{0.5}$ [see Equation (3.25)]. The collector resistance would be proportional to $k_h^{-2} k_v^{-0.5} k_u^{0.95}$. Finally, the collector-base capacitance is expected to scale by k_h^2. The rest of the parameters are modified as in the CIC scaling scheme.

Table 3.5 summarizes the scaling laws of the bipolar transistor for both CIC and CJC schemes.

Table 3.5 Scaling laws of the Bipolar transistor

Parameter	Scaling Factor (CIC)	Scaling Factor (CJC)
Horizontal dimensions	k_h	k_h
Vertical dimensions	k_v	k_v
Collector-current density J_C	k_h^{-2}	1
Collector-current I_C	1	k_h^2
Voltage V_{BC}	k_u	k_u
Base doping N_B	$k_v^{-2}k_m$	$k_v^{-2}k_m$
Epi-layer doping N_{epi}	k_h^{-2}	1
Current gain β_{f0}	$k_\beta^{-1}k_m^{-0.75}$	$k_\beta^{-1}k_m^{-0.75}$
Knee current I_K	1	k_h^2
Intrinsic base resistance R_{Bi0}	$k_v^{0.3}k_m^{-0.65}$	$k_v^{0.3}k_m^{-0.65}$
Collector resistance R_C	$k_h k_v^{-0.9}k_u^{0.95}$	$k_h^{-2}k_v^{-0.9}k_u^{0.95}$
Collector depletion capacitance C_C	k_h	k_h^2
Emitter depletion capacitance C_E	$k_h^2k_v^{-1}k_m^{0.5}$	$k_h^2k_v^{-1}k_m^{0.5}$
Substrate depletion capacitance C_S	k_h	k_h
Base transit time τ_f	$k_v^{1.1}k_m^{0.45}$	$k_v^{1.1}k_m^{0.45}$

3.5. Chapter Summary

This chapter addresses important device aspects that BiCMOS circuit designers need to comprehend for high performance designs. The relation between key device parameters and process parameters has been reviewed. The trade offs involved in BiCMOS device design are also summarized. It has been shown that the N-well, that is shared by the MOS and the bipolar transistors, is the most difficult step in the process. Finally, a scaling theory for MOS and bipolar transistors is derived.

References

1. A.S. Grove, *Physics and Technology of Semiconductor Devices*, John Wiley & Sons, New York (1967).
2. S.M. Sze, *Physics of Semiconductor Devices*, John Wiley & Sons (1981).
3. D.J. Roulston, *Bipolar Semiconductor Devices*, McGraw-Hill Publishing Company (1990).
4. T. Ikeda, A. Watanabe, Y. Nishio, I. Masuda, N. Tamba, M. Okada, and K. Ogiue, ''High-Speed BiCMOS Technology with a Buried Twin Well Structure,'' *IEEE Trans. on Electron Devices*, **Vol. ED-34, No. 6** pp. 1304-1309 (June 1987).
5. J.Y. Chen, ''CMOS- - The Emerging VLSI Technology,'' *IEEE Circuits and Devices*, **Vol. 2** p. 16 (March 1986).
6. R.R. Troutman, *Latchup in CMOS Technology*, Kluwer Academic Pub. (1986).
7. H. Hanaoka, and A. Anzai, ''Perspective of Scaled Bipolar Devices,'' *International Electron Devices Meeting Tech. Dig.*, pp. 512-515 (December 1981).
8. J.E. Brighton, D.P. Verret, T.T. Ten Eyck, M.T. Welch, R.E. MaMann, M.L. Torreno, A.T. Appel, and M.P. Keleher, ''Scaling Issues in the Evolution of ExCL TM Bipolar Technology,'' *Bipolar Circuits and Technology Meeting Tech. Dig.*, pp. 121-124 (1988).

9. P.F. Lu, G.P. Li, and D.D. Tang, ''Lateral Encroachment of Extrinsic-Base Dopant in Submicronmeter Bipolar Transistors,'' *IEEE Electron Device Letters*, **Vol. EDL-8, No. 10** pp. 496-498 (October 1987).

10. L. Deferm, S. Decoutere, C. Claeys, and G. Declerck, ''Latchup in BiCMOS Technology,'' *International Electron Devices Meeting Tech. Dig.*, pp. 130-133 (December 1988).

11. R.H. Dennard, et al., ''Design of Ion Implanted MOSFETs with Very Small Physical Dimensions,'' *IEEE Journal of Solid-State Circuits*, **SC-9** pp. 256-266 (Oct. 1974).

12. P.K. Chatterjee, et al., ''The Impact of Scaling Laws on the Choice of N-Channel or P-Channel for MOS VLSI,'' *IEEE Electron Device Letters*, **EDL-1** pp. 220-223 (Oct. 1980).

13. G. Baccarani, et al., ''Generalized Scaling Theory and its Application to a 1/4 Micrometer MOSFET Design,'' *IEEE Trans. on Electron Devices*, **ED-31, No. 4** pp. 452-462 (Apr. 1984).

14. A. Bellaouar, S.H.K. Embabi, M.I. Elmasry, ''Scaling of Digital BiCMOS Circuit Structures,'' *International Electron Devices Meeting Tech. Dig.*, pp. 437-440 (Dec. 1989).

15. P.M. Solomon, and D.D. Tang, ''Bipolar Circuit Scaling,'' *International Solid-State Circuits Conference Tech. Dig.*, pp. 86-87 (Feb. 1979).

16. T.H. Ning, et al., ''Scaling Properties of Bipolar Devices,'' *International Electron Devices Meeting Tech. Dig.*, pp. 61-64 (Dec. 1980).

17. T.H. Ning, ''Materials and Physics Issues in Scaling Bipolar Devices,'' *J. Vac. Sc. Technol.*, **A 5 (4)** pp. 1394-1395 (Jul/Aug. 1987).

18. D.D-L. Tang, et al., ''Junction Degradation in Bipolar Transistors and the Reliability Imposed Contraints to Scaling and Design,'' *IEEE Trans. on Electron Devices*, **ED-35, No. 12** pp. 2101-2107 (Dec. 1988).

19. T.H. Ning, and R. D. Isaac, ''Effect of Emitter Contact on Current Gain of Silicon Bipolar Devices,'' *IEEE Electron Device Letters*, **ED-27** pp. 2051-2055 (Nov. 1980).

20. B. Soerowirdjoeral, ''The Influence of Surface Treatment on the Electrical Characteristics of Poly-emitter Bipolar Transistors,'' *International Electron Devices Meeting Tech. Dig.*, pp. 668-671 (Dec. 1982).

21. J-C. S. Woo, and J.D. Plummer, ''Optimization of Silicon Bipolar Transistors for High TCurrent Gain at Low Tempuratures,'' *IEEE Trans. on Elecstron Devices*, **ED-35, No. 8** pp. 1311-1321 (Aug. 1988).

22. D. Arora, J.R. Hauser, and D.J. Roulston, ''Electron and Hole Mobilities in Silicon as a Function of Concentration and Temperature,'' *IEEE Trans. on Electron Devices*, **ED-29** pp. 284-291 (Feb. 1982).

23. C.L. Williams, et al., ''Optimization and Tempurature Dependance of Current Gain in Polysilicon-Emitter-Contacted Bipolar Transistors,'' *Bipolar Circuits and Technology Meeting Tech. Dig.*, pp. 50-54 (Sep. 1988).

CHAPTER 4

DEVICE MODELING

The objective of this chapter is two-fold. It is intended to review the basics of the MOS and the bipolar transistors, which are prerequisites for Chapters 5. to 7., and to introduce commonly used models of both devices [Sections 4.1. and 4.2.]. In this chapter we consider simple analytical models which can be used for first order circuit analysis and design. The more sophisticated SPICE device models are also presented to allow the reader to appreciate the meaning of the model parameters as well as the capabilities and limitations of these models. The SPICE parameters for the $0.8\mu m$ BiCMOS process presented in Chapter 2. are included in this chapter for readers who are interested in designing and simulating BiCMOS circuits.

4.1. Modeling of the MOS Transistor

4.1.1. MOSFET Structure and Operation

Fig. 4.1[1] shows cross sections and views of an N-channel MOS transistor. By applying a positive voltage on the gate V_{GS}, a depletion layer is induced in the channel. Further increase in V_{GS} results in a surface inversion layer. The surface charge of the semiconductor (Q_S $coul/cm^2$) is equal in magnitude to the charge of the gate electrode (Q_G $coul/cm^2$). Thus, we have:

---

[1] The effective channel width W_{eff} is equal to $W_{mask}-2W_d-x_w$ and the effective channel length is equal to $L_{mask}-2L_d-x_l$. W_{mask} and L_{mask} are the drawn channel width and length. $2W_d$ ($2L_d$) is the total reduction in the channel width (length) due to the lateral diffusion. x_w and x_l represent the effect of the processes of masking and etching on the channel width and length respectively.

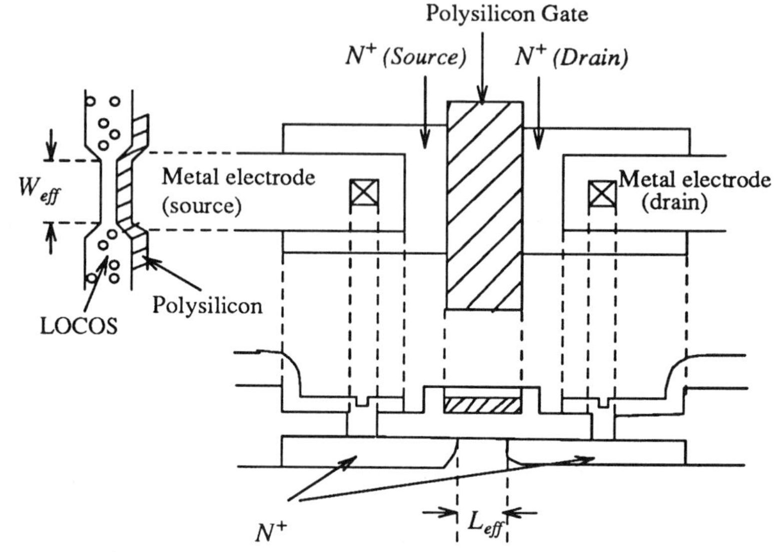

(a)

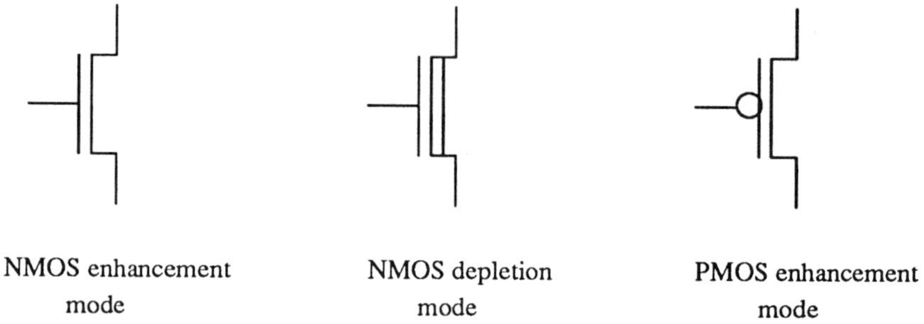

(b)

Figure 4.1 (a)The layout and cross-sectional views of an NMOS transistor. (b) Symbols of different types of MOS transistors.

$$Q_S = -Q_G = -(V_{GS} - V_{FB} - \phi_s)C_{ox} \tag{4.1}$$

where V_{GS} is the gate-source voltage and ϕ_s is the semiconductor surface potential. C_{ox} is the gate oxide capacitance per unit area and is given by:

$$C_{ox} = \frac{\varepsilon_o}{t_{ox}} \tag{4.2}$$

where ε_o is the oxide permittivity and t_{ox} is the gate oxide thickness. The flatband voltage V_{FB} is given by:

$$V_{FB} = \phi_{ms} - \frac{Q_0}{C_{ox}} \tag{4.3}$$

Q_0 is the total of all charges in the oxide and near the interface oxide/silicon. This charge is positive. The work function difference between the gate electrode and the semiconductor ϕ_{ms} depends on the type of the electrode and the doping concentration of the semiconductor. For an aluminum electrode, we have:

$$\phi_{ms} = -0.61 + \phi_f \tag{4.4}$$

For N^+ polysilicon electrode, we have:

$$\phi_{ms} = -0.55 + \phi_f \tag{4.5}$$

The *fermi potential* ϕ_f in Equations (4.4) and (4.5) is given by:

$$\phi_{fp} = -V_t \ln(\frac{N_a}{n_i}) \quad \text{for } P\text{-}type\ Si \tag{4.6}$$

$$\phi_{fn} = +V_t \ln(\frac{N_d}{n_i}) \quad \text{for } N\text{-}type\ Si \tag{4.7}$$

where $V_t = KT/q$. The charge Q_S is the sum of the charge in the depletion layer Q_B and the inversion layer Q_I. Therefore;

$$V_{GS} = V_{FB} + \phi_s - \frac{Q_B + Q_I}{C_{ox}} \tag{4.8}$$

The bulk depletion charge (per unit area) consists of ionized acceptors (P-type substrate) or donors (N-type substrate). The depletion charge of a P-type bulk, with zero bias bulk-source voltage (V_B=0), is given by:

$$Q_{B0} = -qN_a W_D \tag{4.9}$$

where the q is the electron charge and N_a is the donor concentration. The width of the depletion layer in the bulk (W_D) is given by:

$$W_D = \left(2\frac{\varepsilon_s}{qN_a}\phi_s\right)^{1/2} \tag{4.10}$$

The turn-on (or threshold) voltage of an NMOS transistor is defined as the gate-source voltage at which the surface potential ϕ_s is equal to $2|\phi_f|$. This condition also defines what is known as the strong inversion[2]. At the onset of strong inversion we can assume that $Q_S \approx Q_B$. Using Equation (4.8), we can write the following expression of the threshold voltage:

$$V_{T0} = V_{FB} + \phi_s - \frac{Q_{B0}}{C_{ox}} \tag{4.11}$$

Q_{B0} is equal to $-qN_a W_{Dm}$, where $W_{Dm}=W_D(\phi_s=2|\phi_f|)$ [3]. Thus, the threshold voltage can be rewritten as:

$$V_{T0} = V_{FB}+2|\phi_f|+\frac{\sqrt{2q\varepsilon_s N_a(2|\phi_f|)}}{C_{ox}} \tag{4.12}$$

If the bulk-source is reverse biased ($|V_B|>0$), the threshold voltage becomes:

---

[2] The surface is considered to be strongly inverted when the concentration of the inversion charge *at the surface* is equal or greater than the concentration of the ionized acceptors. Based on that definition it can be shown that at the onset of strong inversion $\phi_s=2|\phi_f|$.

[3] The increase of the depletion region width W_D in the bulk is negligible beyond the onset of strong inversion. Hence, we can assume that the depletion region width reaches its maximum (W_{Dm}) at the onset of strong inversion. Therefore, under strong inversion conditions, $W_D=W_{Dm}=\sqrt{2\varepsilon_s(2|\phi_f|)/qN_a}$.

$$V_T = V_{FB} + 2|\phi_f| + \frac{\sqrt{2q\varepsilon_s N_a(|V_B| + 2|\phi_f|)}}{C_{ox}} \tag{4.13}$$

This equation can be rewritten as:

$$V_T = V_{T0} + \gamma(\sqrt{|V_B| + 2|\phi_f|} - \sqrt{2|\phi_f|}) \tag{4.14}$$

where the body effect coefficient γ is given by:

$$\gamma = \frac{\sqrt{2q\varepsilon_s N_a}}{C_{ox}} \tag{4.15}$$

Consider an N-channel MOS transistor with $N_a=10^{16}cm^{-3}$, $t_{ox}=20nm$, $V_{FB}=-1.08V$, then $2|\phi_f|=0.7V$. Equation (4.12) gives $V_{T0}\approx-0.1V$. This value is negative and is not suitable for digital circuits where a positive V_{T0} is required for switching. To get a reasonable V_{T0}, the device surface is implanted with boron. The implanted dose D_I causes V_{T0} to increase by the amount $\frac{qD_I}{C_{ox}}$. The threshold voltage is hence given by:

$$V_{T0}=V_{FB} + 2|\phi_f| + \gamma\sqrt{2|\phi_f|} + q\frac{D_I}{C_{ox}} \tag{4.16}$$

Consider now the previous example, with $D_I=9.7\times10^{11}cm^{-2}$ and $\gamma=0.335V^{1/2}$ we find that V_T is equal to $0.8V$ when $|V_B|=0V$ and is equal to $1.32V$ when $|V_B|=5V$.

The symbols of the NMOS and PMOS transistors are shown in Fig. 4.1(c). Typical values of the V_T are $0.5V$ to $1V$ for enhancement-mode NMOS devices, $-0.5V$ to $-1V$ for enhancement-mode PMOS devices, and $-2.5V$ to $-4V$ for depletion-mode NMOS devices.

When $V_{GS}<V_{T0}$, the transistor is in the *cutoff region*, since no inversion layer exists, as shown in Fig. 4.2(a). The drain current is, therefore, approximately zero. When $V_{GS}>V_{T0}$, the channel is formed and a drain current flows from the drain to the source [Fig. 4.2(b)]. The transistor is in the *linear region* (also called *ohmic region*) when V_{GD} (*i.e.* $V_{GS}-V_{DS}$)$\geq V_T$. When $V_{GS}>V_T$ and $V_{DS}>V_{GS}-V_T$ (i.e. $V_{GD}<V_T$) the channel is pinched off as illustrated in Fig. 4.2(c) and the device enters the *saturation region*. The drain-source voltage which causes the channel to pinchoff at the drain edge is commonly known as the saturation drain-source voltage V_{DSsat} and is equal to

$V_{GS}-V_T$.

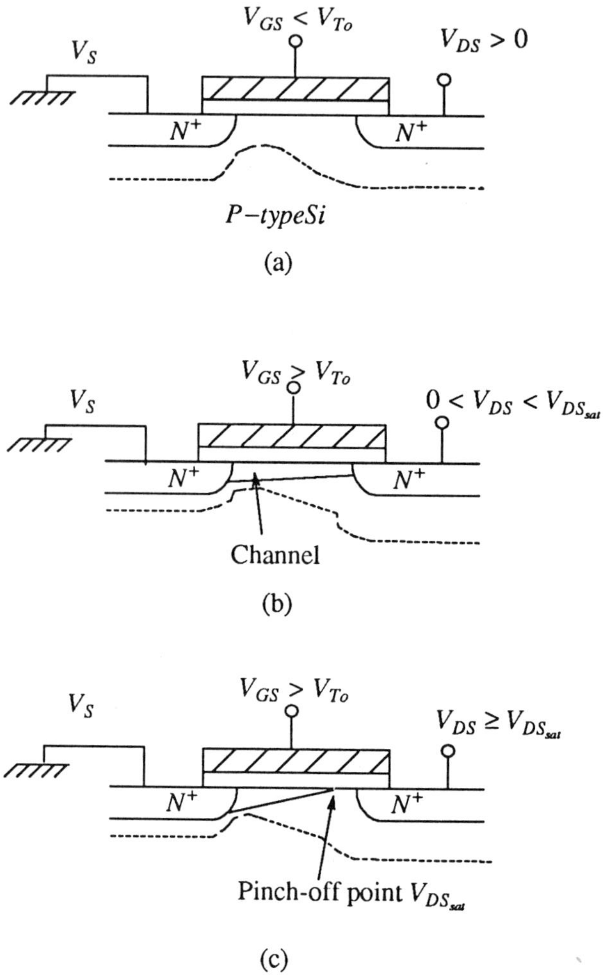

Figure 4.2 Cross-sectional view of an MOS transistor showing the channel in the (a) cut-off, (b) linear, and (c) saturation regions.

The voltage drop between the pinchoff point and the source is V_{DSsat}. Any V_{DS} higher than V_{DSsat} will appear between the pinchoff point and the drain. If we assume that the distance between the pinchoff point and the drain is extremely small compared with the overall length, then for $V_{DS}>V_{DSsat}$ the drain current is constant. The carriers which reach the pinchoff point are swept across to the drain by the potential $(V_{DS}-V_{DSsat})$ between the drain and the end of the channel.

4.1.2. SPICE Models of the MOS Transistor

4.1.2.1. The Simple MOS DC Model

Let us now analyse the simple DC model describing the I-V characteristics of an MOS transistor.

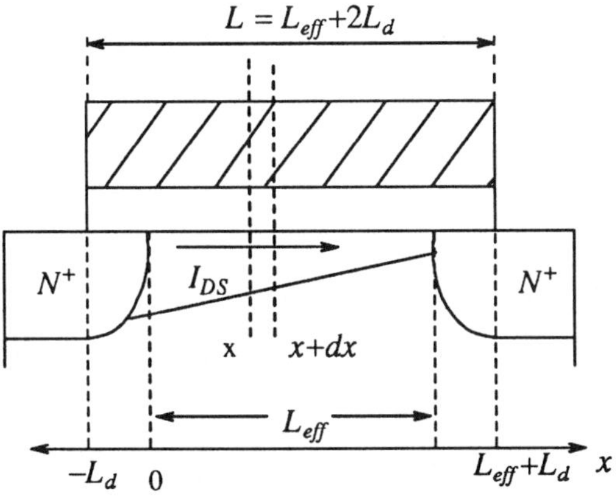

Figure 4.3 Cross-sectional view of an MOS transistor with the channel in the linear region.

From Fig. 4.3 it can be shown that the element dx has a resistance

$$dR = -\frac{dx}{\mu W_{eff} Q_I(x)} \tag{4.17}$$

We assume that the mobility (μ) of the electrons in the channel of an NMOS device is constant. A current I_{DS} crossing the incremental resistance dR causes a voltage drop of

$$dV = I_{DS} dR \tag{4.18}$$

Substituting from Equation (4.17) in Equation (4.18) and integrating from the source to the drain, we obtain:

$$I_{DS} \int_0^{L_{eff}} dx = -\mu W_{eff} \int_0^{V_{DS}} Q_I(x) dV \tag{4.19}$$

To solve this integration, we need to express the electron inversion charge density $Q_I(x)$ in terms of V. From Equation (4.8), we have:

$$Q_I(x) = -\left(V_{GS} - V_{FB} + \frac{Q_{B0}}{C_{ox}} - \phi_s(x) \right) C_{ox} \tag{4.20}$$

The surface potential ϕ_s at any point x along the channel is equal to $2|\phi_f|+V(x)$. By substituting for $V_{FB}-Q_{B0}/C_{ox}+2|\phi_f|$ by V_{T0} [Equation (4.11)] in Equation (4.20) we get:

$$Q_I(x) = -[V_{GS} - V_{T0} - V(x)] C_{ox} \tag{4.21}$$

The surface potential at the drain is larger than that at the source by V_{DS}. Therefore, the magnitude of Q_I decreases with the distance across the channel. This is why the inversion layer is triangular as illustrated in Fig. 4.3. Assuming that Q_{B0} is constant across the channel and substituting for Q_I from Equation (4.21) into Equation (4.19), we obtain:

$$I_{DS} = \mu \frac{W_{eff}}{L_{eff}} C_{ox} \int_0^{V_{DS}} [V_{GS} - V_{TO} - V(x)]dV \tag{4.22}$$

or

$$I_{DS} = \mu \frac{W_{eff}}{L_{eff}} C_{ox}[(V_{GS} - V_{TO})V_{DS} - ½ V_{DS}^2] \tag{4.23}$$

or

$$I_{DS} = k_p \frac{W_{eff}}{L_{eff}} [(V_{GS} - V_{TO})V_{DS} - ½ V_{DS}^2] \tag{4.24}$$

where k_p is a process-dependent parameter defined as $k_p = \mu C_{ox}$. Equation (4.24) is valid only for $V_{DS} \leq V_{DSsat}$ (ohmic region). When V_{DS} exceeds V_{DSsat} the drain-source current saturates. The saturation current can be found by substituting for V_{DS} by V_{DSsat} in Equation (4.24) and is hence given by:

$$I_{DS} = \frac{k_p}{2} \frac{W_{eff}}{L_{eff}} (V_{GS} - V_{TO})^2 \tag{4.25}$$

The characteristics of an MOS transistor based on Equations (4.24) and (4.25) are shown in Fig. 4.4. The current equations (4.24) and (4.25) have to be modified if the bulk-source voltage is greater than zero by replacing V_{TO} by V_T [see Equation (4.14)]. Note that when V_{DS} is small (say 50mV), Equation (4.24) can be approximated by:

$$I_{DS} = k_p \frac{W_{eff}}{L_{eff}} (V_{GS} - V_{TO})V_{DS} \tag{4.26}$$

This equation expresses a linear relationship between I_{DS} and V_{GS}. Using linear extrapolation, V_{TO} and $k_p \frac{W_{eff}}{L_{eff}}$ can be determined as shown in Fig. 4.4(b).

The measured I-V characteristics show that the drain current, in the saturation region, is a weak function of V_{DS}. This is due to the channel length modulation phenomenon which can be explained as follows. Let us define

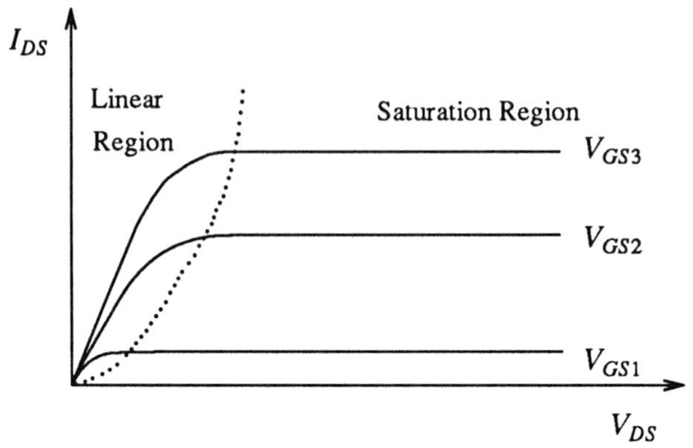

(a)

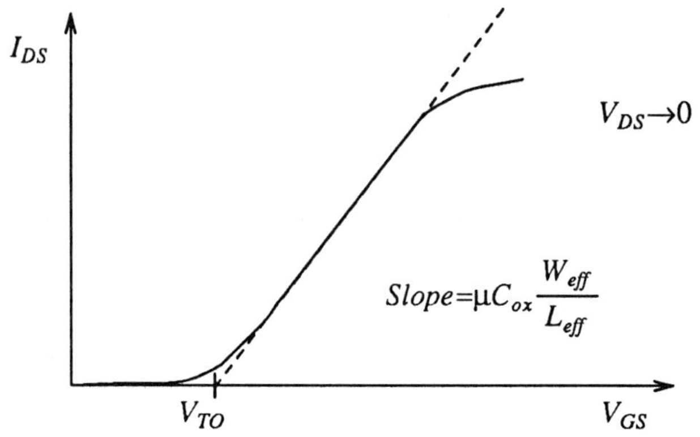

(b)

Figure 4.4 (a) Output characteristic of an MOS transistor (I_D vs V_{DS}). (b) The transfer characteristics of an MOS transistor (I_D vs V_{GS}).

$$L'_{eff} = L_{eff} - \Delta L \tag{4.27}$$

where ΔL is width of the depletion layer between the pinchoff point and the drain as shown in Fig. 4.5. The voltage across this depletion layer is $V_{DS}-V_{DSsat}$, therefore ΔL can be written as:

$$\Delta L = \left\{2\frac{\varepsilon_s}{qN_a}(V_{DS} - V_{DSsat})\right\}^{1/2} \tag{4.28}$$

The corrected saturation current becomes:

$$I_{DS} = \frac{k_p}{2}\frac{W_{eff}}{L'_{eff}}(V_{GS} - V_T)^2 = \frac{k_p}{2}\frac{W_{eff}}{L_{eff}}(V_{GS} - V_T)^2 \frac{1}{\left(1 - \frac{\Delta L}{L_{eff}}\right)} \tag{4.29}$$

If we assume that $\Delta\frac{L}{L_{eff}} \ll 1$, then we can rewrite the current as:

$$I_{DS} = \frac{k_p}{2}\frac{W_{eff}}{L_{eff}}(V_{GS} - V_T)^2 \left(1 + \frac{\Delta L}{L_{eff}}\right) \tag{4.30}$$

The ratio $\frac{\Delta L}{L_{eff}}$ can be related to V_{DS} by the following empirical relation:

$$\frac{\Delta L}{L_{eff}} = \lambda V_{DS} \tag{4.31}$$

The channel modulation factor λ is very small. A typical value of λ is 0.01 V^{-1}.

The drain current model described, so far, is known as the LEVEL 1 (MOS1) model in SPICE[4]. This model is also called the Shichman-Hodges model. However, this model is still very simple[5] to account for state-of-the-art CMOS devices and might lead to a 100% error in the current. However, k_p (or μ) can be used as a fitting parameter to reduce this error. This model is most suitable for preliminary analysis.

---

[4] SPICE2G6 or 3B1 or 3C1

[5] This model was used in the 70's.

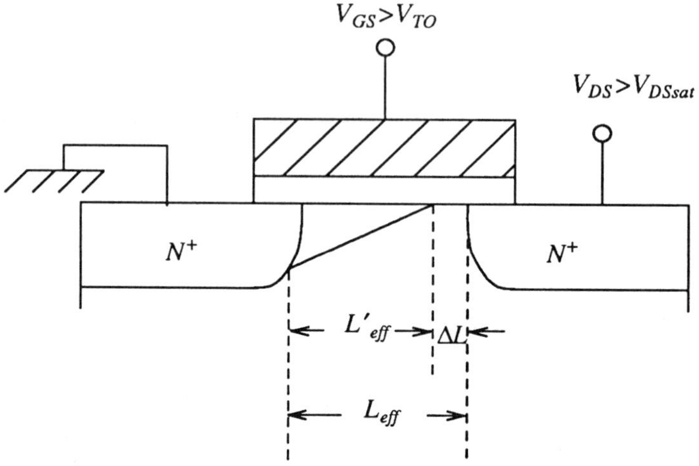

Figure 4.5 Channel length modulation phenomenon in an MOS transistor.

4.1.2.2. MOS Model (LEVEL 2) of SPICE[1]

Let us now include the second order effects in an MOS transistor to improve the previous model (MOS1). These include short-channel effects on threshold voltage, mobility degradation with electric field, and weak inversion conduction.

Correction 1: depletion layer fixed charge (Q_{B0})

Earlier we assumed that Q_{B0} is constant along the channel. Since the surface potential varies in the channel, Q_{B0} can be rewritten as:

$$Q_{B0} = \frac{1}{C_{ox}}\sqrt{2q\varepsilon_s N_a[2|\phi_s| + V(x)]} \qquad (4.32)$$

Correction 2: short- and narrow-channel effects on V_T

Equation (4.13) of V_T can be applied to MOS devices with large W and L. When the channel is short or narrow, two dimensional effects affect the charge Q_B. First, let us examine the effect of short-channel on the threshold voltage. The existence of the source and drain depletion regions at the edges of the channel reduces the charge Q_B. Therefore, the real value of V_T is reduced. Fig. 4.6 shows that the bulk charge Q_B', which contributes to the threshold voltage, occupies a volume with a trapezoidal cross-section[2]. Using the geometrical model of Fig. 4.6(b) it can be shown that the bulk charge Q_B' is given by:

$$Q_B' = Q_B \left(1 - \frac{\Delta L_D}{L_{eff}}\right) \tag{4.33}$$

The correction factor $(1 - \frac{\Delta L_D}{L_{eff}})$ is geometry-dependent. From Fig. 4.6(c) we have:

$$1 - \frac{\Delta L_D}{L_{eff}} = 1 - \frac{x_j}{L_{eff}} \left[\sqrt{1 + 2W_{Dm}/x_j} - 1\right] \tag{4.34}$$

The effect of short-channel on Q_B, which is represented by Equations (4.33) and (4.34), can be introduced to modify the expression of the threshold voltage [Equation (4.14)] by modifying the value of γ as follows:

$$\gamma = \left(1 - \frac{x_j}{L_{eff}} \left[\sqrt{1 + 2W_{Dm}/x_j} - 1\right]\right) \frac{\sqrt{2q\varepsilon_s N_a}}{C_{ox}} \tag{4.35}$$

To demonstrate the effect of using the correct expression of γ on V_T, let us assume the following parameters for an MOS transistor: t_{ox}=20nm, N_a=$10^{16}cm^{-3}$, x_j=0.3μm and L_{eff}=1μm. The threshold voltage, which is calculated from Equation (4.16) using the expression of γ given in Equation (4.15), is 0.8V. If, however, the expression of Equation (4.35) is used, the threshold voltage reduces to 0.725V. It is also interesting to see the effect of the channel length on the threshold voltage. If we use the above data, we find that the threshold voltage of a transistor with L_{eff}>4μm is equal to 0.8V. If the effective channel length of the same device is reduced to 0.5μm, V_T falls to 0.65V.

In deriving Equation (4.35) the effects of V_B and V_{DS} were not accounted for. Fig. 4.7 illustrates the model for the calculation of V_T when the drain and the substrate are biased. Using the same technique employed before, we can demonstrate that

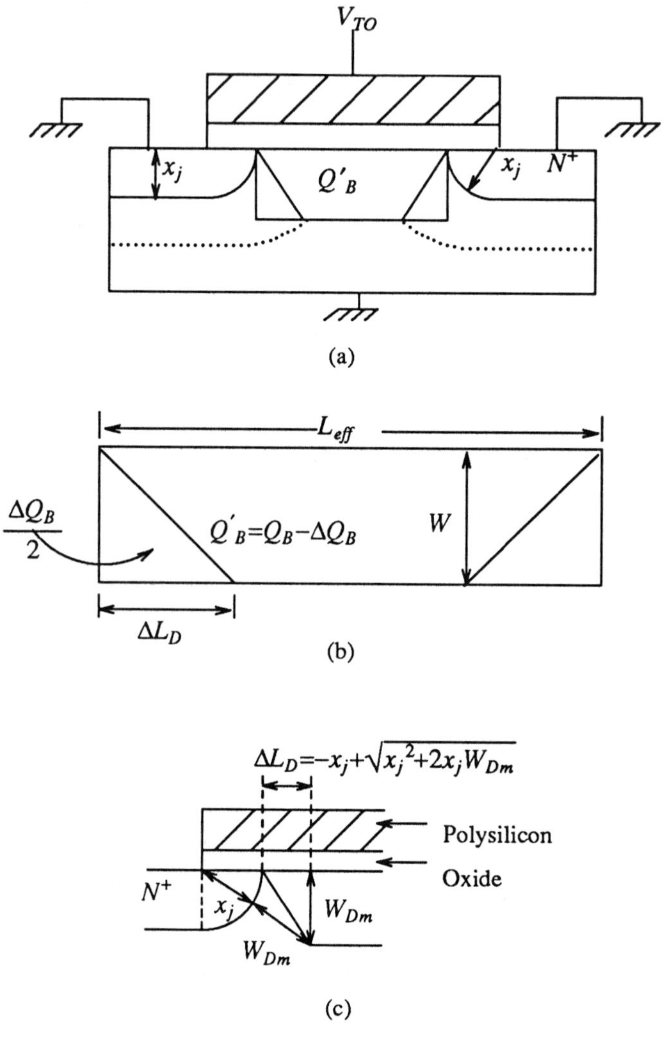

Figure 4.6 Charge model for the calculation of V_{TO}.

$$\gamma = \left[1 - \frac{x_j}{2L_{eff}}\left(\sqrt{1+2W_S/x_j} + \sqrt{1+2W_D/x_j} - 2\right)\right]\frac{\sqrt{2q\varepsilon_s N_a}}{C_{ox}} \tag{4.36}$$

where

$$W_D \approx \sqrt{2\frac{\varepsilon}{qN_a}(2|\phi_f| + |V_B| + V_{DS})} \tag{4.37}$$

and

$$W_S \approx \sqrt{2\frac{\varepsilon}{qN_a}(2|\phi_f| + |V_B|)} \tag{4.38}$$

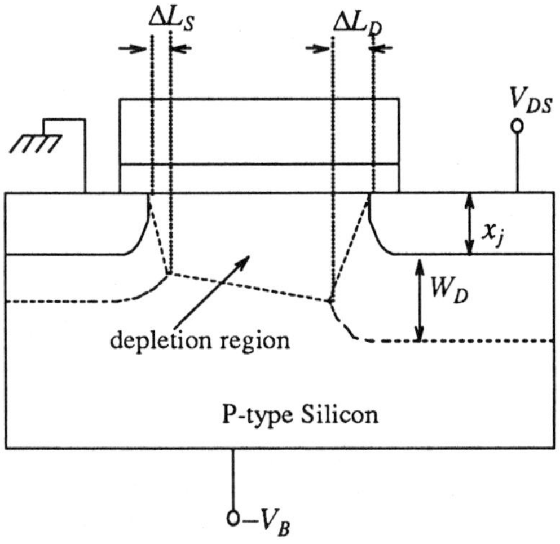

Figure 4.7 The effect of V_B and V_{DS} on the charge Q_B.

Let us now examine the effect of narrow width. So far, we have assumed that Q_B is the depletion charge in the volume with a width W_{eff}. However, when W_{eff} is reduced, the relative magnitude of the charge near the edges of the channel (the bird's beak region of the LOCOS) is significant and causes the actual V_T to be greater than predicted by the simple model. Fig. 4.8 shows a simple model to account for the additional depletion layer charge. The new charge is given by:

$$Q_B' = Q_B(1+\frac{A}{W_{Dm}W_{eff}}) \tag{4.39}$$

The new corrected body effect coefficient is

$$\gamma = \left(1 + \frac{A}{W_{Dm}W_{eff}}\right)\frac{\sqrt{2q\varepsilon_s N_a}}{C_{ox}} \tag{4.40}$$

The area A is a function of the depletion layer shape at the edges. It can be determined empirically by fitting to measured data.

In the MOS LEVEL 2 model, the short- and narrow- channel effects on the threshold voltage are accounted for through the following expression[1]:

$$V_T = V_{TO} + \gamma\left[\sqrt{2|\phi_f|+|V_B|}\right] - \frac{\sqrt{2q\varepsilon_s N_a|\phi_f|}}{C_{ox}} + \frac{\pi\varepsilon_s\delta}{4C_{ox}W_{eff}}(2|\phi_f|+|V_B|) \tag{4.41}$$

Where γ is given by Equation (4.36) which includes the short-channel effect. The effect of a narrow channel is characterized by the parameter δ. A typical value of δ is 0.5.

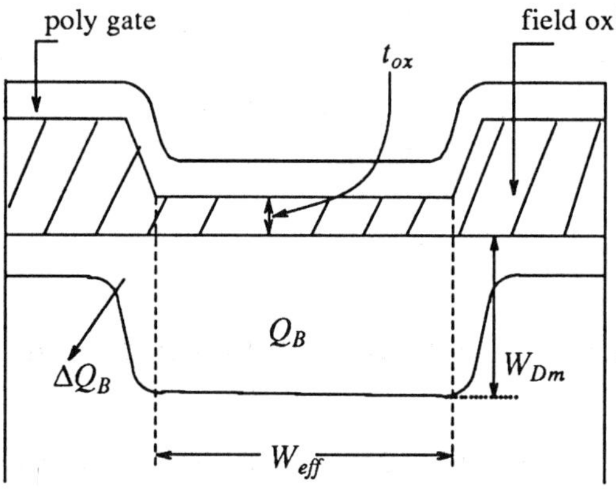

Figure 4.8 The bulk charge distribution in a narrow-channel MOS transistor.

Correction 3: mobility degradation

In the previous model (MOS1), the mobility was assumed independent on the vertical and horizontal electric fields. However, when the dimensions of the MOS transistor are reduced, the electric fields increase and the mobility of the carriers begins to decrease. Fig. 4.9(a) shows the carrier drift velocity for silicon as a function of the horizontal electric field[3]. At low electric field, the mobility μ ($=v/E$), defined as the average drift velocity over the electric field, is constant. The electrons have higher mobility (3-4 times) than that of the holes. At a high electric field, the electron and hole velocities converge towards the same saturation velocity [6] $v_{\max}$. Fig. 4.9(b) illustrates the measured mobility as a function of the average vertical electric field $(V_{GS} - V_T - V_{DS}/2)$[4].

The effective mobility used in the MOS2 model is given by the following empirical expression[5]:

$$\mu_{eff} = \mu_0 \left[\frac{\varepsilon_s}{\varepsilon_{ox}} \frac{U_c}{(V_{GS} - V_T - U_t V_{DS})} \right]^{U_e} \tag{4.42}$$

where the critical field U_c is in the order of $1\text{x}10^4$–10^6 V/cm, U_e is approximately 0.1-0.3, and U_t is an empirical parameter of the order of 0-0.5. The expression $C_{ox}(V_{GS}-V_T-U_t V_{DS})/2\varepsilon_s$ represents the average electric field between the gate and the channel. The term $U_t V_{DS}$ is introduced to account for the inversion layer charge loss due to pinchoff at high V_{DS}. The effective mobility in this model takes into account only the vertical degradation.

Correction 4: subthreshold conduction

In the LEVEL 1 model of SPICE, the drain current in the weak inversion region was assumed zero. The modeling of the subthreshold current in LEVEL 2 is based on the analysis by Swanson and Meindl[6]. The drain current in weak inversion, which is basically a diffusion current, is given by[6]:

---

[6] Sometimes referred to as the maximum carrier velocity.

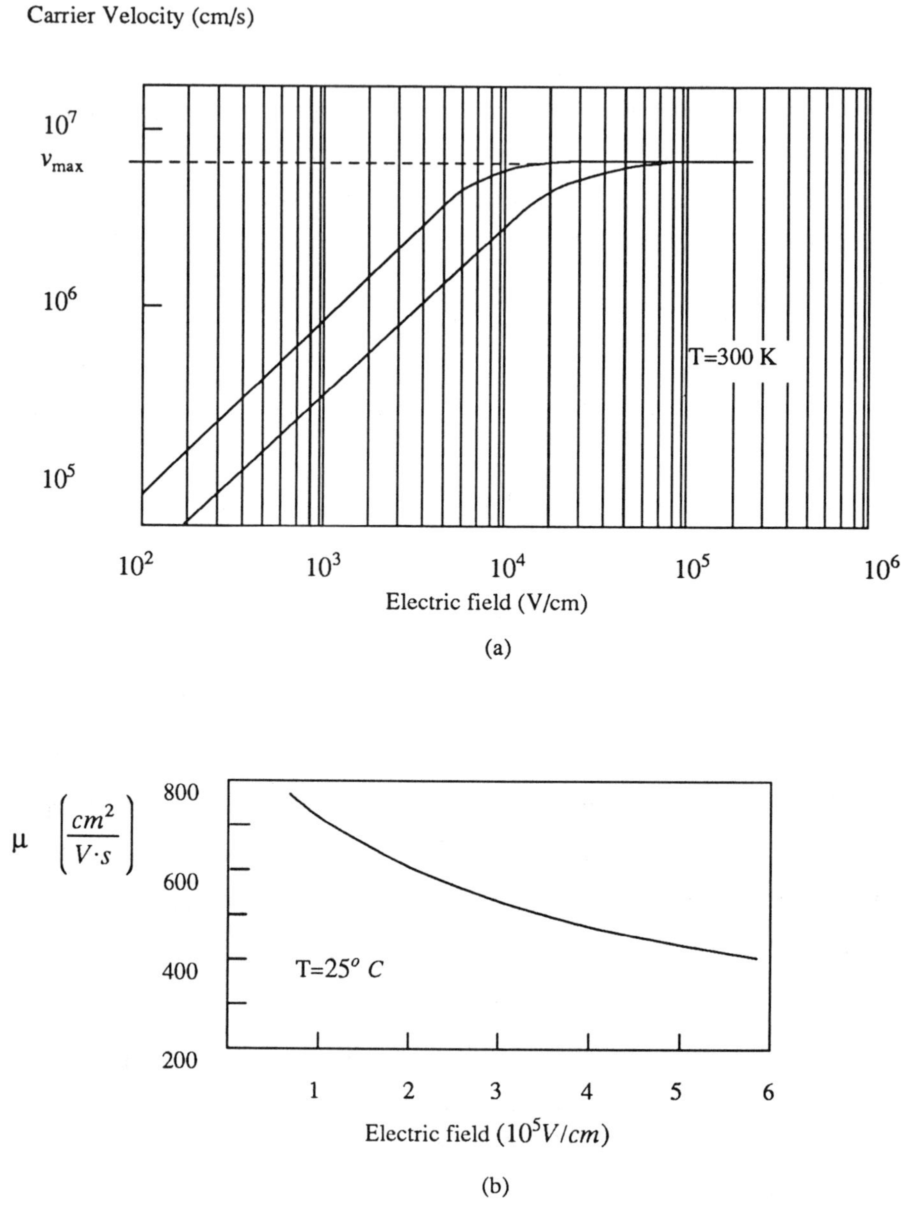

Figure 4.9 (a) Carrier drift velocities vs the horizontal electrical field [3]. (b) The electron mobility as function of the electric field[4].

$$I_{DS} = I_{on} e^{[(V_{GS}-V_{on})/nV_t]} \tag{4.43}$$

where

$$V_{on} = V_T + nV_t \tag{4.44}$$

and

$$n = 1 + \frac{qN_{FS} + C_d}{C_{ox}} \tag{4.45}$$

where

$$C_d = \frac{dQ_B}{dV_B} \tag{4.46}$$

and N_{FS} is a curve fitting parameter. V_{on} marks the point between the weak and strong inversion modes. Typical values of n range from >1.0 to 2.5. I_{on} is related to the current of Equation (4.47) by taking $V_{GS}=V_{on}$.

Fig. 4.10 illustrates the transfer characteristics of the weak inversion and drift model. The voltage V_{on} insures the continuity of the current, but it is clear from the figure that at $V_{GS}=V_{on}$ a discontinuity exists in the derivative. Therefore, the MOS2 model is not precise in simulating the intermediate region where the diffusion and drift currents are comparable.

Modified Drain Current Model:

When $V_{GS}>V_T$, the drain current may be reintegrated to take into account the corrections 1, 2 and 3. First, we obtain the modified V_T by substituting for γ from Equation (4.36) in Equation (4.41). The resulting expression of V_T and the new expression of the mobility [Equation (4.42)] can then be used to derive the following modified drain current equation:

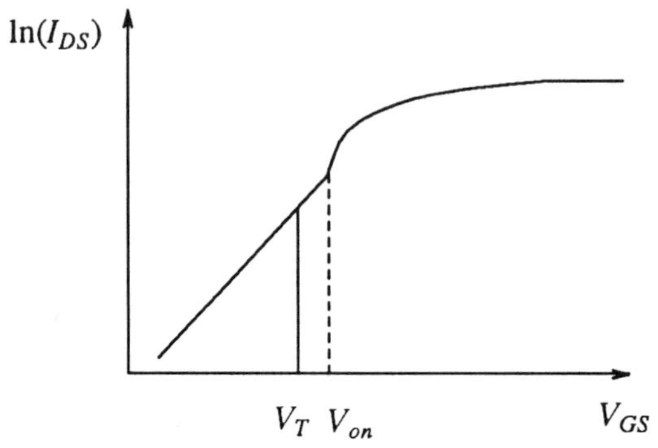

Figure 4.10 Subthreshold and drift current of an MOS transistor vs the gate-source voltage.

$$I_{DS} = \mu_{eff} C_{ox} \frac{W_{eff}}{L_{eff}} \left[(V_{GS} - V_T' - \xi \frac{V_{DS}}{2}) V_{DS} - \frac{2}{3} \gamma \left((V_{DS} + |V_B| + 2|\phi_f|)^{\frac{3}{2}} - (|V_B| + 2|\phi_f|)^{\frac{3}{2}} \right) \right] (1 + \lambda V_{DS}) \quad (4.47)$$

where

$$V_T' = V_{T0} - \frac{\sqrt{2q\varepsilon_s N_a |\phi_f|}}{C_{ox}} + (\xi - 1)(|\phi_f| + |V_B|) \quad (4.48)$$

$$\xi = 1 + \frac{\pi \varepsilon_s \delta}{4 C_{ox} W_{eff}} \quad (4.49)$$

The saturation current is determined by equating $\partial I_{DS} / \partial V_{DS}$ to zero. For short-channel devices, the velocity of the carriers may reach its maximum before pinchoff. Therefore, the drain current saturates before the channel is pinched off. In this case, the parameter λ can be calculated, more accurately, using the Baum model[7] as follows:

$$\lambda = \frac{X_d}{L_{eff} V_{DS}} \left[\left((\frac{X_d v_{max}}{2\mu_{eff}})^2 + V_{DS} - V_{DSsat} \right)^{1/2} - \frac{X_d v_{max}}{2\mu_{eff}} \right] \tag{4.50}$$

where X_d is defined by:

$$X_d = \left(\frac{2\varepsilon_s}{qN_{eff} N_a} \right)^{1/2} \tag{4.51}$$

where N_{eff} is a fitting parameter.

4.1.2.3. Semi-Empirical Short-Channel Model (LEVEL 3)[1]

The MOS2 model, described in section 4.1.2.2, has the following drawbacks:

- The channel length modulation in saturation is overestimated;
- The mobility model is complex;
- The impact of the DIBL effect on the threshold voltage is not accounted for.

As a result, the MOS2 model is not accurate for small geometry transistors. The MOS3 model (or MOS LEVEL 3) has been developed for short- and narrow-channel MOS ($L \leq 2\mu m$, $W \leq 2\mu m$). The MOS3 model has the following features (compared to MOS2):

- A simple and more accurate model for the mobility;
- An improved model for the threshold voltage of short-channel devices (the DIBL effect is accounted for);
- An improved model for the channel length modulation phenomenon.

The threshold voltage expression is given by [1]:

$$V_T = V_{FB} + 2|\phi_F| - \sigma V_{DS} + \gamma F_S \sqrt{2|\phi_F| + |V_B|} + F_N(2|\phi_F| + |V_B|) \tag{4.52}$$

γ in this expression is given by Equation (4.15). This expression includes:

- The static feedback effect coefficient σ (Due to DIBL effect)[8]:

$$\sigma = \eta \frac{8.15\times 10^{-22}}{C_{ox}L_{eff}^{3}} \tag{4.53}$$

where η is an empirical coefficient;

- The correction factor for short-channel effect is based on a modified trapezoidal approach for calculating the charge Q_B [Fig. 4.11]. The correction factor can be obtained from[9]:

$$F_S = 1 - \frac{x_j}{L_{eff}}\left[\frac{L_d + W_c}{x_j}\left(1 - \frac{W_D}{x_j + W_D}\right)^{1/2} - \frac{L_d}{x_j}\right] \tag{4.54}$$

where W_c, the depletion layer width of a cylindrical junction and is given by:

$$W_c = 0.0831353 + 0.8013929\frac{W_D}{x_j} - 0.0111077\left(\frac{W_D}{x_j}\right)^2 \tag{4.55}$$

- The correction factor for narrow-channel MOS is given by [Equation (4.41)]:

$$F_N = \pi\varepsilon_s \frac{\delta}{4C_{ox}W_{eff}} \tag{4.56}$$

Mobility degradation:

The mobility degradation due to the vertical electric field is modeled by the following simple equation[10]:

$$\mu_s = \frac{\mu_0}{[1 + \theta(V_{GS} - V_T)]} \tag{4.57}$$

where θ is an empirical constant which depends on the oxide thickness. A typical value of θ is 0.05. To account for the effect of lateral average electric field, the effective mobility is related to the drain-source voltage and the channel length by[10]:

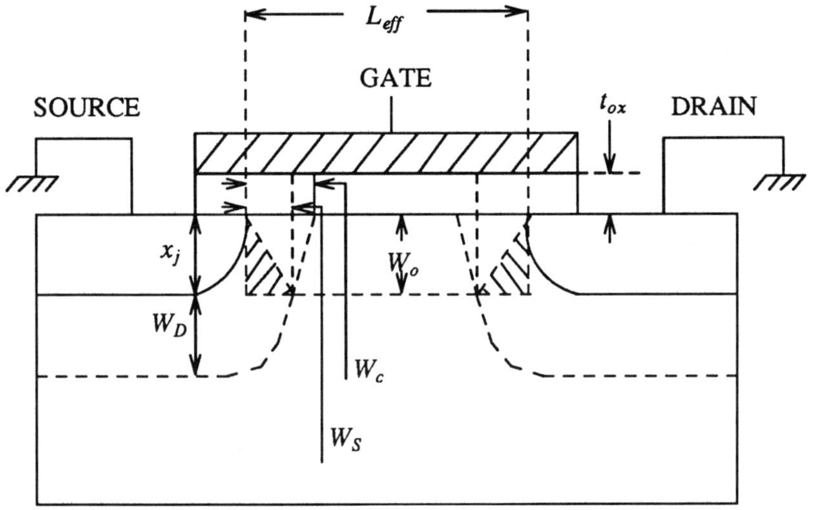

Figure 4.11 The model for the charge distribution in an MOS transistor showing that the bulk charge can be approximated by a trapezoid.

$$\mu_{eff} = \frac{\mu_s}{1 + \mu_s \dfrac{V_{DS}}{v_{\max} L_{eff}}} \tag{4.58}$$

In this expression, when the device operates in the saturation, V_{DS} is replaced by V_{DSsat}.

Channel length modulation:

When $V_{DS} \geq V_{DSsat}$, the channel length is modulated by an amount ΔL. This channel length reduction is formulated in MOS3 by Baum's model[7]. In this model the voltage across the depletion surface of length ΔL is modeled by $\kappa(V_{DS} - V_{DSsat})$. κ is a fitting parameter.

Drain current

The model of drain current in the weak inversion is the same as that of MOS2. In the strong inversion, the drain current can be expressed as:

$$I_{DS} = \mu_{eff} C_{ox} \frac{W_{eff}}{L_{eff}} \int_0^{V_{DS}} [V_{GS} - V_T(x)] dV \tag{4.59}$$

The threshold voltage along the channel is given by:

$$V_T(x) = V_T + \gamma F_S(\sqrt{|2\phi_f| + |V_B| + V(x)} - \sqrt{|2\phi_f| + |V_B|}) + F_N V(x) \tag{4.60}$$

Using Taylor series expansion, we have:

$$V_T(x) = V_T + (1+F_B)V(x) \tag{4.61}$$

where

$$F_B = \frac{\gamma F_S}{4}\left(2|\phi_f| + |V_B|\right)^{-1/2} + F_N \tag{4.62}$$

By substituting for V_T from Equation (4.61) in Equation (4.59), and integrating we obtain the following expression for the drain current:

$$I_{DS} = \mu_{eff} C_{ox} \frac{W_{eff}}{L_{eff}} \left[V_{GS} - V_T - \frac{1+F_B}{2} V_{DS}\right] V_{DS} \tag{4.63}$$

The saturation voltage, which takes into account the carrier velocity saturation effect, is given by:

$$V_{DSsat} = V_{sat} + V_c - \sqrt{V_{sat}^2 + V_c^2} \tag{4.64}$$

where

$$V_{sat} = (V_{GS} - V_T)/(1+F_B) \tag{4.65}$$

$$V_c = v_{max} L_{eff} / \mu_s \tag{4.66}$$

Typical values for parameters of LEVEL 3 are shown in Table 4.1 for MOS devices of the 0.8μm BiCMOS process described in Chapter 2.

Table 4.1
HSPICE MOSFET model parameters (LEVEL=3) (0.8μm BiCMOS process)

Parameter	SPICE Keyword	Description	N-channel	P-channel	Units
-	LEVEL	Model level	3	3	-
V_{TO}	VTO	Zero-bias threshold voltage	0.9	-0.85	V
t_{ox}	TOX	Gate oxide thickness	19×10^{-9}	19×10^{-9}	m
N_a	NSUB	Substrate doping	6.903×10^{16}	5.800×10^{16}	cm^{-3}
N_{FS}	NFS	Surface fast state density	772.0×10^{9}	650.5×10^{9}	$cm^{-2}V^{-1}$
μ_0	UO	Surface mobility	431	145	$cm^2/(V.s)$
v_{max}	VMAX	Maximum drift velocity of carriers	167.5×10^{3}	835.5×10^{3}	m/s
η	ETA	Static feedback on threshold voltage	36.0×10^{-3}	14.0×10^{-3}	-
κ	KAPPA	Saturation field factor	21.0×10^{-18}	133.5	$1/V$
θ	THETA	Mobility degradation factor	53.3×10^{-3}	0.14	$1/V$
δ	DELTA	Width effect on threshold voltage	0.496	0.614	-
x_j	XJ	Junction depth	16.06×10^{-9}	14.08×10^{-9}	m
C_j	CJ	Zero-bias bulk junction capacitance	550.0×10^{-6}	450.0×10^{-6}	F/m^2
M_j	MJ	Bulk junction grading coefficient	0.5	0.5	-
ϕ_j	PB	Junction potential	0.95	0.94	V
C_{jsw}	CJSW	Zero-bias side wall capacitance	500.3×10^{-12}	450.6×10^{-12}	F/m
M_{jsw}	MJSW	Sidewall capacitance grading coef.	0.333	0.330	-
C_{GD0}	CGDO	Gate-drain overlap capacitance	230×10^{-12}	128×10^{-12}	F/m
C_{GS0}	CGSO	Gate-source overlap capacitance	230×10^{-12}	128×10^{-12}	F/m
C_{GB0}	CGBO	Gate-bulk overlap capacitance	1.25×10^{-9}	1.25×10^{-9}	F/m
-	RD	Drain ohmic resistance	554	870	Ω
-	RS	Source ohmic resistance	554	870	Ω
L_d	LD	Lateral diffusion from drain or source	64×10^{-9}	0.	m
W_d	WD	Lateral diffusion along the width	0.	0.	m
x_l	XL	Masking and etching effects on W	0.15×10^{-6}	0.15×10^{-6}	m

x_w	XW	Masking and etching effects on L	-0.25×10^{-6}	0.25×10^{-6}	m
-	ACM	Area calculation method	2	2	-
-	LDIF	Lateral diffusion beyond the gate	933×10^{-9}	1×10^{-6}	m

The effects of some important parameters such as v_{max}, θ, and η on the characteristics of a short channel NMOS transistor are illustrated in Fig. 4.12. It is interesting to note that the velocity saturation phenomenon has a major effect particularly when the device is operating in saturation. The LEVEL 3 model approximates the device physics and relies on the proper choice of the empirical parameters to accurately reproduce the device characteristics.

4.1.2.4. BSIM Model (LEVEL 4)[11].

BSIM (Berkeley Short-Channel IGFET Model) is a simple and accurate short channel MOS transistor model[11]. It is implemented in SPICE as LEVEL 4. The model was tested for effective channel length down to $1\mu m$. This model includes:

- Vertical field dependence of carrier mobility;
- Carrier velocity saturation;
- Drain-induced barrier lowering effect;
- Non-uniform doping in the channel surface and sub-surface regions effect;
- Depletion charge sharing by the drain and source;
- Channel-length modulation;
- Dependance of some electrical parameters on drain and substrate biases;
- Better modeling of weak-, medium-, and strong- inversion regions and elimination of the discontinuity problem in the drain-current;

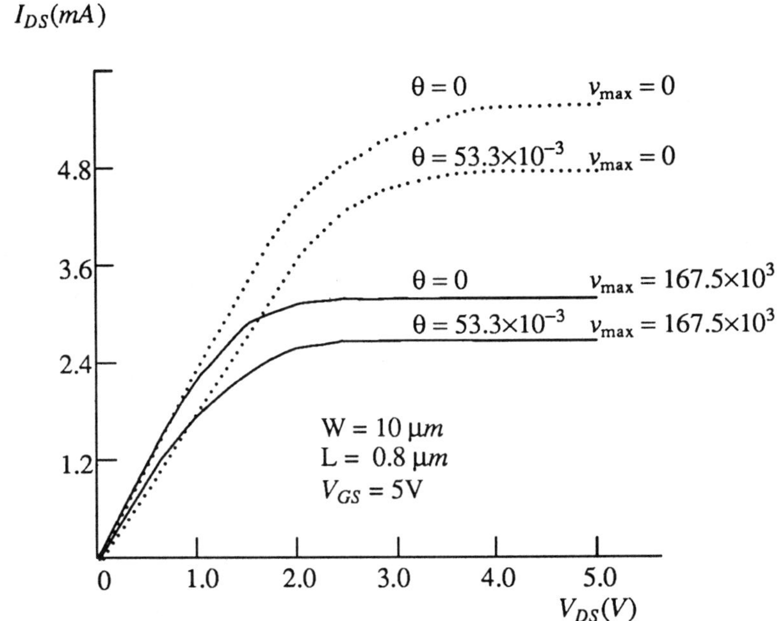

(a)

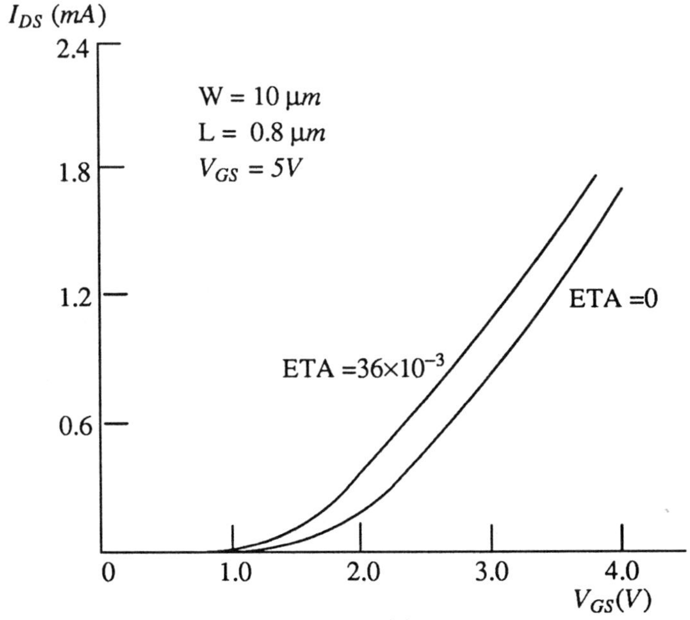

(b)

Figure 4.12 The effect of v_{max}, θ and η on the characteristics of an NMOS transistor.

Geometric dependencies;

Threshold voltage:

The threshold voltage is given by:

$$V_T = V_{FB} + \phi_s + K_1\sqrt{\phi_s + |V_B|} - K_2(\phi_s + |V_B|) - \eta V_{DS} \tag{4.67}$$

The two parameters, K_1 and K_2, model the effect of non-uniform doping of the substrate on the threshold voltage. Typical values for K_1 and K_2 are $1V^{1/2}$ and 0.12 respectively. The factor η models the DIBL effect and accounts for the channel-length modulation effect. It is a function of V_{DS} and V_B.

Drain current:

When $V_{DS} \leq V_{DSsat}$ we have:

$$I_{DS} = \frac{\mu_0}{1 + U_0(V_{GS} - V_T)} \frac{C_{ox}\frac{W_{eff}}{L_{eff}}}{(1 + \frac{U_1}{L_{eff}}V_{DS})} \left[(V_{GS} - V_T)V_{DS} - \frac{a}{2}V_{DS}^2 \right] \tag{4.68}$$

where

$$a = 1 + g\frac{K_1}{2}\left(\phi_s + |V_B|\right)^{-1/2} \tag{4.69}$$

and

$$g = 1 - \frac{1}{1.744 + 0.836(\phi_s + |V_B|)} \tag{4.70}$$

The parameters $U_0 = U_0(V_B)$, $U_1 = U_1(V_B)$ and $\mu_0 = \mu_0(V_{DS}, V_B)$ are bias sensitive. For $V_{DS} > V_{DSsat}$, the drain current is given by:

$$I_{DS} = \frac{\mu_0}{1 + U_0(V_{GS} - V_T)} C_{ox} \frac{W_{eff}}{L_{eff}} \frac{(V_{GS} - V_T)^2}{2aK'} \tag{4.71}$$

where

$$K' = \frac{1+v_c+\sqrt{1+2v_c}}{2} \tag{4.72}$$

and

$$v_c = \frac{U_1}{L_{eff}} \frac{(V_{GS} - V_T)}{a} \tag{4.73}$$

The drain-source saturation voltage is given by:

$$V_{DSsat} = \frac{V_{GS} - V_T}{a\sqrt{K'}} \tag{4.74}$$

"a" is called body-effect coefficient.

Subthreshold current:

In BSIM, the total drain current is modeled as the linear sum of a strong-inversion component and a weak-inversion component I_w. I_w is expressed as:

$$I_w = \frac{I_{\exp} I_{\lim}}{I_{\exp}+I_{\lim}} \tag{4.75}$$

where

$$I_{\exp} = \mu_0 C_{ox} \frac{W_{eff}}{L_{eff}} V_t^2 e^{1.8} e^{(V_{GS}-V_T)/nV_t} \left(1 - e^{-V_{DS}/V_t}\right) \tag{4.76}$$

and

$$I_{\lim} = \mu_0 \frac{C_{ox}}{2} \frac{W_{eff}}{L_{eff}} (3V_t)^2 \tag{4.77}$$

The factor $e^{1.8}$ is empirical to achieve the best fit. The subthreshold parameter n is a function of V_{DS} and V_B.

Sensitivity Factors of Model Parameters:

BSIM uses the following formula to account for the sensitivity of each parameter to the width and length of the channel:

$$P = P_0 + \frac{LP_0}{L_{eff}} + \frac{WP_0}{W_{eff}} \tag{4.78}$$

where P_0 is an arbitrary parameter, LP_0 and WP_0 are the L and W sensitivity factors of P_0.

4.1.2.5. MOS Capacitances

In transient simulation, MOS capacitances are very important for CMOS and BiCMOS circuits analysis. The MOS capacitances can be divided into two types of lumped capacitors:

(i) the depletion capacitors of the bulk-drain and bulk-source pn junctions (C_{BD} and C_{BS}) [Fig. 4.13].

(ii) the capacitors associated with the gate (C_{GS}, C_{GD}, C_{GB}, C_{GSov}, C_{GDov} and C_{GBov}) [see Fig. 4.13, except for C_{GBov}].

Junction Depletion Capacitances

The bulk-source and the bulk-drain junctions have a bottom area A_S and A_D respectively and a sidewall with a perimeter P_S and P_D respectively. Each of the bottom area and the sidewall contributes to the total depletion capacitance. The bottom area capacitance is measured per unit area, while the sidewall capacitance is measured per unit perimeter. Both of these components are voltage dependent. As these junctions are normally reverse biased, we will consider the case when the bulk-source and bulk-drain voltages (V_{BS} and V_{BD}) are less than or equal to $0.5\phi_j$ (ϕ_j is the junction built-in potential).

The total bulk-source and bulk-drain capacitances can be expressed by the following relations[1]:

$$C_{BD} = \frac{C_j A_D}{\left(1 - \frac{V_{BD}}{\phi_j}\right)^{M_j}} + \frac{C_{jsw} P_D}{\left(1 - \frac{V_{BD}}{\phi_j}\right)^{M_{jsw}}} \tag{4.79}$$

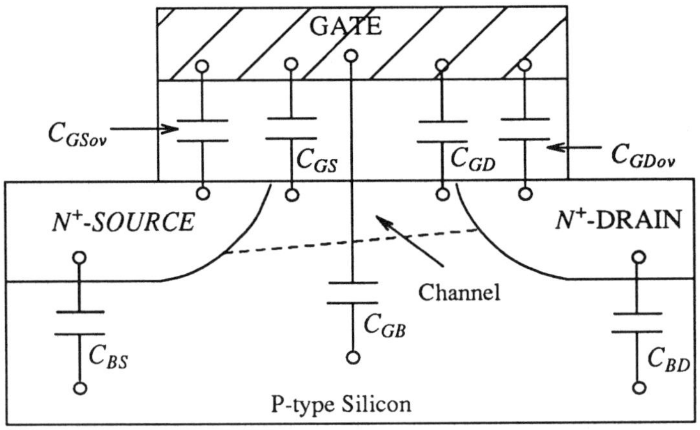

Figure 4.13 MOSFET capacitance model.

$$C_{BS} = \frac{C_j A_S}{\left(1 - \frac{V_{BS}}{\phi_j}\right)^{M_j}} + \frac{C_{jsw} P_S}{\left(1 - \frac{V_{BS}}{\phi_j}\right)^{M_{jsw}}} \tag{4.80}$$

The exponential factors M_j and M_{jsw} are in the order of 0.3-0.5. C_j is the zero-bias capacitance of the bottom junction per unit area and C_{jsw} is the zero-bias capacitance per unit perimeter.

Gate Capacitances

The gate capacitances can be divided into two categories:

- *The fixed overlap capacitances*: gate-drain (C_{GDov}), gate-source (C_{GSov}) , and gate-bulk (C_{GBov}) overlap capacitances. Both C_{GSov} and C_{GDov} exist due to the lateral diffusion of the source and drain under the gate. They are usually given per unit width as C_{GS0} and C_{GD0}. The total gate-source and gate-drain overlap capacitance is given by:

$$C_{GSov} = C_{GS0}\, W_{eff} \tag{4.81}$$

$$C_{GDov} = C_{GD0}\, W_{eff} \tag{4.82}$$

where C_{GS0} and C_{GD0} are equal to $C_{ox}L_d$. The capacitor C_{GBov} is due to the overlap of the gate oxide and the bulk along the channel length at both ends of the active area of the transistor. This capacitance is typically normalized to the effective channel length, the total C_{GBov} is hence given by:

$$C_{GBov} = C_{GB0}\, L_{eff} \tag{4.83}$$

where C_{GB0} is equal to $C_{ox}W_d$.

The nonlinear capacitance due to the charge of the bulk or the channel. This capacitance is actually distributed but can be modeled by lumped capacitances. In the case when the channel does not exist the capacitance can be expressed as:

$$C_{GB} = C_{ox}W_{eff}L_{eff} \tag{4.84}$$

When the device is in the linear region the channel is extending uniformly from the source to the drain. The channel shields the bulk and the capacitance exists only between the gate and the channel. The gate-bulk capacitance goes to zero. The gate-channel capacitance can be expressed in terms of two equal lumped capacitances, a gate-source and a gate-drain capacitance, which are denoted C_{GS} and C_{GD} and are given by:

$$C_{GS} = C_{GD} = \frac{1}{2}C_{ox}W_{eff}L_{eff} \tag{4.85}$$

Finally, when the device enters saturation, the channel at the drain pinches off and hence the gate-drain capacitance component becomes zero while the gate-source capacitance can be expressed by:

$$C_{GS} = \frac{2}{3}C_{ox}W_{eff}L_{eff} \tag{4.86}$$

Fig. 4.14 depicts the change of the capacitance components as a function of the gate-source voltage (assuming that the source-bulk voltage is zero). The total gate-source capacitance is given by the summation of the C_{GSov} and C_{GS}, and similarly, the total gate-drain capacitance is given by the summation of C_{GDov} and C_{GD}.

The above described capacitance model can be used for circuit analysis and circuit design. SPICE uses a charge-control model, which was developed by Ward and Dutton[12]. This model is based on the actual distribution of charge in the MOS structure and its conservation.

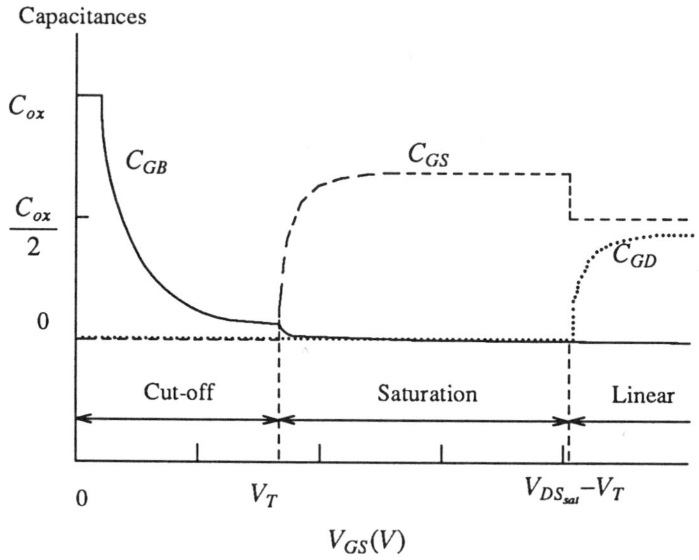

Figure 4.14 MOSFET capacitances as function of gate to source voltage.

4.1.3. Analytical Model for Short-Channel MOS Devices[13]

The MOS models discussed previously have been developed for circuit simulators. These models (e.g. BSIM) involve large numbers of parameters whose values must be derived from device measurements. With these models it is difficult to develop an intuitive understanding of the device behavior. Therefore, an analytical drain current model valid for micrometer and submicrometer MOSFETs is needed for hand calculation and first order circuit analysis, with reasonable accuracy. This section is based on the model proposed by Sodini[13].

For long-channel devices, the carrier drift velocity v is related to the horizontal electric field E by a simple linear relation ($v=\mu E$) where the carrier mobility is constant. For short-channel devices, the mobility is no longer a constant and is a function of the vertical electric field in the inversion layer. At this point we prefer to use the symbol μ_v for the mobility to denote its dependence on the vertical electric field. Also, the velocity (v) is no longer proportional to E but is given by the following two-region piecewise empirical model[14]:

$$v = \begin{cases} \mu_v E/(1+E/E_c) & \text{if } E \le E_c \\ v_{max} & \text{if } E > E_c \end{cases} \tag{4.87}$$

where

$$E_c = \frac{2v_{max}}{\mu_v} \tag{4.88}$$

The drain current at any point x along the channel is given by:

$$I_{DS} = W_{eff} Q_I(x) v(x) \tag{4.89}$$

Recall from Equation (4.21) that:

$$Q_I(x) = C_{ox}[V_{GS} - V_T - V(x)] \tag{4.90}$$

where V_T is the extrapolated threshold voltage which is given by the following relation [see Equations (4.13) and (4.36)]:

$$V_T = V_T(long\text{-}channel)$$

$$- qN_a W_D \frac{x_j}{2C_{ox}L_{eff}} \left[\sqrt{1+2W_S/x_j} + \sqrt{1+2W_D/x_j} - 2\right] \quad (4.91)$$

The depletion layer width under the inversion layer, W_D, can be calculated for buried-channel MOS devices from the model in[15]. From the velocity expression [Equation 4.88] and Equations (4.87) and (4.90), we can write:

$$E(x) = \frac{dV(x)}{dx} = \frac{I_{DS}}{W_{eff}\mu_v C_{ox}(V_{GS} - V_T - V(x)) - \frac{I_{DS}}{E_c}} \quad (4.92)$$

Separating variables and integrating Equation (4.92), we get:

$$I_{DS}\int_0^{L_{eff}} dx = \int_0^{V_{DS}} [\mu_v C_{ox} W_{eff}(V_{GS} - V_T - V(x)) - \frac{I_{DS}}{E_c}]dV(x) \quad (4.93)$$

or

$$I_{DS} = \mu_v C_{ox} \frac{W_{eff}}{L_{eff}} \frac{1}{(1 + \frac{V_{DS}}{E_c} L_{eff})} (V_{GS} - V_T - \frac{V_{DS}}{2})V_{DS} \quad (4.94)$$

Equation (4.94) is valid for the triode region only. In saturation (i.e. $v(x)=v_{max}=\mu_v E_c/2$), the drain current can be derived by substituting from Equations (4.90) and (4.87) into Equation (4.89) to get:

$$I_{DSsat} = \frac{\mu_v}{2} C_{ox} W_{eff}(V_{GS} - V_T - V_{DSsat})E_c \quad (4.95)$$

By equating (4.95) and (4.94) we can derive the following expression for V_{DSsat}:

$$V_{DSsat} = (1-K)(V_{GS} - V_T) \quad (4.96)$$

where

$$K = \frac{1}{1+E_c\dfrac{L_{eff}}{(V_{GS}-V_T)}} \tag{4.97}$$

The drain current in the saturation can be rewritten as:

$$I_{DSsat} = Kv_{max}C_{ox}W_{eff}(V_{GS}-V_T) \tag{4.98}$$

The effective mobility due to the vertical electric field is approximated by the following empirical form[16]:

$$\mu_v = \mu_0\left(\frac{E_0}{E_{eff}}\right)^{C_1} \tag{4.99}$$

where the effective field in the inversion layer is given by[4]:

$$E_{eff} = \frac{1}{\varepsilon_s}(Q_B + \alpha Q_I) \tag{4.100}$$

α=1/2 for electrons and α=1/3 for holes. Q_B and Q_I can be calculated for buried-channel PMOS devices using the model given in[15]. The empirical parameter C_1 is 1/3 for electrons and holes provided that the gate oxide thickness is between 10 and 40 nm and the surface doping is between 10^{16} and $1.5\times10^{17}cm^{-3}$. The μ_0 is defined at an arbitrary field E_0. For E_0=0.1MV/cm, μ_0 for electrons is 620$cm^2/V.s$ and for holes is 160$cm^2/V.s$[17]. This analytical model can be used for gate length down to 0.5μm. A modification and extension of that model has been discussed in[18].

4.2. Modeling of the Bipolar Transistor

4.2.1. BJT Structure and Operation

Fig. 4.15 shows a cross-sectional view of a NPN bipolar junction transistor with geometrical layout and the corresponding symbols for NPN and PNP. To understand the basic operation of the bipolar transistor, one dimensional representation of the active region can be used. Fig. 4.16(a) illustrates a typical profile of the one-dimensional section of the active region [Fig. 4.16(b)]. The N^+PN^- sandwich forms the heart of BJT.

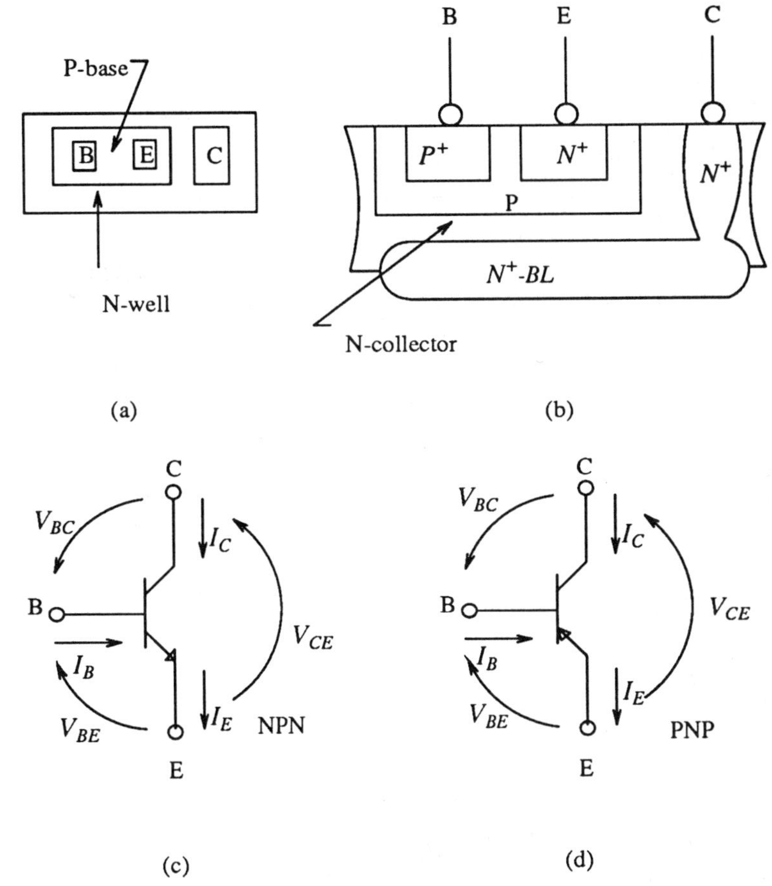

Figure 4.15 **(a) The layout, (b) The cross-sectional view and (c) The symbol of an NPN BJT, (d) The symbol of an PNP BJT.**

Consider an NPN transistor with $V_{BE}>0.5V$ and $V_{BC}<0V$ (forward-active mode). The corresponding energy band diagram is shown in Fig. 4.16(c). When the N^+P (emitter-base) junction is forward-biased, electrons are injected from the emitter into the base (current I_{nE}). A small fraction of these electrons recombine in the neutral base (I_{rB})[7]. The rest of the electrons, of which the current I_{nC} is constituted, diffuse

[7] In order to reduce I_{rB}, the width of the neutral base should be narrow (i.e. $W_B \ll L_p$, where L_p is the diffusion length of the electrons in the base).

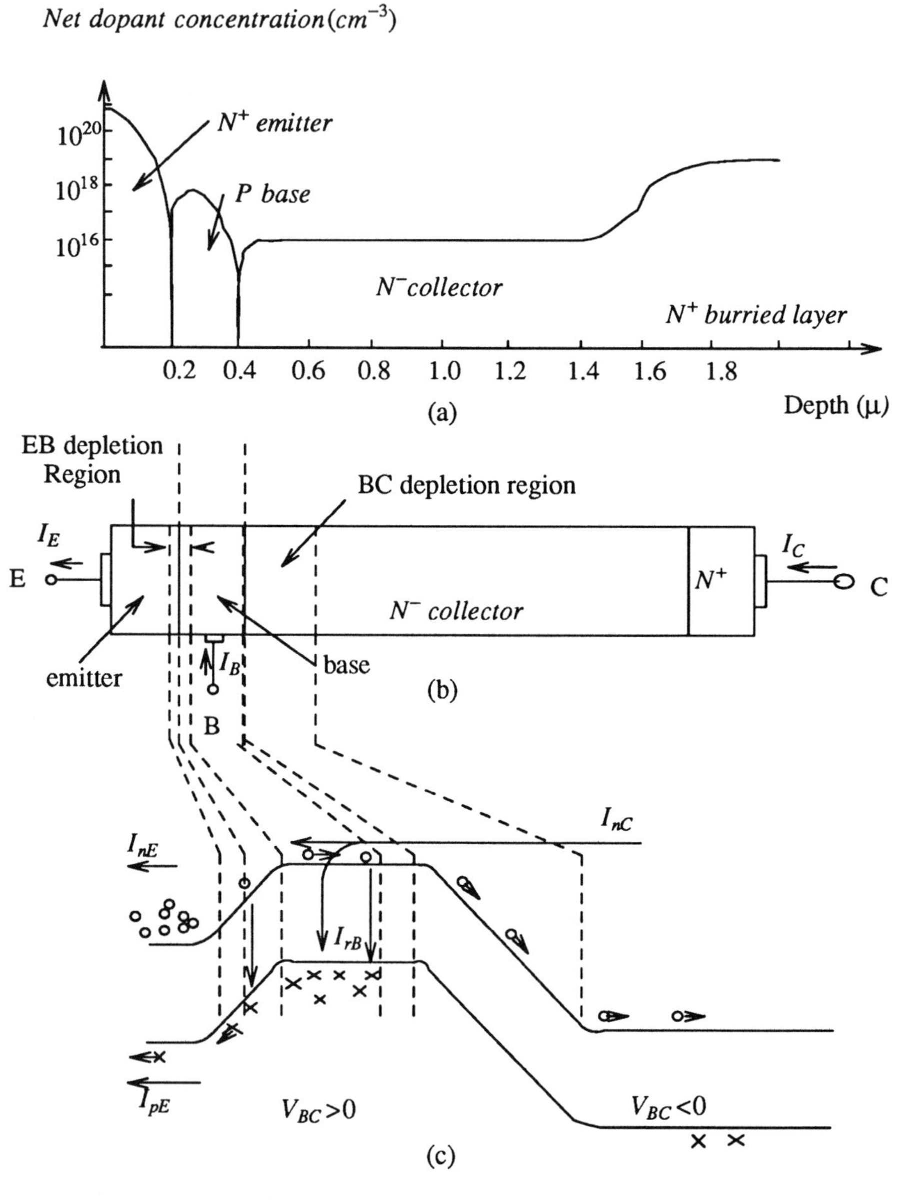

Figure 4.16 (a) Doping profile of a BJT, (b) One dimensional cross-sectional view of a BJT in the active region, (c) Energy band diagram of a BJT in the active region.

through the base towards the reverse-biased base-collector junction where they are swept by the electric field into the base-collector depletion layer. On the other hand, some of the holes in the base are injected into the N^+ emitter region resulting in a current I_{pE}. This component is small compared to I_{nE} because the holes' concentration in the base is much smaller than the electron concentration in the emitter. The emitter-base depletion layer can be a site for the recombination between the injected electrons and holes resulting in a current I_{rscl}. Moreover, some holes are swept into the base due to the generation in the base-collector depletion region, but this component is very small ($\approx 10^{-17} A/\mu m^2$). The terminal currents can be written as follows:

$$I_C = I_{nC} \tag{4.101}$$

$$I_B = I_{pE} + I_{rscl} + I_{rB} \tag{4.102}$$

$$I_E = I_{nE} + I_{rscl} + I_{pE} \tag{4.103}$$

Note that it has been assumed that the base and collector currents are flowing in the device, while the emitter current is flowing out of it [Fig. 4.16]. The emitter injection efficiency, which is defined as the ratio of the electron's current injected into the base to the total emitter current, is given by:

$$\gamma_E = \frac{I_{nE}}{I_{nE}+I_{rscl}+I_{pE}} \tag{4.104}$$

This ratio γ_E has to be near unity; that is, the emitter current should mostly be due to electrons for an NPN transistor. The ratio

$$\beta = \frac{I_C}{I_B} \tag{4.105}$$

is defined as the DC current gain.

When the emitter-base junction is reverse-biased and the collector-base junction is forward-biased, the transistor is in the inverse region where the emitter and collector may be exchanged. When both junctions are reverse-biased the transistor is in the *cutoff* region. But when they are forward-biased, the device is said to be in the saturation region. In this situation, both junctions are injecting into the base, the small electric fields in the two depletion regions sweep the carriers into the emitter and collector regions. Both junctions collect as well as emit.

4.2.2. Ebers-Moll Model

In this section, we present the Ebers-Moll (EM) model, which is a simple DC model of the bipolar transistor. The Ebers-Moll model can be used for hand calculations and first order circuit analysis. The derivation of the model equations, in this section, is based on the analysis by Roulston[19]. In Section 4.2.1, we have discussed the device operation in the forward active region only. For a general analysis, we assume that the base-emitter and the base-collector junctions are forward biased. In the following discussion we will neglect the currents due to recombination in the space charge layers and in the base. This implies that $I_{nC} = I_{nE}$[8], hence, Equation (4.103) reduces to:

$$I_E = I_{nC} + I_{pE} \tag{4.106}$$

The current due the holes injected from the base into the emitter is given by[19]:

$$I_{pE} = \frac{q\,A_E\,D_{pE}\,p_{nE0}}{W_E}\left[e^{V_{BE}/V_t} - 1\right] \tag{4.107}$$

where p_{nE0} is the equilibrium hole concentration in the emitter and W_E is the neutral emitter width. The current I_{nC} is dominated by the diffusion current in the base and is proportional to the gradient of the minority carriers (electrons) in the neutral base. Because the neutral base width (W_B) is very thin, this gradient is approximately a constant. Therefore, we can write I_{nC} as[19]:

$$I_{nC} = q\,A_E\,D_{nB}\left[\frac{n_B(0) - n_B(W_B)}{W_B}\right] \tag{4.108}$$

where $n_B(0)$ and $n_B(W_B)$ are the electron concentrations at the edges of the emitter-

---

[8] By applying KCL (i.e. $I_C + I_B - I_E = 0$), we can easily see that I_{rB} is the difference between I_{nE} and I_{nC}. If the recombination in the base is neglected ($I_{rB} = 0$), we can assume that $I_{nE} \approx I_{nC}$.

base and collector-emitter depletion regions respectively [see Fig. 4.17]. Note that the slope of the electrons in the base is given by the term between the brackets as demonstrated by Fig. 4.17.

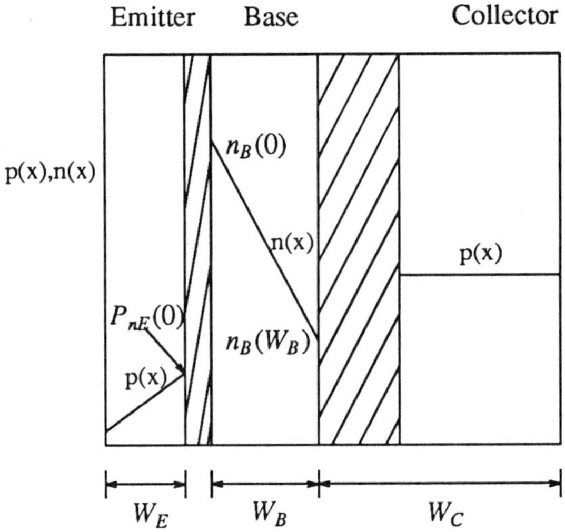

Figure 4.17 Minority carrier distribution in a BJT with both junctions forward biased.

Using the junction law, the electron concentrations $n_B(0)$ and $n_B(W_B)$, can be expressed in terms of V_{BE} and V_{BC} respectively. The current I_{nC} can hence be given by[19]:

$$I_{nC} = \frac{q\, A_E\, D_{nB}}{W_B\, N_B} \left[e^{V_{BE}/V_t} - e^{V_{BC}/V_t} \right] \tag{4.109}$$

where N_B is the base impurity concentration.

The collector current is given by:

$$I_C = I_{nC} - I_{pC} \tag{4.110}$$

The current I_{pC} is due to the holes injected from the base to the collector[9]. The

[9] Note that I_{pC} was not included in Equation (4.101) because in deriving Equation (4.101) we have assumed that the collector-base junction was reverse biased.

base-collector junction is basically a P^+NN^+ structure as shown in Fig. 4.16(a). An expression for I_{nC} can be derived from the analysis of a P^+NN^+ diode. The reader is adviced to consult with reference [19] for the details of this analysis. The current I_{pC} is given by[19]:

$$I_{pC} = \frac{q\, A_E\, p_{nC0}\, W_C}{\tau_{epi}} \left[e^{V_{BC}/V_t} - 1 \right] \tag{4.111}$$

where p_{nC0} is the equilibrium hole concentration in the collector, W_C is the epitaxial thickness under the base and τ_{epi} is the hole lifetime in the epitaxial layer. By substituting from Equations (4.107) and (4.109) in Equation (4.106) and from Equations (4.109) and (4.111) in Equation (4.110) we get the following equations for I_E and I_C:

$$I_E = I_f - \alpha_r I_r \tag{4.112}$$

$$I_C = -I_r + \alpha_f I_f \tag{4.113}$$

where

$$I_f = qA_E \left(\frac{D_{pE}\, p_{nE0}}{W_E} + \frac{D_{nB}\, n_{B0}}{W_B} \right) (e^{V_{BE}/V_t} - 1) \tag{4.114}$$

$$I_r = qA_E \left(\frac{D_{nB}\, n_{B0}}{W_B} + \frac{p_{nC0}\, W_C}{\tau_{epi}} \right) (e^{V_{BC}/V_t} - 1) \tag{4.115}$$

$$\alpha_f I_f = \alpha_r I_r = \frac{q\, A_E\, D_{nB}\, n_{B0}}{W_B} \tag{4.116}$$

Equations (4.112) and (4.113) are called the Ebers-Moll equations. Fig. 4.18 shows the equivalent circuit of the BJT based on the Ebers-Moll equations.

The Ebers-Moll model described above is general and can be used for any region of operation by substituting for V_{BE} and V_{BC} by the appropriate values. In the forward active region, assuming that V_{BE}=0.8V and V_{BC}<0.3V the emitter and collector current of Equations (4.112) and (4.113) reduce to:

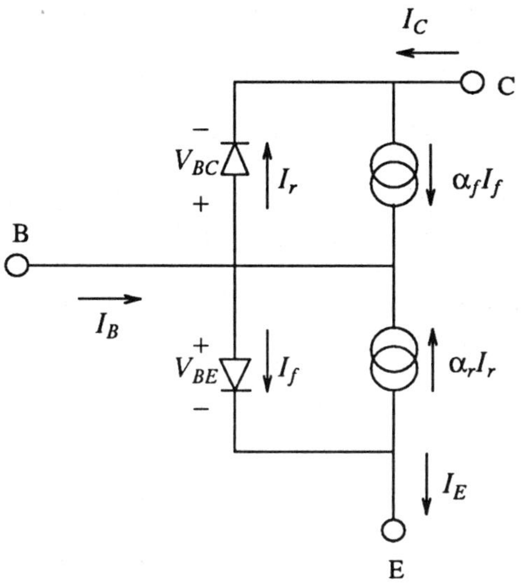

Figure 4.18 Equivalent DC circuit of the BJT based on the Ebers-Moll model.

$$I_E = I_f \approx I_{Es}\, e^{V_{BE}/V_t} \tag{4.117}$$

and

$$I_C = \alpha_f I_f \approx \alpha_f I_{Es}\, e^{V_{BE}/V_t} = \alpha_f I_E \tag{4.118}$$

where the reverse saturation current of the base-emitter junction I_{Es} can be derived from Equation (4.114) and is given by:

$$I_{Es} = qA_E \left(\frac{D_{pE}\, p_{nE0}}{W_E} + \frac{D_{nB}\, n_{B0}}{W_B} \right) \tag{4.119}$$

It can easily be shown that the base current can be expressed as

$$I_B = \frac{1-\alpha_f}{\alpha_f} I_C \tag{4.120}$$

Equations (4.117), (4.118) and (4.120) are the well-known current equations of a forward biased bipolar transistor. Note that Equation (4.120) yields the famous relation between α_f and the DC forward current gain β [$\beta=\alpha_f/(1-\alpha_f)$].

The simple Ebers-Moll model lacks accuracy for the following three reasons:

1- It does not account for the parasitic resistors of the emitter, base and collector.

2- It does not account for the Early effect, which causes the collector current to increase as the collector-emitter voltage increases.

3- It does not account for the effect of the high collector currents on the current gain.

Next, we will discuss the modeling of each phenomena separately.

The Parasitical Resistors of a Bipolar Transistor

Fig. 4.19 shows the modification of the EM model by the addition of the base resistance R_B, the collector resistance R_C and the emitter resistance R_E. These extrinsic components represent the transistor's parasitic resistances from their active region to their base, collector and emitter terminals, respectively.

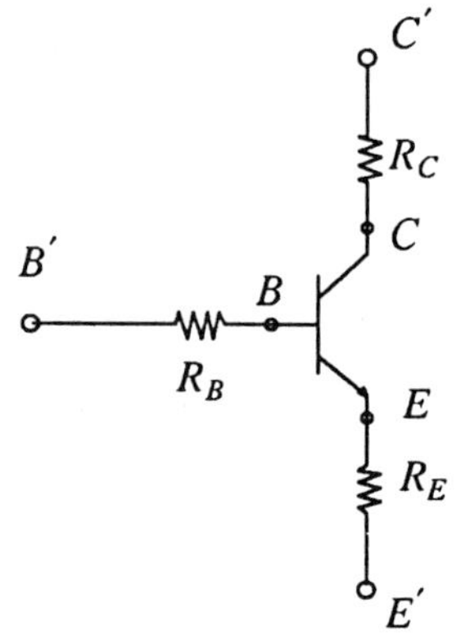

Figure 4.19 The extrinsic resistances of a bipolar transistor.

The effect of the parasitic resistances is important because the voltage drop across them contribute to the external base-emitter and collector-emitter voltages $V_{B'E'}$ and $V_{C'E'}$ respectively, as shown by the following two equations.

$$V_{B'E'} = V_{BE} + R_B I_B + R_E I_E \tag{4.121}$$

$$V_{C'E'} = V_{CE} + R_C I_C + R_E I_E \tag{4.122}$$

The drop across the parasitic resistors has to be accounted for to get more accurate results from the EM model. Neglecting these drops may even lead to erroneous results. For example, if the external collector-emitter voltage is found to be equal to $2V$ one may deduce that the BJT operates in the active region. However, if R_C=$1.8K$ and R_E=$0.1K$ and $I_C \approx I_E = 1mA$, then the intrinsic collector-emitter voltage (V_{CE}) is $0.1V$. This implies that the bipolar transistor is actually saturated. This phenomenon is known as *Quasi-Saturation.*

The Early Effect

The Early effect refers to the base width modulation due to the change of the collector base reverse voltage (in the forward active region). As the collector-base reverse voltage increases, the base-collector depletion layer widens. The resulting reduction in the neutral base width causes the current gain to increase which, in turn, leads to an increase in the collector current [see Fig. 4.20]. This effect can be modeled by introducing the Early voltage (V_{Af}) in the expression of the collector current as follows:

$$I_C = I_s e^{V_{BE}/V_t} \left(1 + \frac{V_{CE}}{V_{Af}}\right) \tag{4.123}$$

The inverse of the forward Early voltage $1/V_{Af}$ is analogous to the coefficient λ in an MOS transistor. A typical value of V_{Af} is 50 V. The AC output resistance of the BJT in the forward active region is related to the Early voltage and is given by:

$$r_o = \frac{V_{Af}}{I_C} \tag{4.124}$$

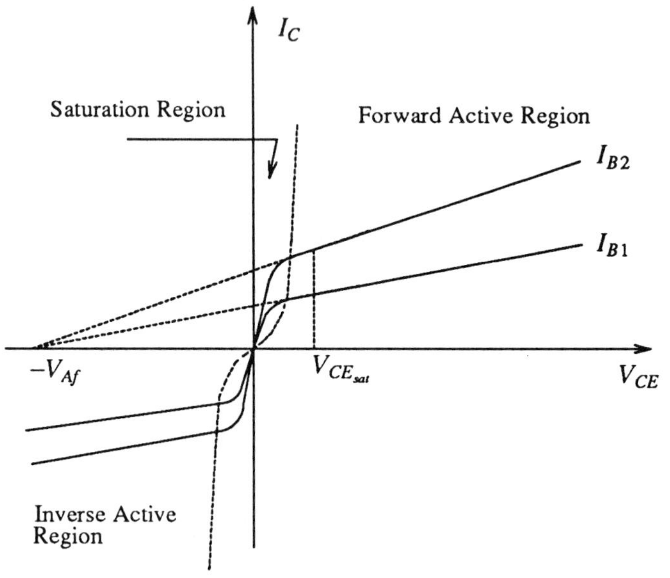

Figure 4.20 The I-V characteristics of a BJT.

The Early effect in the inverse active region can be modeled by using the reverse Early voltage (V_{Ar}) which characterizes the slope of the collector current in that region (inverse active region).

High Current Effects

The current gain and the cut-off frequency are degraded due to high collector current. Fig. 4.21 shows the effect of the collector current on the gain. This degradation can be referred to the high level injection in the base (Webster effect) and/or the base pushout (Kirk effect). For a detailed discussion on these phenomenon, the reader is advised to consult reference[19]. In the case, where the injection level in the base is high (Webster effect) the collector current can be expressed as[19]:

$$I_C = \sqrt{I_s I_{Kf}}\, e^{V_{BE}/2V_t} \tag{4.125}$$

where the knee current I_{Kf} is defined as the collector current at which its slope in the

[9] A typical value of I_{Kf} per unit area is $1 mA/\mu m^2$.

Gummel plot changes from 1 to ½ [see Fig. 4.22]. This current marks the onset of high level injection. The degradation of the current gain, when $I_C > I_{Kf}$, can be described by the following relation[19]:

$$\beta = \frac{I_C}{I_B} = \beta_0 \frac{I_{Kf}}{I_C} \tag{4.126}$$

where β_0 is the value of the gain when $I_C < I_{Kf}$. The modeling of the Kirk effect is very complex. However, a simple model for the current gain, which can be used in first order circuit analysis, is given below[20]:

$$\beta = \frac{\beta_0}{1 + \dfrac{I_C}{I_{Kf}}} \tag{4.127}$$

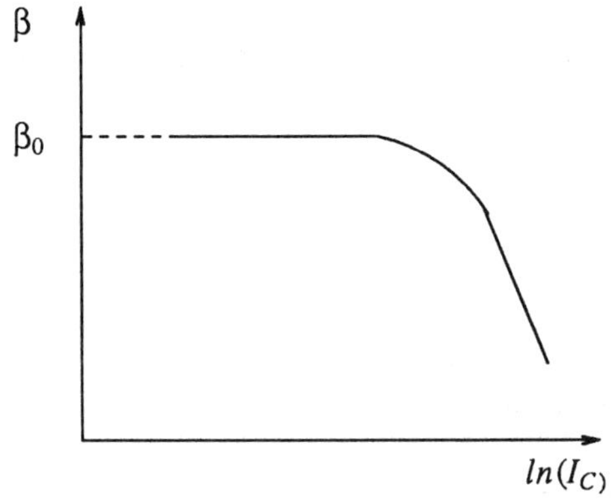

Figure 4.21 Current gain vs collector current at medium and high currents.

The accuracy of the simple EM model can be enhanced by accounting for the parasitic resistors, the Early effect and high current effect which can be modeled by simple analytical expressions as shown above.

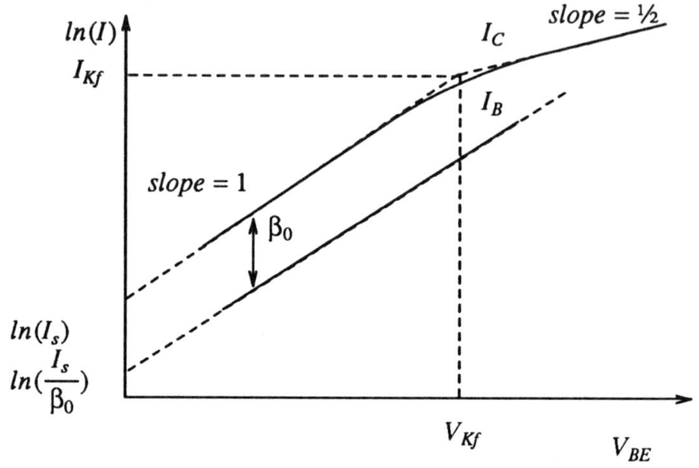

Figure 4.22 Variation of $ln(I_C)$, $ln(I_B)$ as function of V_{BE}.

4.2.3. Bipolar Models in SPICE

Two BJT models are implemented in SPICE. The Ebers-Moll model and a more sophisticated one, which is based on the Gummel-Poon (GP) model[21]. The second model includes the following second order effects:

- Very low current Effect on the gain.
- Base width modulation effect.
- High-level injection effects (the Kirk effect is not included)
- Base resistance variation with current.

The GP model is based on one-dimensional analysis. It is valid for all regions of operation: cutoff, forward-active, inverse-active, and saturation. The GP-based bipolar model is illustrated by the equivalent circuit shown in Fig. 4.23. The two back-to-back diodes on the right represent the intrinsic base-emitter and base-collector junctions and their currents are given by[22]:

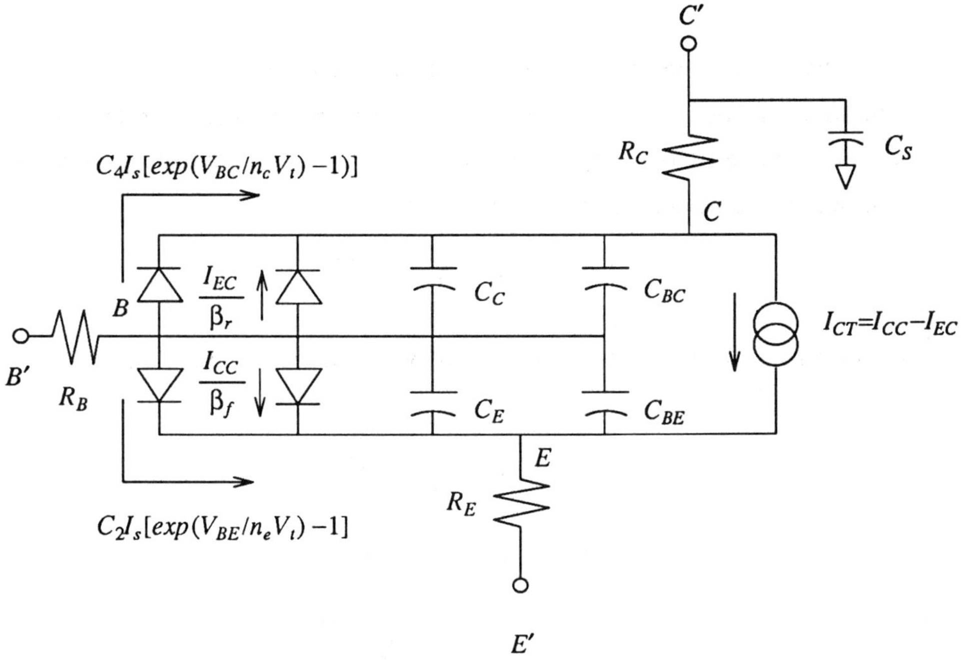

Figure 4.23 The GP-based model of a bipolar transistor.

$$I_{CC} = \frac{I_s}{q_b}(e^{V_{BE}/n_f V_t} - 1) \tag{4.128}$$

$$I_{EC} = \frac{I_s}{q_b}(e^{V_{BC}/n_r V_t} - 1) \tag{4.129}$$

where I_s is given by[22]:

$$I_s = \frac{q\, D_{nB}\, n_i^2\, A_E}{\int_0^{W_B} N_B(x)dx} \qquad (4.130)$$

The forward and reverse current emission coefficient (n_f and n_r), which are introduced in Equations (4.128) and (4.129), are used to model the low currents. The parameter q_b (base charge factor) accounts for the high current and base width modulation effects. It is given by[22]:

$$q_b = \frac{q_1}{2}[1 + \sqrt{1 + 4q_2}] \qquad (4.131)$$

q_1 models the effects of base width modulation and can be expressed as:

$$q_1 = \frac{1}{(1 - \frac{V_{BC}}{V_{Af}} - \frac{V_{BE}}{V_{Ar}})} \qquad (4.132)$$

q_2 models the high-level injection and is expressed as:

$$q_2 = \frac{I_s}{I_{Kf}}(e^{V_{BE}/n_f V_t} - 1) + \frac{I_s}{I_{Kr}}(e^{V_{BC}/n_r V_t} - 1) \qquad (4.133)$$

The general expression of q_b [Equation (4.130)] can be simplified for low-level and high-level injection conditions.

$$q_b = \begin{cases} q_1 & \text{if } q_2 \ll q_1^2/4 \quad (low-level\ injection) \\ \sqrt{q_2} & \text{if } q_2 \gg q_1^2/4 \quad (high-level\ injection) \end{cases} \qquad (4.134)$$

The two back-to-back diodes on the left [Fig. 4.23] account for the currents caused by the recombination of carriers in the emitter-base and the collector-base space-charge layers and other recombinations. These currents can be modeled by[22]:

$$C_2 I_s(e^{V_{BE}/n_e V_t} - 1) \qquad (4.135)$$

$$C_4 I_s(e^{V_{BE}/n_c V_t} - 1) \qquad (4.136)$$

where C_2, C_4, n_e and n_c have been introduced to fit the measured currents.

Further improvements to this model are possible by the inclusion of three parasitic resistances (R_C, R_B, R_E); three junction capacitances (C_E, C_C, C_S); and two diffusion capacitances (C_{dE}, C_{dC}) as shown in Fig. 4.23.

The model of the base resistance takes into account the effect of the current (current crowding) through the following expression[23]:

$$R_B(z) = R_{Bm} + 3(R_B - R_{Bm})\frac{\tan(z) - z}{z\ \tan(z)^2} \tag{4.137}$$

where the variable z is given by:

$$z = \frac{-1 + \sqrt{1 + 144\ I_B/(\pi^2 I_{rB})}}{\frac{24}{\pi^2}\sqrt{(I_B/I_{rB})}} \tag{4.138}$$

R_B represents the low-current maximum resistance and R_{Bm} high-current minimum resistance.

The junction depletion capacitance is a function of the junction voltage (V). This function can be approximated by the following two expressions:

$$C_{jdep} = C_j(1 - \frac{V}{\phi_j})^{-M_j} \qquad \text{if } V<FC\phi_j \tag{4.139}$$

$$C_{jdep} = C_j\frac{1 - FC(1 + M_j) + M_j\frac{V}{\phi_j}}{(1 - FC)^{1+M_j}} \qquad \text{if } V>FC\phi_j \tag{4.140}$$

The empirical factor FC has a value between 0 and 1. Its default value in SPICE is 0.5. Note that Equations (4.139) and (4.140) apply for a reverse and forward biased junction respectively.

The diffusion capacitances model the charge associated with injected carriers. For example, the electrons injected in the base have a corresponding storage charge

$$Q_{DE} = \tau_f I_{CC} \tag{4.141}$$

The forward transit time τ_f is current-dependent and is given by[23]:

$$\tau_f = \tau_{f0}\left(1 + XTF\left(\frac{I_{CC}}{I_{CC}+ITF}\right)^2 e^{V_{BC}/1.44VTF}\right) \tag{4.142}$$

Where VTF is a fitting parameter to model the change of τ_f as a function of V_{BC} (or V_{CE}), ITF models the change due to I_C and XTF controls the increase of τ_f.

The diffusion capacitance (associated with the injected electrons from the emitter into the base, when the base-emitter junction is forward biased) is given by:

$$C_{DE} = \frac{\partial Q_{DE}}{\partial V_{BE}} \tag{4.143}$$

Similarly, the base-collector junction has a diffusion capacitance, which is given by:

$$C_{DC} = \frac{\partial Q_{DC}}{\partial V_{BC}} \tag{4.144}$$

where

$$Q_{DC} = \tau_r I_{EC} \tag{4.145}$$

Although the SPICE models account for most of the first and second order effects, they are not highly accurate. This originates from some weaknesses in the theory on which the models are based. As the device features are scaled down the currently available models become less accurate. The physics and the theory of the scaled devices is more complex. Hence, accurate modeling becomes very difficult. One way around that problem is to chose the model parameters such that simulated device characteristics agree with measurements. In practice, the models' parameters are extracted automatically using parameter analyzers with software tools to obtain the best fit. As a result, the values of the extracted parameters may not correspond to their actual values. For example, it is common to find a discrepancy of 20% between the measured current gain of a bipolar transistor and that listed in the SPICE file. Another approach, which is equivalent to tweaking the parameters, is to use empirical models (e.g. BISM model), in which the empirical (fitting) parameters can be optimized to get the best fit between simulation and measurements.

Typical GP parameters, for the 0.8μm BiCMOS presented in Chapter 2., are shown in Table 4.2.

Table 4.2

SPICE BJT model parameters (0.8μm BiCMOS process)

Parameter	SPICE Keyword	Description	Value	Units
I_s	IS	Saturation current	3×10^{-18}	A
β_f	BF	Ideal maximum forward gain	100	-
β_r	BR	Ideal maximum reverse gain	4	-
n_f	NF	Forward current-emission coefficient	1	-
n_r	NR	Reverse current-emission coefficient	1	-
V_{Af}	VAF	Forward early voltage	50	V
V_{Ar}	VAR	Reverse early voltage	4.7	V
I_{Kf}	IKF	Forward-knee current	2×10^{-3}	A
I_{Kr}	IKR	Reverse-knee current	9×10^{-3}	A
I_{se}	ISE	Base-emitter leakage saturation current	1×10^{-16}	A
I_{sc}	ISC	Base-collector leakage saturation current	1×10^{-16}	A
n_e	NE	Base-emitter leakage emission coefficient	1.5	-
n_c	NC	Base-collector leakage emission coefficient	2.0	-
R_E	RE	Emitter resistance	10	Ω
R_C	RC	Collector resistance	100	Ω
R_B	RB	Base resistance at zero current	450	Ω
I_{rB}	IRB	Base current where R_B=$R_B(0)/2$	0.	A
R_{Bm}	RBM	Minimum high-current base resistance	1×10^{-4}	Ω
C_E	CJE	Base-emitter zero-bias depletion cap.	1.21×10^{-14}	F
ϕ_{je}	VJE	Base-emitter built-in potential	0.988	V

M_{je}	MJE	Base-emitter junction grading factor	0.33	-
C_C	CJC	Base-collector zero-bias depletion cap.	0.96×10^{-14}	F
ϕ_{jc}	VJC	Base-collector built-in potential	0.808	V
M_{jc}	MJC	Base-collector junction grading factor	0.33	-
C_S	CJS	Collector-substrate zero-bias cap.	2.52×10^{-14}	F
ϕ_{js}	VJS	Collector-substrate built-in potential	0.60	-
M_{js}	MJS	Collector-substrate junction grading factor	0.33	-
-	XCJC	Internal base fraction of base-collector cap.	1.0	-
FC	FC	Coefficient for forward-bias depletion cap.	0.5	-
τ_f	TF	Forward transit time	1.2×10^{-12}	s
XTF	XTF	TF bias-dependent coefficient	7.2	-
VTF	VTF	TF base-collector voltage dependence coef.	1.3	-
ITF	ITF	TF high current parameter	4.5×10^{-3}	s
τ_r	TR	Reverse transit time	4×10^{-9}	s
-	XTB	Forward and reverse β temp. exponent	2.0	-
-	XTI	Saturation current temperature exponent	6.0	-
-	EG	Energy gap	1.11	eV
-	KF	Flicker noise coefficient	0.0	-
-	AF	Flicker noise exponent	1.0	-

4.3. Chapter Summary

In this Chapter, we have reviewed the fundamentals of the MOS and bipolar devices. The most common device models used in SPICE have been presented. The key device parameters of each model have been defined and explained, so that the reader is familiar with the details of these models and can appreciate the importance of the different model parameters. The reader is given a list of model parameters, for a typical 0.8 μm BiCMOS process, that can be used for circuit simulations.

References

1. A. Vladimirescu, and S. Liu, *The Simulation of MOS Integrated Circuits using SPICE2*, Memo. No. UCB/ERL M80/7, Univ. California, Berkeley (Oct. 1980).

2. L.D. Yau, "A Simple Theory to Predict the Threshold Voltage of Short-Channel IGFET's," *Solid State Electronics*, **17** pp. 1059-1063 (1974).

3. P. Smith, M. Inoue, and J. Frey, "Electron Velocity in Si GaAs at Very High Electric Fields," *Appl. Phys. Lett.*, **Vol. 37** pp. 797-798 (1980).

4. A.G. Sabnis and J.T. Clemens, "Characterization of Electron Mobility in the Inverted <100> Si Surface," *International Electron Devices Meeting Tech. Dig.*, pp. 18-21 (1979).

5. D. Frohman-Bentchkowsky, "On the Effect of Mobility Variation on MOS Device Characteristics," *Proc. IEEE*, **Vol. 56**(1968).

6. R.M. Swanson and J.D. Meindl, "Ion-Implanted Complementary MOS Transistors in Low-Voltage Circuits," *IEEE Journal of Solid-State Circuits*, **Vol. SC-7** pp. 146-153 (1972).

7. G. Baum and H. Beneking, "Drift Velocity Saturation in MOS Transistors," *IEEE Trans. on Electron Devices*, **Vol. ED-17** pp. 481-482 (1970).

8. T. Ikeda, A. Watanabe, Y. Nishio, I. Masuda, N. Tamba, M. Okada, and K. Ogiue, "High-Speed BiCMOS Technology with a Buried Twin Well Structure," *IEEE Trans. on Electron Devices*, **Vol. ED-34, No. 6** pp. 1304-1309 (June 1987).

9. R.L.M. Dang, "A Simple Current Model for Short-Channel IGFET and its Application to Circuit Simulation," *IEEE Journal of Solid-State Circuits*, **Vol. SC-14** pp. 358-367 (1979).

10. G. Merkel, J. Borel and N.Z. Cupcea, "An Accurate Large Signal MOS Transistor Model for Use in Computer-Aided Design," *IEEE Trans. on Electron Devices*, **Vol. ED-19**(1972).

11. B.J. Sheu, D.L. Scharfetter, P.-K. Ko, and M.C. Jeng, "BSIM: Berkeley Short-Channel IGFET Model for MOS Transistors," *IEEE Journal of Solid-State Circuits*, **Vol. SC-22** pp. 558-566 (1987).

12. D.E. Ward and R.W. Dutton, "A Charge-Oriented Model for MOS Transistors Capacitances," *IEEE Journal of Solid-State Circuits*, **SC-13** pp. 703-707 (1978).

13. C.G. Sodini, P. Ko, and J.L. Moll, "The Effect of High Fields on MOS Device and Circuit Performance," *IEEE Trans. on Electron Devices*, **Vol. ED-31, No. 10** pp. 1386-1393 (1984).

14. B. Hoefflinger, H. Sibbert, and G. Zimmer, "Model and Performance of Hot-Electron MOS Transistor for VLSI," *IEEE Trans. on Electron Devices*, **Vol. ED-26** p. 513 (1979).

15. J.S.T. Huang, J.W. Schrankler, and J.S. Kueng, "Short-Channel Threshold Model for Buried-Channel MOSFET," *IEEE Trans. on Electron Devices*, **Vol. ED-31** pp. 1889-1894 (1984).

16. G. Baccarani, et al., "Generalized Scaling Theory and its Application to a 1/4 Micrometer MOSFET Design," *IEEE Trans. on Electron Devices*, **ED-31, No. 4** pp. 452-462 (Apr. 1984).

17. M. Kakumu, M. Kinugawa, and K. Hashimoto, "Choice of Power-Supply Voltage for Half-Micrometer and Lower Submicrometer CMOS Devices," *IEEE Trans. on Electron Devices*, **Vol. ED-37** pp. 1334-1342 (1990).

18. K-Y. Toh, "An Enginnering Model for Short-Channel MOS Devices," *IEEE Journal of Solid-State Circuits*, **23 No. 4** pp. 950-958 (Aug. 1988).

19. D.J. Roulston, *Bipolar Semiconductor Devices,* McGraw-Hill Publishing Company (1990).

20. K. Nakazato, et al., "Characteristics and Scaling Properties of n-p-n Transistors with a Sidewall Base Contact Structure," *IEEE Trans. on Electron Devices*, **ED-32, No 2** pp. 328-332 (Feb. 1985).

21. H.K. Gummel and H.C. Poon, ''An Integral Charge Control Model of Bipolar Transistors,'' *Bell Syst. Tech. J.*, **Vol. 49**(1970).

22. I. Getreu, *Modeling the Bipolar Transistor*, Tektronix, Inc. (1976).

23. P. Antognetti and G. Massobrio, ''Semiconductor Device Modeling with SPICE,'' *McGraw Hill*, (1988).

CHAPTER 5

MOS DIGITAL INTEGRATED CIRCUITS[1]

The dual purpose of this chapter is to review the design and to analyze basic MOS digital circuits using first-order dc and transient design equations. Digital NMOS circuits are covered for completeness. They also highlight commonly used circuit analysis and design techniques. Circuit techniques as well as circuit structures are presented including CMOS transmission gate logic, dynamic CMOS, domino logic, NORA and Zipper circuit structures, CMOS nonthreshold logic and cascade voltage switch logic.

5.1. The Static NMOS Inverter

Fig. 5.1(a) shows an NMOS inverter consisting of an enhancement type driver transistor and a load. The load is one of the following: (a) a saturated enhancement-type NMOS device, (b) a nonsaturated enhancement-type NMOS device, (c) a depletion-type NMOS device, or (d) a polysilicon resistor.

Fig. 5.1(b) shows the load lines of the above four loads super-imposed on the I_{DS} vs. V_{DS} of the driver. The intersection of the load line with the driver's characteristics for $V_{GS} = V_{in} = V_0$, gives $V_{DS} = V_{out} = V_1$ where V_0 is the low voltage level representing logical '0' [2], $V_0 < V_{TD}$, V_{TD} is the threshold voltage of the driver, and V_1 is the high voltage level representing logical '1'. Similarly, the intersection of the load line with the driver's characteristics for $V_{GS} = V_{in} = V_1$, gives $V_{DS} = V_{out} = V_0$.

---

[1] This chapter is based on a tutorial paper by M.I. Elmasry[1].

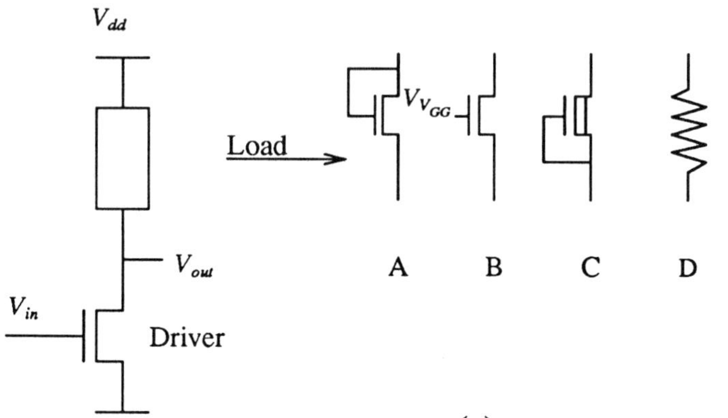

(a)

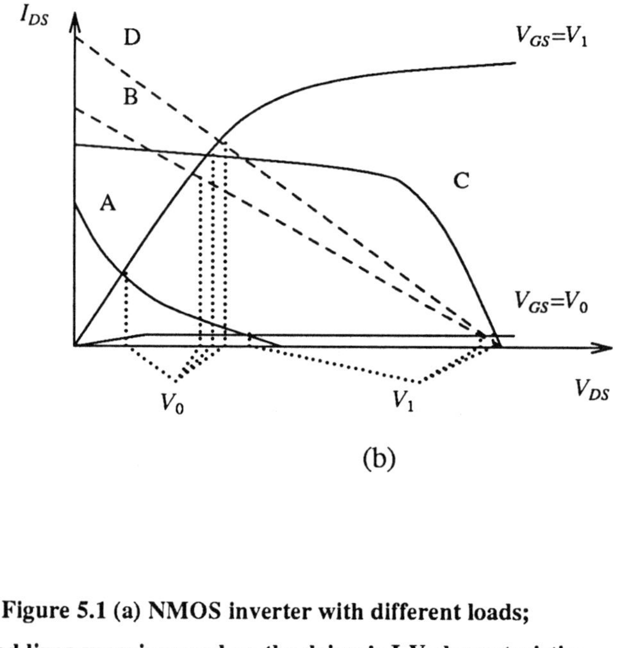

(b)

Figure 5.1 (a) NMOS inverter with different loads; (b) load lines superimposed on the driver's I-V characteristics.

5.1.1. DC Analysis

5.1.1.1. Saturated Enhancement-Type Load

The saturated enhancement-type load was used in the early digital NMOS integrated circuits. Fig. 5.2 shows an inverter using NMOS devices. The load device has its gate connected to its drain (i.e., $V_{DS} = V_{GS}$), and operates in the saturation region when it conducts, since $V_{DS} > (V_{GS} - V_{TL})$, where V_{TL} is the threshold voltage of the load device.

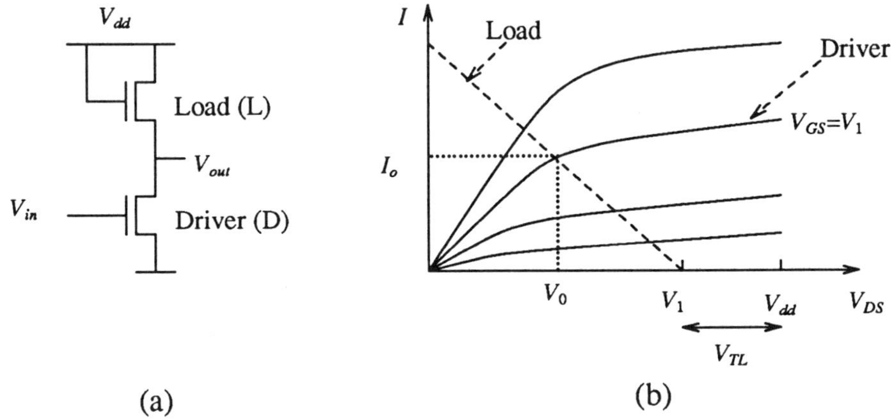

Figure 5.2 (a) NMOS inverter with saturated enhancement-type load; (b) load line superimposed on the driver I_{DS} vs V_{DS} characteristic, showing the operating current I_o when $V_{out}=V_o$.

The dc circuit operation is explained as follows:

(a) when $V_{in} = V_0 < V_{TD}$, the driver is off, $I \approx 0$ and

$$V_{out} = V_1 = V_{dd} - V_{TL} \tag{5.1}$$

(b) when $V_{in} = V_1$, $V_1 > V_{TD}$, the driver is operating in the nonsaturation region, the load device is in saturation and $I = I_0$ where:

$$I_0 = \frac{K'(\frac{W}{L})_L}{2} [(V_{dd} - V_0) - V_{TL}]^2 \tag{5.2.a}$$

$$= K' (\frac{W}{L})_D [(V_1 - V_{TD}) V_0 - ½ V_0^2] \tag{5.2.b}$$

If $V_0 \ll V_{dd}$ and $V_0 \ll V_1 - V_{TD}$ then Equation (5.2) can be simplified:

$$I_0 \approx \frac{K' (\frac{W}{L})_L}{2} (V_{dd} - V_{TL})^2 = \frac{K' (\frac{W}{L})_L}{2} V_1^2 \tag{5.3.a}$$

$$\approx K' (\frac{W}{L})_D (V_1 - V_{TD}) V_0 \tag{5.3.b}$$

From Equation (5.3) we obtain:

$$\frac{(\frac{W}{L})_D}{(\frac{W}{L})_L} \geq \frac{V_1}{2V_0} \tag{5.4}$$

and

$$V_0 \approx \frac{I_0}{K'(\frac{W}{L})_D (V_1 - V_{TD})} \tag{5.5}$$

Thus from Equation (5.4) if V_1/V_0 is taken to be 10, $(W/L)_D/(W/L)_L$ should be ≥5. Because the ratio of (W/L) of the driver to that of the load must be greater than unity, the circuit is referred to as "a ratioed circuit." To summarize: the voltage levels of the NMOS inverter V_1, V_0, V_t, where V_t is the logic threshold of the inverter, and the dc power dissipation are given by:

$$V_1 = V_{dd} - V_{TL}$$

$$V_0 \approx \frac{I_0}{K' \left(\frac{W}{L}\right)_D (V_1 - V_{TD})}$$

$$V_t \geq V_{TD}$$

$$P_{DC}(@\, V_{out} = V_1) \approx 0$$

$$P_{DC}(@\, V_{out} = V_0) = I_0 V_{dd} \tag{5.6}$$

where I_0 are given by Equation (5.3.a).

In Equation (5.6) I_0 is determined by transient considerations or the allowable power dissipation. V_{dd} is usually fixed by sub-system design considerations and today it is typically = 5 V. It is clear that V_1 increases as the power supply voltage V_{dd} increases. The maximum allowable V_{dd} must be less than the junction breakdown voltage of the drain-substrate junction. It should also be less than the voltage at which the parasitic field-oxide MOS transistors start to conduct.

Because the substrate-source junction of the load device is more reverse biased than the driver, V_{TL} is larger than V_{TD} due to the back-gate bias effect. The higher the V_{BB} value of the load, the higher the value of V_{TL}, and the lower the value of V_1. Because $V_1 = V_{dd} - V_{TL}$, the saturated enhancement-type MOS load is said to introduce "threshold losses" in the value of V_1. As we explain in the following sections, depletion-type nonsaturated enhancement type and bootstrapped MOS loads do not introduce threshold losses.

The low logical voltage level, V_0, is a function of the operating current I_0. As I_0 increases, the dc power dissipation increases and the speed also increases as we explain in Section 5.2.2. V_0 can be reduced (hence increasing the logic swing and the noise margin NM_o) by increasing the geometrical ratio $(W/L)_D$ of the driver transistor, hence increasing its size, with respect to $(W/L)_L$.

The threshold voltage V_t of the inverter[2] is $\geq V_{TD}$. The higher the ratio $[(W/L)_D/(W/L)_L]$ is, the closer V_t to V_{TD} becomes. If the driver is replaced with a number of stacked transistors connected in a series, as in the case of a MOS NAND gate, then V_{TD} and hence V_T of the upper input transistors are higher than that of the lower transistors because of the back-gate bias effect. This is a drawback of using NAND gates in NMOS logic design.

5.1.1.2. Nonsaturated Enhancement-Type Load

One drawback of using a saturated enhancement-type load is the threshold voltage losses caused by the load, resulting in $V_1 = V_{dd} - V_{TL}$.

This situation can be rectified if the load is operating in the nonsaturation region by connecting its gate to V_{GG} where $V_{GG} > (V_{dd} + V_{TL})$. In this case V_1 can be made close to V_{dd} by increasing V_{GG} as shown in Fig. 5.1(b). The inverter dc circuit operation is explained as follows:

(a) when $V_{in} = V_0$, $V_0 < V_{TD}$ the driver is off, the load is nonsaturated and $V_{out} = V_1 = V_{dd} - V_{DS}|_{load}$, where $V_{DS}|_{load}$ is the voltage drop across the load device. If the inverter is driving only MOS gates, which is usually the case with high input impedances, then $V_{DS}|_{load} \approx 0$ and $V_1 \approx V_{dd}$.

It should be noted that $V_{DS}|_{load}$ can be reduced for a given load current by increasing $(W/L)_L$ or $(V_{GG} - V_{TL})$ at the expense of increasing the gate power dissipation.

(b) when $V_{in} = V_1$, $V_1 > V_{TD}$, the driver is operating in the nonsaturation region, the load is nonsaturated and $I = I_0$ where

$$I_0 = K' \left(\frac{W}{L}\right)_L [(V_{GG} - V_{TL} - V_0)(V_{dd} - V_0) - \tfrac{1}{2}(V_{dd} - V_0)^2] \tag{5.7.a}$$

$$= K' \left(\frac{W}{L}\right)_D [(V_1 - V_{TD}) V_0 - \tfrac{1}{2} V_0^2] \tag{5.7.b}$$

If $V_0 \ll V_{dd}$, $V_0 \ll V_1 - V_{TD}$, $V_{GG} - V_{TL} \gg V_{dd}$ then Equation (5.7) can be simplified:

$$I_0 \approx K' \left(\frac{W}{L}\right)_L (V_{GG} - V_{TL}) V_{dd} \tag{5.8.a}$$

$$\approx K' \left(\frac{W}{L}\right)_D (V_1 - V_{TD}) V_0 \tag{5.8.b}$$

From Equation (5.8) we obtain

$$\frac{(\frac{W}{L})_D}{(\frac{W}{L})_L} \geq \frac{V_{GG} - V_{TL}}{V_0} \tag{5.9}$$

and

$$V_0 \approx \frac{I_0}{K' \, (\frac{W}{L})_D \, (V_1 - V_{TD})} \tag{5.10}$$

where I_0 is given by Equation (5.8.a).

In summary, using a nonsaturated enhancement-type load offers a higher V_1, and hence higher logic swing and higher noise margins. Moreover, it offers an improvement in transient performance, as shown in Section 5.2.2. This improved performance is obtained at the expense of adding the extra power supply V_{GG}.

5.1.1.3. Depletion-Type Load (E/D NMOS Logic)

Depletion-type devices are used as loads in NMOS E/D logic circuits. They offer a voltage logic level $V_1 \approx V_{dd}$ as shown in Fig. 5.1(b). In addition, the constant current characteristic improves the transient performance of the circuit[3]. The inverter dc circuit operation is explained as follows:

(a) when $V_{in} = V_0$, $V_0 < V_{TD}$, the driver is off, the load is nonsaturated and similar to the nonsaturated enhancement-type case:

$$V_{out} = V_1 = V_{dd} - V_{DS}|_{load} \approx V_{dd}$$

It should be noted that $V_{DS}|_{load}$ can be reduced for a given load current by increasing $(W/L)_L$ or $|V_{TL}|$.

(b) when $V_{in} = V_1$, $V_1 > V_{TD}$, the driver is operating in the nonsaturation region, the load is saturated and $I = I_0$ where

$$I_0 = \frac{K'(\frac{W}{L})_L}{2} V_{TL}^2 \tag{5.11.a}$$

$$= K'(\frac{W}{L})_D [(V_1 - V_{TD}) V_0 - \tfrac{1}{2} V_0^2] \tag{5.11.b}$$

If $V_0 \ll V_1 - V_{TD}$, then Equation (5.11) can be simplified:

$$I_0 = \frac{K'(\frac{W}{L})_L}{2} V_{TL}^2 \tag{5.12.a}$$

$$\approx K'(\frac{W}{L})_D (V_1 - V_{TD}) V_0 \tag{5.12.b}$$

From Equation (5.12) we obtain:

$$\frac{(\frac{W}{L})_D}{(\frac{W}{L})_L} \geq \frac{V_{TL}^2}{2V_0(V_1 - V_{TD})} \tag{5.13}$$

$$V_0 \approx \frac{I_0}{K'(\frac{W}{L})_D (V_1 - V_{TD})} \tag{5.14}$$

where I_0 is given by Equation (5.12.a). It should be noted here that because of the back-gate bias effect on the depletion load, V_{TL}, and hence I_0, should be either taken as an average or a worst case value.

5.1.1.4. Resistive Load

Polyresistors can be used in the design of NMOS digital integrated circuits, especially static memories. As shown in Fig. 5.1(b), a resistive load provides $V_1 \approx V_{dd}$. In fact, the nonsaturated enhancement-type MOS load dc characteristic approaches that of a resistance when $V_{GG} \gg V_{dd} + V_{TL}$ as shown in Fig. 5.1(b), case (b_2). The inverter dc circuit operation is explained as follows:

(a) when $V_{in} = V_0$, $V_0 < V_{TD}$, the driver is off and similar to the nonsaturated enhancement-type case:

$$V_{out} = V_1 = V_{dd} - V|_{load} \approx V_{dd}$$

Similarly $V|_{load}$ can be reduced for a given load current by decreasing the value of the load resistance R.

(b) when $V_{in} = V_1, V_1 > V_{TD}$, the driver is operating in the nonsaturation region and $I = I_0$ where

$$I_0 = \frac{V_{dd} - V_0}{R} \tag{5.15.a}$$

$$= K'(\frac{W}{L})_D \left[(V_1 - V_{TD})\, V_0 - ½\, V_0^2\right] \tag{5.15.b}$$

If $V_0 \ll V_{dd}$, then Equation (5.15) can be simplified

$$I_0 \approx \frac{V_{dd}}{R} \tag{5.16.a}$$

$$\approx K'\,(\frac{W}{L})_D\,(V_1 - V_{TD})V_0 \tag{5.16.b}$$

From Equation (5.16) we obtain

$$RK'\,(\frac{W}{L})_D \geq \frac{V_{dd}}{V_0(V_1 - V_{TD})} \tag{5.17}$$

$$V_0 \approx \frac{I_0}{K'\,(\frac{W}{L})_D\,(V_1 - V_{TD})} \tag{5.18}$$

$$= I_0\, R_D$$

where I_0 is given by Equation (5.16.a) and R_D is the ON resistance of the driver device.

5.1.2. Transient Analysis

The transient performance of a MOS inverter is a function of two times [see Fig. 5.3]:

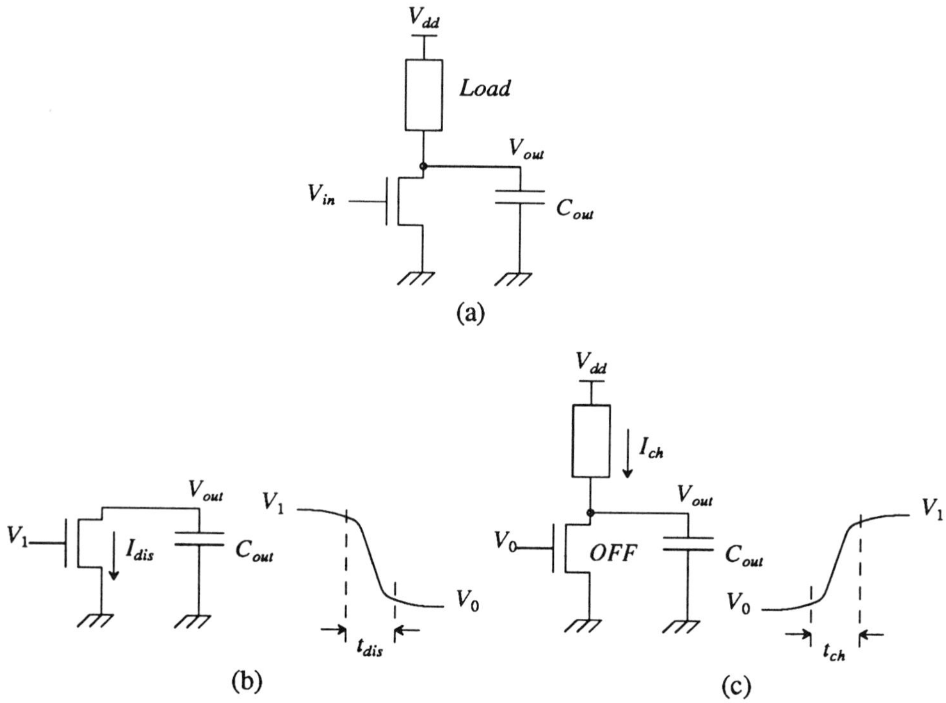

Figure 5.3 (a) NMOS inverter showing output node capacitance, (b) simple circuit for discharging, (c) simple circuit for charging.

1. The discharging time t_{dis}, which is the time taken by the output node capacitance C_{out} to discharge through the driver transistor, from V_1 to V_0, where V_1 and V_0 are chosen to be 0.9 and 0.1 of their dc values respectively.

2. The charging time t_{ch}, which is the time taken by C_{out} to charge, through the load from V_0 to V_1, where V_1 and V_0 are chosen to be 0.9 and 0.1 of their dc values respectively.

In calculating t_{dis} and t_{ch} we will direct our attention to simplified expressions for the purpose of using them in the first phase of circuit design. The approach follows that of Crawford[4]. It is important to remember here that the time (t) taken by a capacitance (C) to change its voltage by ΔV, using a constant current (I), is given by the simple relation $t = C\Delta V/I$. In the following analysis we shall assume that the input voltage to the MOS inverter is an ideal voltage step.

5.1.2.1. The Discharging Time t_{dis}

As shown in the dc analysis of the MOS inverter, the ratio of

$$(\frac{W}{L})_D \ to \ (\frac{W}{L})_L$$

is large, hence the on resistance of the driver transistor is much less than the on resistance of the load transistor. As a result, during the discharging of the output node capacitance, when $V_{in} > V_{TD}$ and the driver is on, it is reasonable to neglect the current through the load. Hence, the discharging circuit can be simplified, as shown in Fig. 5.3(b), to C_{out} with initial output voltage $= V_1$ and an on driver transistor with $V_{in} = V_1$. The final output voltage $= V_0$.

During the discharging time, the driver is initially in the saturation region and the constant drain current:

$$I_D = \frac{\beta_D}{2}(V_1 - V_{TD})^2$$

will partially discharge C_{out} in a time given by

$$(t_{dis})_1 = \frac{C_{out}\Delta V}{I_D} \tag{5.19}$$

where

$$\Delta V = V_1 - V_{out}''$$

and V_{out}'' is the output voltage at which the driver begins operating in the nonsaturation region (i.e., $V_{out}'' = V_1 - V_{TD}$).

Equation (5.19) can be written in the form

$$(t_{dis})_1 = 2\frac{C_{out}}{g_{mD}}\frac{V_{TD}}{V_1 - V_{TD}} \tag{5.20}$$

where

$$g_{mD} = \beta_D(V_1 - V_{TD})$$

When the driver transistor begins to operate in the nonsaturation region at $V_{out} = V''_{out}$, the available current to discharge the output capacitance is not a constant and the discharging time $(t_{dis})_2$ for C_{out} to discharge from V''_{out} to V_0 can be obtained from the solution of the differential equation which results from equating the capacitance current through C_{out} to the current through the driver:

$$C_{out}\frac{dV_{out}}{dt} = -\beta_D(V_1 - V_{TD})\,V_{out} + \frac{\beta_D}{2}V_{out}^2$$

which gives

$$V_{out} = (V_1 - V_{TD})\frac{2e^{-t/\tau_D}}{1 + e^{-t/\tau_D}} \tag{5.21}$$

where

$$\tau_D = C_{out}/g_{mD}$$

$$g_{mD} = \beta_D(V_1 - V_{TD})$$

Solving Equation (5.21) for $(t_{dis})_2$ as a difference between the time at which $V_{out} = V''_{out}$ and the time at which $V_{out} = V_0$ gives

$$(t_{dis})_2 = \tau_D\, ln\left[\frac{2(V_1 - V_{TD})}{V_0} - 1\right] \tag{5.22}$$

From Equations (5.20) and (5.22):

$$t_{dis} = \tau_D \left(\frac{2V_{TD}}{V_1 - V_{TD}} + ln \left[\frac{2(V_1 - V_{TD})}{V_0} - 1 \right] \right) \tag{5.23}$$

As shown from Equation (5.23), t_{dis} is reduced if (V_1/V_{TD}) is increased, hence the higher the logical '1' voltage level, the lower the discharging time.

Because $(t_{dis})_2 \gg (t_{dis})_1$, $V_1 \gg V_{TD}$, and $V_1 \gg V_0$, Equation (5.23) can be simplified:

$$t_{dis} = \tau_D \, ln \frac{2V_1}{V_0} \approx \tau_D \, ln \frac{V_1}{V_0} = \frac{C_{out}}{\beta_D \, (V_1 - V_{TD})} \, ln \frac{V_1}{V_0} \tag{5.23.a}$$

i.e., t_{dis} is linearly proportional to C_{out}, β_D^{-1} and $(V_1 - V_{TD})^{-1}$, and is logarithmically proportional to V_1/V_0. Thus t_{dis} can be reduced for a given C_{out}, by increasing $(W/L)_D$ or increasing the driving voltage factor $(V_1 - V_{TD})$.

5.1.2.2. The Charging Time t_{ch}

During the charging time the driver is off and C_{out} will charge through the load as shown in Fig. 5.3(c).

Case 1: Saturated Enhancement-Type Load

Equating the capacitive current through C_{out} to that through the load we obtain the differential equation for the output voltage V_{out}:

$$C_{out} \frac{dV_{out}}{dt} = \frac{\beta_L}{2} \, [V_{dd} - V_{out} - V_{TL}]^2$$

With the initial condition:

$$at \; t = 0, \; V_{out} = V_0$$

the solution is:

$$V_{out} = (V_1 - V_0) \frac{t/\tau_L}{2 + \dfrac{t}{\tau_L}} + V_0$$

hence

$$t_{ch} = \tau_L \left(\frac{V_1 - V_0}{V_0} \right) = \tau_L \left(\frac{V_1}{V_0} - 1 \right) = \tau_L \frac{V_l}{V_0} \tag{5.24}$$

where

$$\tau_L = \frac{C_{out}}{g_{mL}}$$

$$g_{mL} = \beta_L (V_{dd} - V_{TL})$$

From (5.24) it is clear that the charging time is proportional to the time constant τ_L. Reducing C_{out}, increasing β_L (i.e., the size of the load device), or the voltage factor $(V_{dd} - V_{TL})$ will decrease t_{ch}. Moreover, reducing the logic swing V_l (or reducing V_1/V_0), at the expense of reducing the noise margins, will also reduce t_{ch}. As the output voltage rises during t_{ch}, the substrate-source bias of the load increases and hence V_{TL} increases. Thus, in calculating g_{mL}, V_{TL} may be taken as an average value in the voltage range $V_0 \le V_{out} \le V_1$. However, a pessimistic value of t_{ch} can be obtained if V_{TL} is taken as the maximum value where $V_{out} = V_1$.

Case 2: Nonsaturated Enhancement-Type Load

Similarly, the differential equation describing the output voltage V_{out} is given by:

$$C_{out} \frac{dV_{out}}{dt} = \frac{\beta_L V_{dd}^2}{2m} \left(1 - \frac{V_{out}}{V_{dd}} \right) \left(1 - m \frac{V_{out}}{V_{dd}} \right)$$

Calculating t_{ch} requires calculating

$$t_{ch} = \int_{t@V_{out}=V_0}^{t@V_{out}=V_1} dt = C_{out} \frac{2m}{\beta_L V_{dd}} \int_{V_1/V_{dd}}^{V_0/V_{dd}} \frac{d(V_{out}/V_{dd})}{\left(1 - \frac{V_{out}}{V_{dd}} \right) \left(1 - m \frac{V_{out}}{V_{dd}} \right)}$$

which gives

$$t_{ch} = \tau_c \frac{1}{1-m} \, ln \left[\frac{\left(1 - m \dfrac{V_1}{V_{dd}}\right)\left(1 - \dfrac{V_0}{V_{dd}}\right)}{\left(1 - m \dfrac{V_0}{V_{dd}}\right)\left(1 - \dfrac{V_1}{V_{dd}}\right)} \right]$$

$$= 2 \frac{C_{out}}{\beta_L V_{dd}} \frac{m}{1-m} \, ln \left[\frac{\left(1 - m \dfrac{V_1}{V_{dd}}\right)\left(1 - \dfrac{V_0}{V_{dd}}\right)}{\left(1 - m \dfrac{V_0}{V_{dd}}\right)\left(1 - \dfrac{V_1}{V_{dd}}\right)} \right]$$

$$0 \le m < 1 \tag{5.25}$$

where

$$m = \frac{V_{dd}}{2(V_{GG} - V_{TL}) - V_{dd}}$$

The factor m is a measure of the biasing conditions on the load. If $V_{GG} \gg V_{dd}$ and $V_{GG} \gg V_{TL}$, m tends to zero and the MOS load transistor acts like a resistance. As m approaches unity, i.e., $(V_{GG} - V_{TL})$ approaches V_{dd}, the nonsaturated load approaches the saturated load. Note that Equation (5.25) is not valid for $m = 1$.

It is clear that as m approaches zero, t_{ch} approaches zero. Reducing C_{out}, increasing β_L, or increasing V_{dd} will decrease t_{ch}. Because of the back-gate bias effect during charging, in calculating m, V_{TL} may be taken as a pessimistic value as is done in Case 1 or can be taken as an average value for the voltage range $V_0 \le V_{out} \le V_1$. For example, it can be taken at:

$$V_{out} = \frac{V_1 + V_0}{2}$$

Case 3: Depletion-Type Load

Assuming that the load current is constant during the charging time as the output voltage rises from V_0 to V_1 then

$$t_{ch} = \frac{C_{out} V_l}{I_0} \tag{5.26}$$

where I_0 is given by Equation (5.11.a) and an average value of V_{TL} is used. The charging time can be reduced by decreasing $V_l\, C_{out}$ or increasing I_0, where I_0 can be increased by increasing V_{TL} or β_L.

Case 4: Resistive Load

From the analysis of a simple RC circuit:

$$t_{ch} = R_L C_{out} ln \frac{V_{dd} - V_0}{V_{dd} - V_1} \tag{5.27}$$

5.1.3. NMOS Delay-Power Tradeoffs

It has been shown previously that the NMOS static inverter consumes dc power only when the output voltage is low. Thus,

$$P_{DC}|_{av} = \frac{1}{2} I_0 V_{dd} \tag{5.28}$$

where I_0 is the inverter dc current when the output voltage is low. In addition to this component of power dissipation, there is a transient power component due to the switching of the output node capacitance C_{out}. This is given by:

$$P_t = C_{out} V_l^2 f \tag{5.29}$$

where V_l is the logic swing and f is the switching frequency. The maximum switching frequency is given by: $f_{\max} = 1/(t_{ch} + t_{dis})$.

In order to *highlight* the tradeoff that exists in the design of MOS digital circuits between delay and power dissipation, let us consider the case of a static inverter with a depletion-type load and assume that $t_{dis} \ll t_{ch}$ and the inverter is on half of the time. Thus, the total power dissipation at the maximum switching frequency is given by:

$$P = \frac{1}{2} I_0 V_{dd} + (C_{out} V_l^2 / t_{ch})$$

and the inverter average delay time is given by:

$$\tau_D = \frac{t_{ch} + t_{dis}}{2} \approx \frac{t_{ch}}{2} = \frac{C_{out} V_l}{2 I_0}$$

Thus, the delay-power produced is given by:

$$\tau_D P = \frac{1}{4} C_{out} V_l V_{dd} + \frac{1}{2} V_l^2 C_{out}$$

$$\approx \frac{3}{4} C_{out} V_{dd}^2 \tag{5.30}$$

This shows that the delay-power product is proportional to $C_{out} V_{dd}^2$.

5.2. NMOS Circuit Configurations

5.2.1. NMOS Source-Followers and Push-Pull Drivers

In addition to the basic NMOS inverter discussed in Section 5.2.1. and Section 5.2.2, other circuit configurations are also used in the design of NMOS digital circuits. Two of these are the source-follower and the push-pull driver (super buffer). Source-followers are used to reduce the capacitive loading at a given node while push-pull drivers are used to drive high capacitive loads, e.g., off-chip loads. Because of the close relationship between the two circuits, they are discussed here with the aid of Fig. 5.4.

The static inverter in Fig. 5.4 consists of a driving transistor Q and a load. The output node of that inverter (node 2) is driving the source-follower transistor Q_1. The load of that source-follower is the switchable transistor Q_2. The capacitive loading of node 2 is the gate-source and the gate-drain capacitances of Q_1, rather than C_{out}. Thus, Q_1 isolates node 2 from the load capacitance C_{out}.

Q_1 and Q_2 are operating in the push-pull mode to charge and discharge C_{out}, thus the name "push-pull driver." As Q_1 is turned on, Q_2 is turned off and vice versa. It can be assumed that the circuit consists of a source-follower transistor Q_1 with a

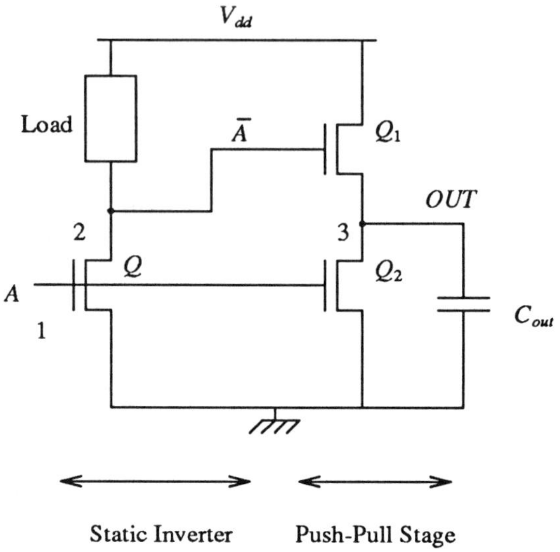

Figure 5.4 NMOS push-pull driver.

switchable load Q_2, or a driving transistor Q_2 with a switchable load Q_1. The push-pull operation allows high charging and discharging currents resulting in low switching times. The transistor Q_1 could be an enhancement or a depletion-type. If it is an enhancement-type, the push-pull driver does not consume dc power since either Q_1 or Q_2 is on during the steady state. Thus, there is no dc requirement on the ratio $[(W/L)_2/(W/L)_1]$ and the circuit configuration is "ratioless". However, the voltage at node 3 suffers from threshold losses. In the case of a depletion-type load, threshold losses are eliminated at the expense of dc power consumption. The circuit in this case must be ratioed.

5.2.2. Bootstrapped Loads

It has been shown in Section 5.2.1.2. that when operating the NMOS load transistor in the nonsaturation region by tying its gate to V_{GG}, $V_{GG} > V_{dd} + V_{TL}$, threshold losses are eliminated and the transient performance is improved. These two results can be obtained without the need for the extra power supply, V_{GG}, by bootstrapping the load.

Bootstrapping allows the gate of the NMOS load transistor to rise above V_{dd} *during switching*, thus allowing the load to operate in the nonsaturation region. The basic circuit is shown in Fig. 5.5(a) and consists of the load transistor Q_1, a biasing transistor Q_2, and a MOS capacitor C. The biasing transistor Q_2 allows node 1 to float at $(V_{dd} - V_{T2})$, where V_{T2} is the threshold voltage of Q_2. As the input voltage drops, the output voltage rises by an amount $= \Delta V$. Because of the capacitive coupling, node 1 rises by an amount $= \Delta V'$, $\Delta V' = \Delta V C/(C_1 + C)$, i.e., if $C \gg C_1$, where C_1 is the parasitic capacitance of node 1, then $\Delta V' \approx \Delta V$. In the limit $\Delta V = V_{dd}$ and the voltage at node 1 can reach up to $(2V_{dd} - V_{T2})$ allowing Q_1 to operate in the nonsaturation region.

It should be noted that as the voltage of node 1 rises above V_{dd}, Q_2 shuts off and isolates node 1 from the rest of the circuit.

It should also be noted that bootstrapping is a dynamic operation occurring when the output voltage switches. If the output node capacitance C_{out} is large and the output voltage rises slowly, then the bootstrap coupling between the output node and node 1 occurs at a slower rate. This can be resolved by isolating the output node capacitance C_{out} from the bootstrapped load using a source-follower as shown in Fig. 5.5(b). The isolating stage consists of a bootstrapped transistor Q_3 and a driver transistor Q_4. The output node is now isolated from the bootstrapped load.

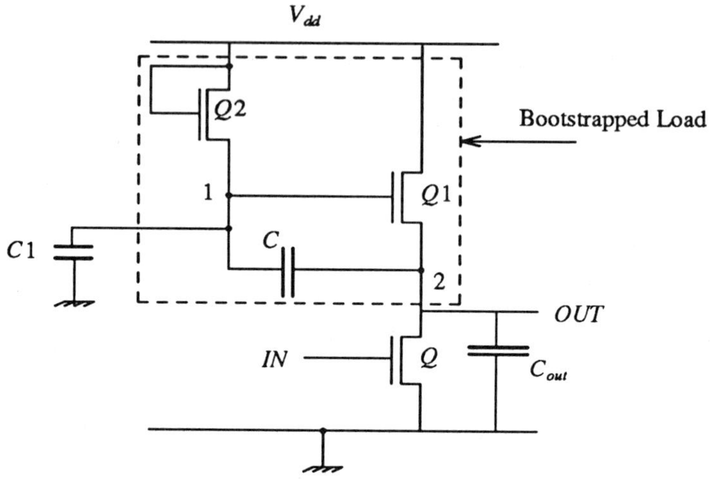

(a)

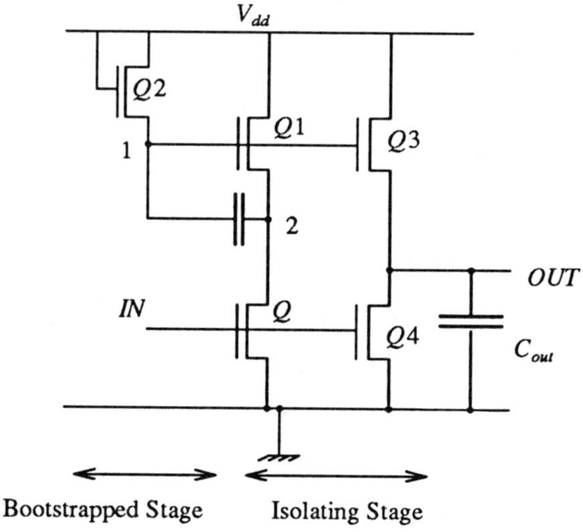

(b)

Figure 5.5 (a) Using Bootstrapped load; (b) Using isolated Bootstrapped load.

5.2.3. NMOS Transmission Gates

The NMOS transmission gate is used in the design of static and dynamic NMOS digital circuits. It consists of a single transistor as shown in Fig. 5.6. The gate is simply a symmetrical switch controlled by the voltage at node 1. If that voltage is greater than V_T, then nodes 2 and 3 are connected through the on resistance of the NMOS device. If that voltage is less than V_T then the NMOS device is off and nodes 2 and 3 are separated by an open circuit. Thus, a transmission gate can be used as a two-input AND gate where the inputs are applied to nodes 1 and 2 (or 3) and the output to node 3 (or 2).

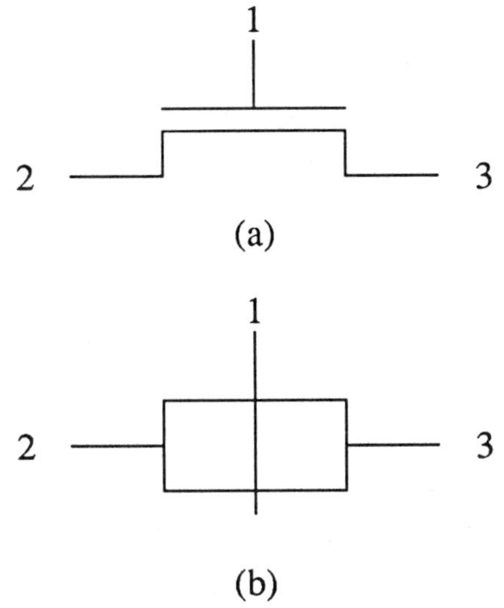

Figure 5.6 NMOS transmission gate, (a) circuit schematic, (b) logic symbol.

The back-gate bias effect should be considered in the design of NMOS transmission gates. Thus V_T is a function of the voltage at nodes 2 and 3. Nodes 2 and 3 take the role of a source or drain depending on the values of their respective voltages. The on resistance of the NMOS device can be reduced by increasing (W/L) or the voltage driving factor $[(V\ at\ node\ 1) - V_T)]$.

Because of the capacitive coupling between node 1 and nodes 2 and 3, a feed-through voltage appears at nodes 2 and 3 as the voltage at node 1 changes. The effect of the feed-through voltages on the overall circuit performance, where the transmission gate is used, should be evaluated.

5.2.4. NMOS Current Mode Logic

Since the introduction of MOS devices in the early 1960's, the devices have been used mainly in a pull-up-pull-down (PUD) circuit configuration. The configuration in single channel static NMOS logic has serious shortcomings[5]: high delay-power product, high sensitivity to interconnections, unsymmetrical static and dynamic operation (unequal noise margins and unequal charging and discharging times) and a logic swing which is highly dependent on the value of the power supply and difficult to change by design to meet different circuit requirements. The most important shortcoming is the configuration's inability to deliver bipolar-like performance with delays in the nanosecond range even with high-performance VLSI NMOS devices.

One circuit configuration which can be used for MOS logic is the current-mode logic (CML) configuration. CML offers symmetrical high-speed operation, equal noise margins, and equal rise and fall times. The symmetrical operation allows operation at lower logic swing, thus enhancing the speed of operation even further. Moreover, the availability of the output logic function and its complementary (the NOR and the OR functions) offers flexibility in logic design and provides two complementary output signals to drive long interconnections.

One basic MOS CML gate is shown in Fig. 5.7[5]. It consists of a differential stage and two source-follower stages. The reference voltage V_r is established inherently by Q_R and Q_{IN}[5]. V_r is chosen to be $V_{dd}/2$, i.e., in a 5 V design $V_r = 2.5\ V$ where

$$V_r \approx |\, V_T \,|_{Q_R} + V_T \,|_{Q_{IN}}$$

where $V_T|_{Q_R}$ and $V_T|_{Q_{IN}}$ are the threshold voltages of Q_R and Q_{IN}, respectively, at the operating values of the substrate bias.

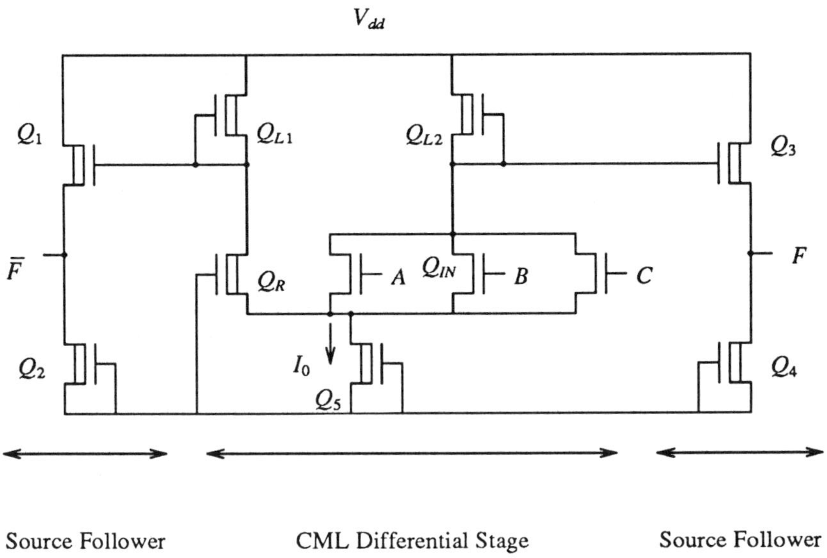

Figure 5.7 Basic NMOS CML gate.

The design of a MOS CML gate starts with choosing a value for V_r and determining the allowable tolerances on this value. The tighter these tolerances are, the lower the logic swing of the gate, the higher the speed, and the lower the logic swing of the gate. We assume here that reasonable tolerances on V_r, which is assumed to be 2.5 V, are ± 500 mV. If we further assume noise margins of 500 mV, then the logic levels of the gate are $V_1 \geq 3.5\ V$ and $V_0 \leq 1.5\ V$. Thus, the minimum logic swing is 2 V, as shown in Fig. 5.8.

Because the operation of the MOS CML gate relies on the gate-source voltage drops (V_{GS}) of the different transistors, it is important to examine analytically the value of V_{GS} of a MOS device.

This is given by[2]:

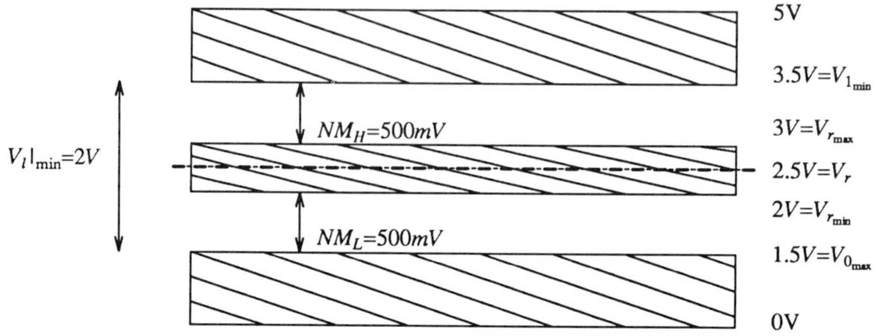

Figure 5.8 Logic levels V_1, V_0, noise-margins (NM) and logic swing (V_l).

$$V_{GS} = V_T + \left(\frac{2I}{n(2-n)\, K'(W/L)} \right)^{1/2} \tag{5.31}$$

where I is the operating drain-source current

$$n = \frac{V_{DS}}{V_{GS} - V_T}$$

$= 1$, if *the transistor is saturated*

< 1, if *the transistor is nonsaturated*

$$V_T = V_{T0} + \Delta V_T(V_{BB}) - \Delta V_T(V_{DS}).$$

V_{T0} is that part of the threshold voltage which is only a function of processing parameters and not a function of operating voltages. $\Delta V_T(V_{BB})$ is the shift in the threshold voltage due to the substrate source voltage bias V_{BB}. $\Delta V_T(V_{DS}) = 0$ for MOS devices which have no short-channel effects, i.e., devices with proper scaled processing parameters[6].

Equation (5.31) shows that V_{GS} depends on the operating current and this function dependency is stronger if the transistor is operating in the nonsaturation region. The ratio $[I/(W/L)]$ controls this current-dependent term.

It has been shown[5] that NMOS CML gates offer nanosecond performance at few picojoules delay-power products. The gates offer two logic output functions. For any added logic input only one transistor is added. This is in contrast with CMOS where only one logic output function is available and for any added logic input two transistors are needed, one of them being a bulky PMOS device.

The NMOS CML gate reported here uses a simple realization for the reference voltage, the current source, the loads, and the source followers, e.g., no biasing chain for V_r or the current source is used.

The NMOS CML circuit configuration could offer the answer to high-performance nanosecond and subnanosecond MOS VLSI, avoiding the complicated clock circuitry required for dynamic circuits and the high area and the complex processing associated with CMOS. NMOS CML bipolar-like performance is obtained at approximately half the area and half the delay-power product of bipolar CML.

5.3. Dynamic NMOS Circuits

The delays of static MOS circuits can be reduced by using advances in MOS technologies. However, it is possible to increase the throughput (the number of processed computations per unit time) of a MOS logic system by pipelining the circuit operation, where the delays of the system building blocks do not accumulate. The key feature of this approach is that the delay contribution of each block is with respect to a master clock, and that delay does not transfer from one block to another. Thus, the maximum frequency of operation of a pipelined system is determined by the delay of a *single*, although the slowest, block. Dynamic MOS logic circuits lead inherently to a pipeline operation, and can realize logic and arithmetic functions with a high throughput rate. Because today's CMOS circuits dominate MOS circuit design, dynamic NMOS circuits are rarely used. In this section only simple circuits are discussed for completeness and to illustrate the principle of operating a dynamic MOS circuit, as this will help in understanding CMOS dynamic circuits as discussed in Section 5.4.

In dynamic MOS circuits, a master clock is used to generate different timing clocks, which are used to control the dynamic operation of the circuit. These clocks are referred to as multiphase clocks [2 phase (2ϕ), 3 phase (3ϕ), 4 phase (4ϕ), etc.].

Although, in principle, the number of clocks can be increased to any number, four clocks is a practical compromise and only 2ϕ circuits are discussed here.

Two-Phase (2ϕ) Circuits

The basic operations in 2ϕ dynamic circuits are:

1. *Charging* a capacitance through a MOS transistor during a first time slot (precharge time).

2. *Logically discharging* that capacitance (discharging it or not depending on the logical state(s) of the input(s) through input transistor(s)) during a second time slot (evaluation time).

3. *Transferring* the logical state (the voltage level on the capacitance) to the input of the next gate during the proper time slot (sampling time).

Fig. 5.9 shows two 2ϕ inverters connected in cascade. Each inverter consists of three transistors Q_1,Q_2 and Q_3. The charging transistor Q_1 permits the output capacitance C_{out} to charge. The input transistor Q_2 provides a discharging path for C_{out}. The transferring transistor Q_3 allows a charge transfer between C_{out} and the input capacitance C_i of the next stage.

The above three basic operations are performed under the control of two nonoverlapping clocks ϕ_1 and ϕ_2 as shown in the timing diagram of Fig. 5.10. We assume that the input sequence is logical '1' - '0' - '1'. The logical inputs are presented by voltage signals which maintain their levels during the ϕ_1 clock cycles. For the case when the input is '1' the following sequence of operations occurs. During the time interval t_{11} (precharge time), C_{out1} is precharged through Q_{11}. During t_{12} (evaluation time), because the input is high, C_{out1} is discharged through Q_{21}. During t_{13} (sampling time), Q_{31} conducts and a charge sharing occurs between C_{out1} and C_{i2}. If $Cout1 \gg C_{i2}$ the voltage level at node 3 will be approximately the same as that of node 2: (V at node 3) = (V at node 2) ($C_{out1}/C_{out1} + C_{i2}$). Note that the input information which was available at node 1 during t_{11} has been inverted at node 3 and is available during t_{13}, i.e., the input information has been inverted with a half-bit delay (a half clock cycle). During the same time interval t_{13},C_{out2} is

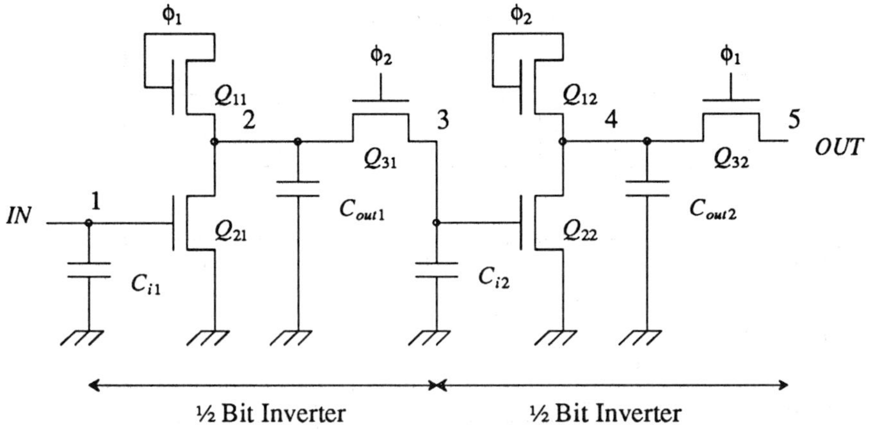

Figure 5.9 Two 2ϕ inverters connected in cascade.

precharged through Q_{12}. During t_{14}, because the input at node 3 is low, node 4 stays high. During t_{21}, Q_{32} conducts and the input capacitance of the next state shares the charge of C_{out2} and a high voltage results at node 5, i.e., the input information at node 3 has been inverted with another half-bit delay.

The sequence of operations for a logical '0' input at node 1 is shown in Fig. 5.10 and can be similarly explained.

Fig. 5.11 shows a logic symbol of the two inverters of Fig. 5.9. The first inverter type (ϕ_1/ϕ_2) accepts an input at ϕ_1 clock cycles and provides an output at ϕ_2 clock cycles. The second inverter type (ϕ_2/ϕ_1) accepts an input at ϕ_2 clock cycles and provides an output at ϕ_1 clock cycles. As a result, two inverters of the same type cannot be successively connected.

Logic gates can be realized by replacing the input transistor Q_2 of the two-phase inverter with a combination of transistors. Parallel transistors provide the NOR function, serial transistors provide the NAND function and mixed serial/parallel arrangements provide complex logic functions. As in the case of NMOS static gates, NOR gates take less area than NAND gates. Flip-flops can be realized using such 2ϕ gates. Fig. 5.12 shows some examples where 2ϕ AND, OR, NOR and noninverting

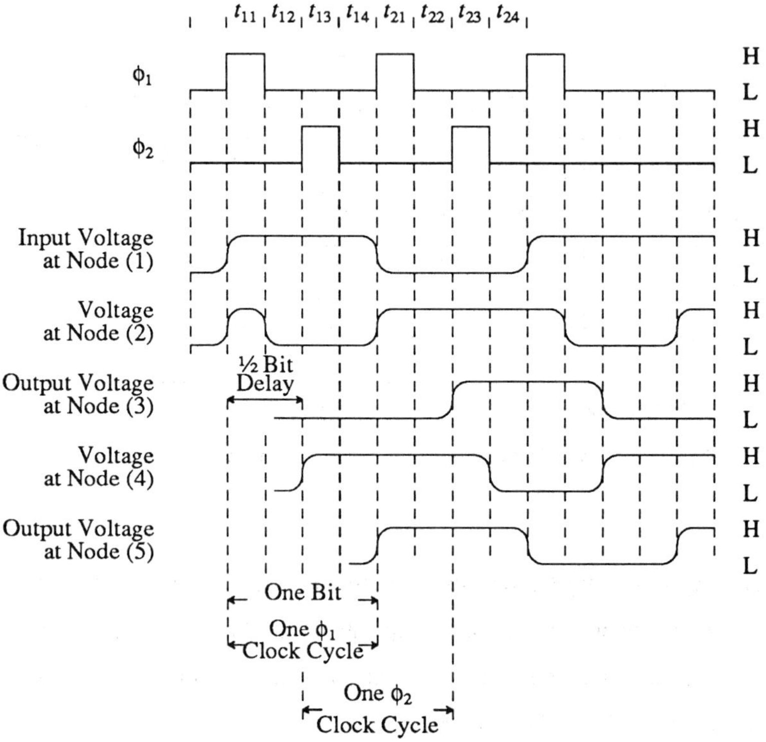

Figure 5.10 Timing diagram of Fig. 5.9.

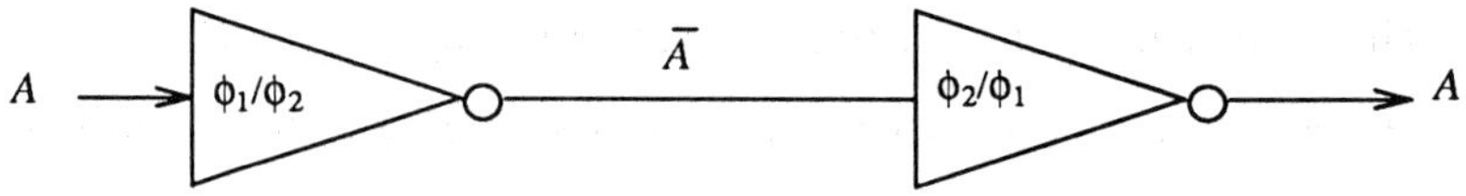

Figure 5.11 Logic symbol of 2ϕ inverters.

gates are used.

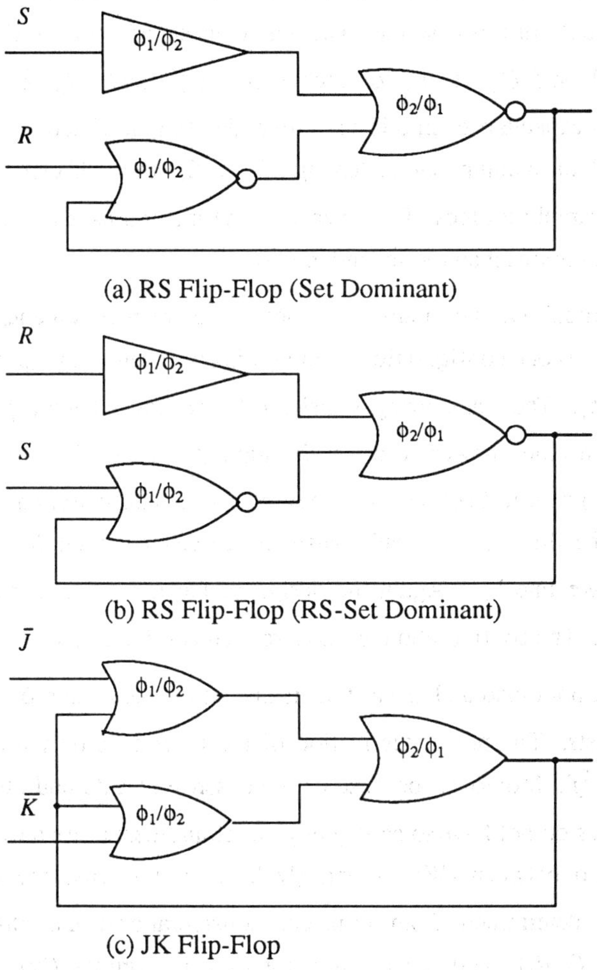

Figure 5.12 Dynamic 2φ flip-flops.

Fanning-out in 2ϕ circuits can be done at node 3. In this case the basic gate consists of Q_1 and Q_2 followed by the transferring transistor Q_3. The interconnection capacitances in this case are added to that of node 3. Alternatively, fanning-out can be done at node 2. In this case the basic gate consists of a transferring transistor Q_3 followed by Q_1 and Q_2. Thus, for each input, a Q_3 and a Q_2 are required. The interconnection capacitances in this case are added to that of node 2. The advantage of this method of fanning out is having $C_{out} > C_i$, thus efficient charge sharing results during sampling times. However, the first method uses fewer transistors and fewer diffusion regions than the second method.

The basic 2ϕ circuit can be modified to meet different requirements. For example, Fig. 5.13(a) shows one configuration where Q_1 is operating from V_{dd}, while its gate is clocked with ϕ_1. The transferring transistor Q_3 is clocked with ϕ_2. The inverter (Q_1 and Q_2) consumes power when ϕ_1 is high. Its design is based on the static inverter design given in Section 5.2. Note that in the configuration of Fig. 5.9 both the gate and the drain of the load transistor are clocked with ϕ_1. This results in saving the power line V_{dd}. Again the design of the inverter is based on the static inverter design. The configuration is also a ratioed configuration.

In Fig. 5.13(b), both load and driver transistors are clocked using ϕ_1 and in this case no dc path exists. The power dissipation of the inverter is only a transient power given by $C_{out}V_i^2f$. Moreover, because of the absence of a dc path, the load and the driver transistors do not have to satisfy any dc requirements. As a result, there is no need for a ratio between $(W/L)_D$ and $(W/L)_L$. In this case the (W/L) of both transistors are determined from transient considerations: the charging time is determined by $(W/L)_L$ and the discharging time determines $(W/L)_D$. It has the advantage of not using a V_{dd} or a ground line, no dc power dissipation, and using relatively small size devices for the load and the driver. However, it has the disadvantage of contributing large load capacitance to the clock lines.

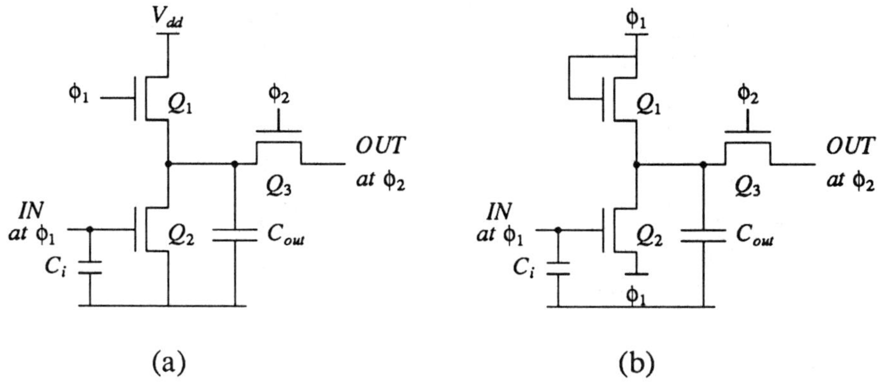

Figure 5.13 Two other 2ϕ configurations: (a) ratioed configuration; (b) ratioless configuration with overlapping clocks.

5.4. Complementary MOS (CMOS) Circuits

The use of a p-channel MOS as a load for an n-channel MOS driver provides the basic complementary CMOS inverter, as shown in Fig. 5.14(a). Logic gates and flip-flops can be realized using the complementary pair. CMOS digital circuits offer many advantages, including high noise immunity, operation at a wide range of power supply voltages, low power dissipation, relatively high speed, and compatibility to other logic families. The main disadvantage of the family is the relatively high silicon area consumed because of the need of complementary MOS pairs for each added logic input as shown in Section 5.5.2.

CMOS circuits can be fabricated using P-well, N-well or twin-tub CMOS technologies[7,8]. In a P-well CMOS technology the starting material is N-silicon substrate and is used to fabricate the PMOS devices, while a diffused or ion implanted P-well is used to fabricate the NMOS devices. N-well CMOS technology is widely used today, where the starting material is a P-silicon substrate and is used to fabricate NMOS devices and an ion implanted N-well is used to fabricate the PMOS devices. In a twin-tub CMOS technology, the starting material is a high resistivity silicon and two ion implanted N and P tubs are used to fabricate the PMOS and NMOS devices respectively.

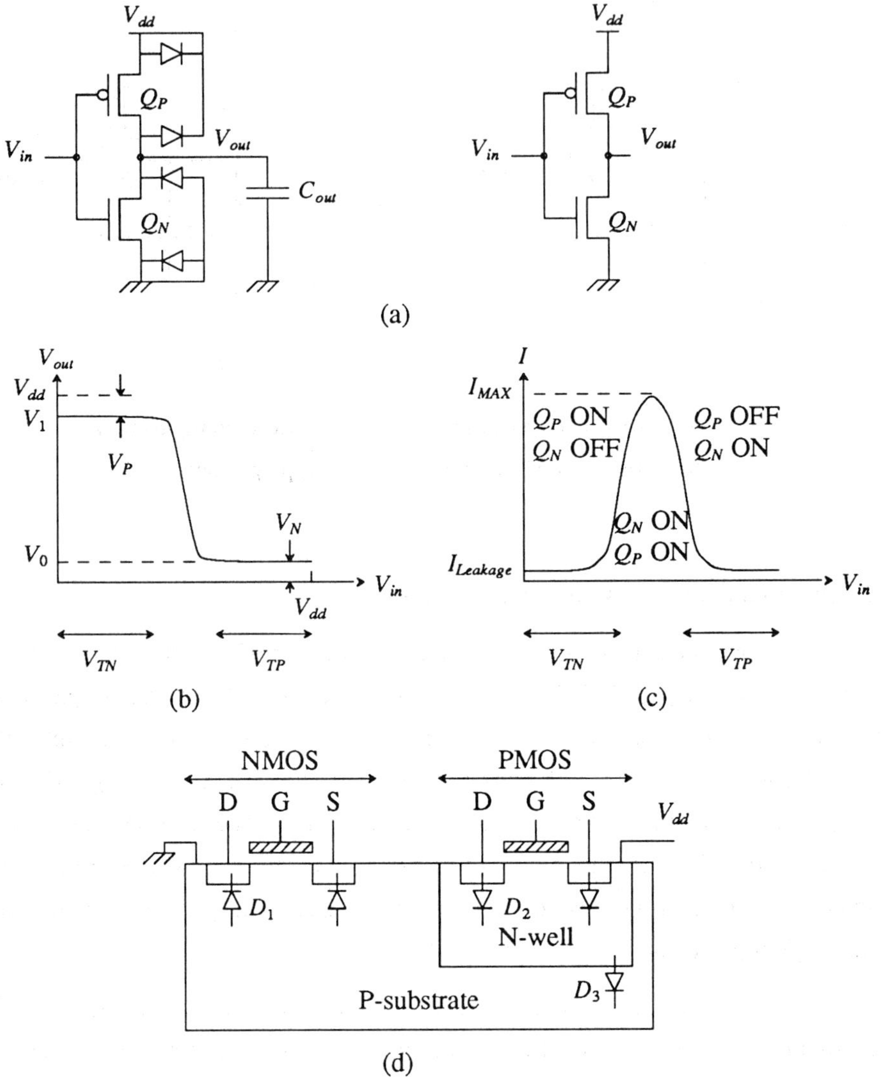

Figure 5.14 CMOS inverter (a) circuit diagram; (b) transfer characteristic; (c) I vs. V_{in}; (d) cross-section in an N-well CMOS technology.

Fig. 5.14(d) shows a cross-section in NMOS and PMOS devices in an N-well CMOS technology. The following parasitic diodes are shown: D_1 is the substrate-source/drain junction diode, D_2 is the well-source/drain junction diode, and D_3 is the well-substrate junction diode. The three diodes make up a three junction thyrister (SCR) and during circuit operation it may latch-up. Techniques are used to minimize this hazardous circuit operation[8].

5.4.1. The CMOS Static Inverter

5.4.1.1. DC Analysis

The basic inverter operation can best be explained with the aid of the transfer characteristic of Fig. 5.14(b).

1. If V_{in} is a logic '0' (i.e., $V_{in} = V_0$, $V_0 < V_{TN}$, where V_{TN} is the threshold voltage of Q_N), Q_N is off. If $|V_0 - V_{dd}| > |V_{TP}|$, where V_{TP} is the threshold voltage of Q_P, then Q_P is on. The output voltage in this case is given by

$$V_{out} = V_1 = V_{dd} - V_P \tag{5.32}$$

where V_P is the voltage drop across Q_P and is given by

$$V_P = I_P R_P \approx I_P \frac{1}{\beta_P[(V_{dd} - V_0) - |V_{TP}|]}$$

where I_P is the current supplied by Q_P to Q_N and to the loads. If the inverter is loaded with MOS gates, as is often the case, then I_P is negligible, in the order of the leakage current of the substrate-source and drain junctions, and $V_1 \approx V_{dd}$. If I_P is not negligible, then by increasing β_P (i.e., the size of Q_P) it is possible to achieve $V_P \ll V_{dd}$ and, hence, $V_1 \approx V_{dd}$.

2. If V_{in} is a logical '1' (i.e., $V_{in} = V_1$, $V_1 > V_{TN}$) then Q_N is on. If $|V_1 - V_{dd}| < |V_{TP}|$ then Q_P is off. The output voltage in this case is given by

$$V_{out} = V_0 = V_N \tag{5.33}$$

where V_N is the voltage drop across Q_N and is given by

$$V_N = I_N R_N \approx I_N \frac{1}{\beta_N (V_1 - V_{TN})}$$

where I_N is the current sunk by Q_N from Q_P and from the loads. If the loads are MOS gates then I_N is a leakage current and $V_N \approx 0$. However, if the loads are bipolar circuits then I_N is not negligible. In this case V_0 can be reduced by increasing β_N.

Thus, the logic levels of a CMOS inverter can be made close to V_{dd} and ground; and a logic swing V_l of the order of V_{dd} results. This is a main feature of CMOS gates. The threshold voltage of CMOS logic gate V_t is close to $V_{dd}/2$ if symmetrical devices are used, i.e., $\beta_N = \beta_P$. The $\beta_N = \beta_P$ condition can be achieved if

$$\left[\left(\frac{W}{L}\right)_P / \left(\frac{W}{L}\right)_N\right] = \frac{\mu_N}{\mu_P} \approx 2.5.$$

Thus, equal noise margins result which are $\approx (V_{dd}/2)$. This is another feature of CMOS gates.

As shown in Fig. 5.14(c), the gate current I passes between V_{dd} and ground only when both Q_P and Q_N are on, i.e., during the transition region. During the two steady states, $I \approx I_{leakage}$ and the dc power dissipation of the gate is given by

$$P_{DC} = V_{dd} I_{leakage} \tag{5.34}$$

This power dissipation is a negligible component of the total power dissipation of the gate as will be explained.

From Fig. 5.14(b) it is clear that the minimum value of V_{dd} is given by:

$$V_{dd}|_{min} = V_{TN} + |V_{TP}| \tag{5.35}$$

If V_{dd} is lower than that value, the gate demonstrates a hysteresis transfer characteristic as shown in Fig. 5.15, and it can not be used as a logic gate.

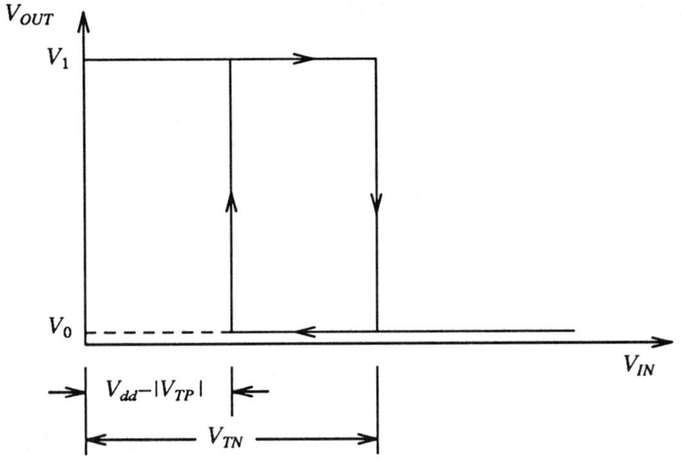

Figure 5.15 Hysteresis transfer characteristic of a CMOS inverter, $V_{dd}< (V_{TN}+|V_{TP}|)$.

5.4.1.2. Transient Analysis

If we consider a CMOS inverter with an ideal step input voltage from 0 to V_{dd}, the charging and discharging times can be calculated following the approach outlined in Section 5.2.2. The discharging time, t_{dis}, which is the time taken for the output capacitance C_{out} to discharge through Q_N from an output logic level of $V_1 \approx V_{dd}$ to an output logic level of V_0, is given by

$$t_{dis} = \tau_N \left[\frac{2}{\dfrac{V_{dd}}{V_{TN}} - 1} + ln \left(\frac{2(V_{dd} - V_{TN})}{V_0} - 1 \right) \right] \tag{5.36}$$

where

$$\tau_N = \frac{C_{out}}{g_{mN}} = \frac{C_{out}}{\beta_N(V_{dd} - V_{TN})}$$

Equation (5.36) is similar to Equation (5.23). Because of the symmetry of the problem, the charging time t_{ch} which is the time taken for C_{out} to charge through Q_P from V_0 to V_1, is given by

$$t_{ch} = \tau_P \left[\frac{2}{\left(\frac{V_{dd}}{|V_{TP}|} \right) - 1} + ln \left(\frac{2(V_{dd} - |V_{TP}|)}{V_{dd} - V_1} - 1 \right) \right] \quad (5.37)$$

where

$$\tau_P = \frac{C_{out}}{g_{mp}} = \frac{C_{out}}{\beta_p(V_{dd} - |V_{TP}|)}$$

It should be noted that Equations (5.36) and (5.37) can be simplified as in Equation (5.23.a).

Another useful transient parameter is the delay through a CMOS inverter τ_D, defined as the delay time between the input and the output waveforms measured at the ($V_{dd}/2$) points for a chain of CMOS inverters or a ring oscillator. This is given by[9]:

$$\tau_D \approx \frac{0.9\ C_{out}}{V_{dd}\beta_N} \left[\frac{1}{\left(1 - \frac{V_{TN}}{V_{dd}}\right)^2} + \frac{1}{\frac{\beta_P}{\beta_N}\left(1 - \frac{|V_{TP}|}{V_{dd}}\right)^2} \right] \quad (5.38)$$

If $(V_{TN}/V_{dd}) \ll 1$ and $|V_{TP}|/V_{dd} \ll 1$ then:

$$\tau_D \approx \frac{0.9\ C_{out}}{V_{dd}\beta_N} \left[1 + \frac{\beta_N}{\beta_P} \right] \approx \frac{2\ C_{out}}{V_{dd}\ \beta_N} \quad (5.38.a)$$

i.e., under these conditions V_{TN} and V_{TP} have only a small effect on the delay time and τ_D is linearly proportional to (C_{out}/V_{dd}) and β_N^{-1} ($\beta_N = \beta_P$ by design). This result is demonstrated in Fig. 5.16.

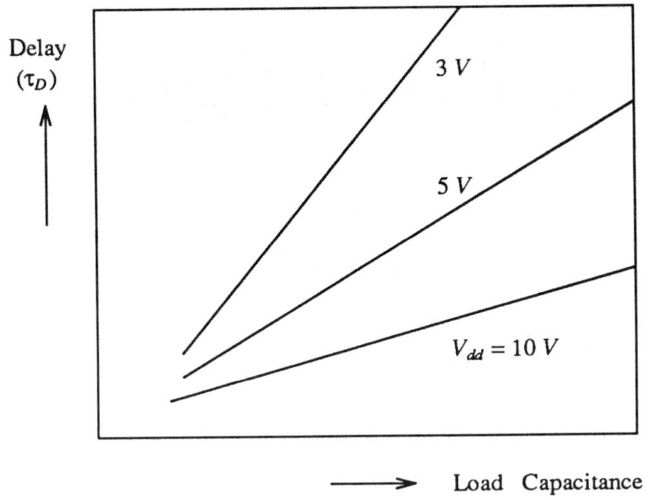

Figure 5.16 Delay vs. load capacitance for different values of V_{dd} for a given CMOS inverter.

5.4.1.3. Transient Power Dissipation

The transient power dissipation of a CMOS inverter has two components. The first results from charging and discharging the output capacitance through the finite on resistance of Q_N and Q_P. This component is given by:

$$P_{t1} = C_{out} V_l^2 f \tag{5.39}$$

where C_{out} is the total output-node capacitance including the capacitive effect of the interconnections. V_l is the logic swing of the gate ($V_l = V_1 - V_0 \approx V_{dd}$), and f is the switching frequency. The second transient power dissipation component arises from the fact that the input (and hence the output) voltage waveforms have finite rise and fall times. The waveform of the current supplied by V_{dd} to a CMOS gate is shown in Fig. 5.17, and as a result a power dissipation P_{t2} occurs:

$$P_{t2} = V_{dd} I_{av},$$

$$I_{av} = \frac{1}{2} I_{\max} \left[\frac{V_{dd} - \left(V_{TN} + |V_{TP}| \right)}{V_{dd}} \right] \left(\frac{t_1 + t_2}{\tau} \right) \tag{5.40}$$

where

$$I_{\max} = \frac{V_{dd}}{R_N + R_p}$$

If t_1 and $t_2 \ll \tau$ then $P_{t2} \ll P_{t1}$.

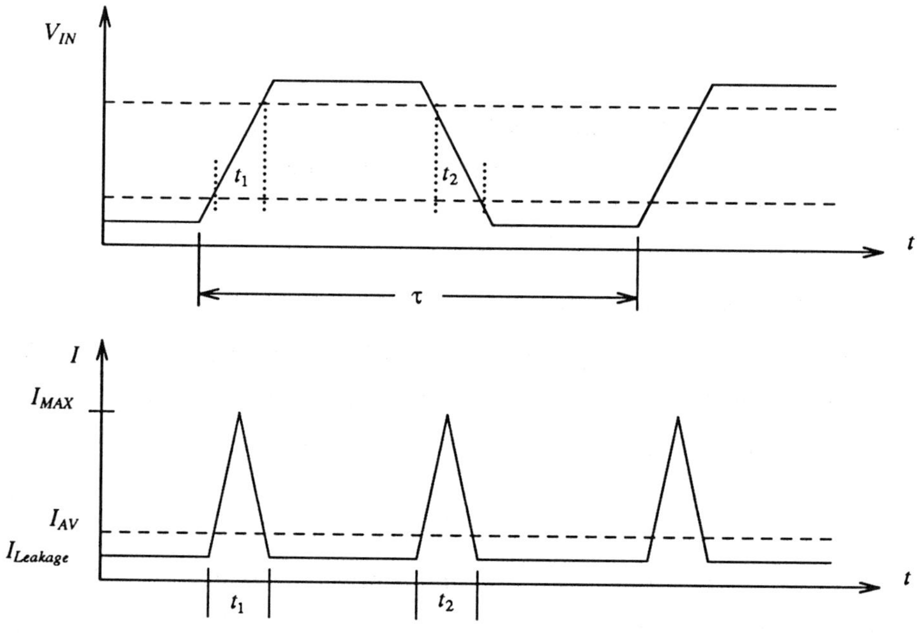

Figure 5.17 Input voltage and current waveforms for a CMOS inverter.

CMOS Delay-Power Trade-Offs

In CMOS circuits the main component of power dissipation is P_{t1}. Thus, the total power dissipation at the maximum frequency of operation is given by:

$$P \approx C_{out} V_l^2 / 2\tau_D \tag{5.41}$$

and the delay-power product is given by:

$$\tau_D P = 0.5\, C_{out} V_l^2 = 0.5\, C_{out} V_{dd}^2. \tag{5.42}$$

This shows that the delay-power product in CMOS digital circuits, as in the case of NMOS circuits, is proportional to $C_{out} V_{dd}^2$.

5.4.2. CMOS Static Gates

Logic gates (and hence flip-flops) can be realized using the complementary pairs. Fig. 5.18(b) and (c) show a two-input NAND and a two-input NOR. Each input requires a complementary pair. In the case of the NAND gate the p-channel devices are connected in parallel while the n-channel devices are connected in series. But in the case of the NOR gate the n-channel devices are connected in parallel while the p-channel devices are connected in series.

In CMOS logic gates the standby dc current of the gate is ≈ 0 because there is no dc path between V_{dd} and ground for any logic combination of the input. For example, for the two-input NOR and NAND gates shown in Fig. 5.18 for any input logic combination ($x_1 x_2 = 00, 01, 11$ *or* 10) the dc current of the gate is zero.

5.4.2.1. NAND/NOR CMOS Gates

The design of a CMOS logic gate follows that of an inverter. First, an inverter is designed to meet a given dc and transient performance, and $(W/L)_N$ and $(W/L)_P$ are determined. Then $(W/L)_N$ and $(W/L)_P$ of the devices of a logic gate are determined as follows. If a CMOS m-input NAND gate is to be designed to have the same dc and transient performance as that of the inverter, then for the same values of C_{out}, $(W/L)_P$ of the NAND gate devices should be $\geq (W/L)_P$ of the inverter while $(W/L)_N$ should be $\geq m(W/L)_N$ of the inverter. On the other hand, if a CMOS m-input

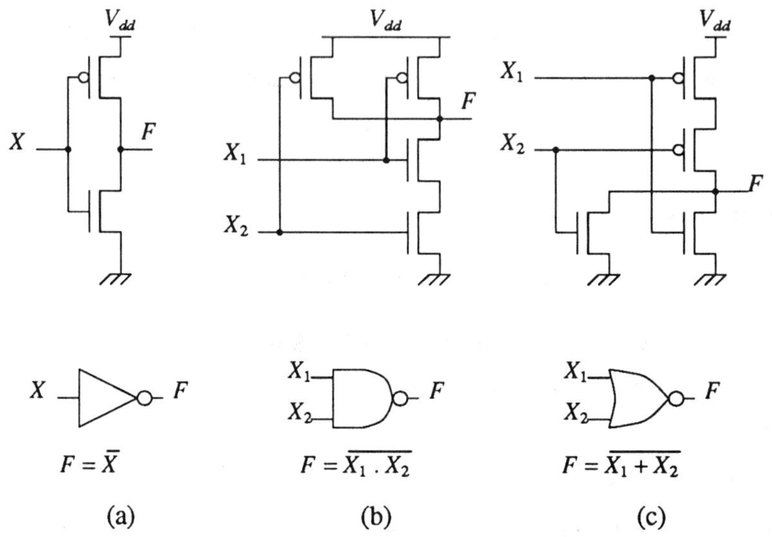

Figure 5.18 CMOS logic gates (a) inverter; (b) two-input NAND; (c) two input NOR.

NOR gate is to be designed to have the same dc and transient performance as that of the inverter, then for the same values of C_{out}, $(W/L)_P$ of the NOR gates should be $\geq m(W/L)_P$ of the inverter, while $(W/L)_N$ should be $\geq (W/L)_N$ of the inverter. Note that the increase in the size of the NMOS devices in the case of the NAND gate and the increase in the size of the PMOS devices in the case of the NOR gates allows the logic levels V_0 and V_1 to be the same as that of the inverter if the dc currents are nonzero. The size of these transistors should be further increased because C_{out} is larger than that of the inverter. Note also that for the same performance and for the same number of inputs the NAND gate consumes less silicon area than that of a NOR gate because of the smaller area taken by the NMOS devices. Hence, CMOS NAND gates are more widely used than NOR gates. This is the opposite of NMOS logic gates where NOR gates are more widely used.

It should be noted that the back-gate bias effect has to be taken into consideration in the design of the NMOS devices of the NAND gate and in the design of the PMOS devices of the NOR gate. In the case of a CMOS inverter, no back-gate bias effect has to be considered because the sources of the n- and p-channel devices are connected to their corresponding substrates.

5.4.2.2. Complex CMOS Logic Gates

Complex logic functions can be realized by connecting several NAND, NOR and Inverter gates. However, they can also be efficiently realized using a single CMOS logic gate, referred to as a "Complex CMOS Gate", with two sets of transistors connected in a series/parallel combination. These transistors form the P and the N logic blocks as shown in Fig. 5.19(a). The two blocks have the same number of transistors. There is no direct dc path between V_{dd} and ground for any of the input logic combination. The topology of one block is the dual of the other block, i.e. parallel connections become series and vice versa. Fig. 5.19(b) shows a two-input complex CMOS gate and its logic gate equivalent. In either the P or the N logic blocks, the series parallel combination of transistors are arranged such that the output capacitance is minimized. For example, the contribution of the P block to the output capacitance in Fig. 5.19(d) is less than that of Fig. 5.19(c). In practice, the complex CMOS gates are used for a maximum number of inputs of 5-6.

5.4.2.3. CMOS Transmission Gates (CMOS TGs)

As shown in Fig. 5.20, a CMOS transmission gate consists of a complementary pair connected in parallel. It acts as a switch, with the logical variable A as the control input. Let us assume that A is connected to the n-channel device and $\bar{A}$ is connected to the p-channel device. If A is high, the gate is on and acts as a switch with an on resistance of R_N and R_p in parallel. If A is low, the gate is off and presents a high resistance between the terminals. It should be noted that in designing CMOS transmission gates to meet specific dc and transient requirements, the back gate bias effect on both the NMOS and PMOS devices must be taken into account.

The advantage of using a complementary pair, rather than a single NMOS or PMOS device, to realize a transmission gate in CMOS can be explained with the aid of Fig. 5.21. Although in Fig. 5.21 we assume that there is no time delay between A and $\bar{A}$, the conclusion of the following discussion is general. Fig. 5.21 shows A and $\bar{A}$ waveforms and also shows the times at which the NMOS and PMOS devices turn on.

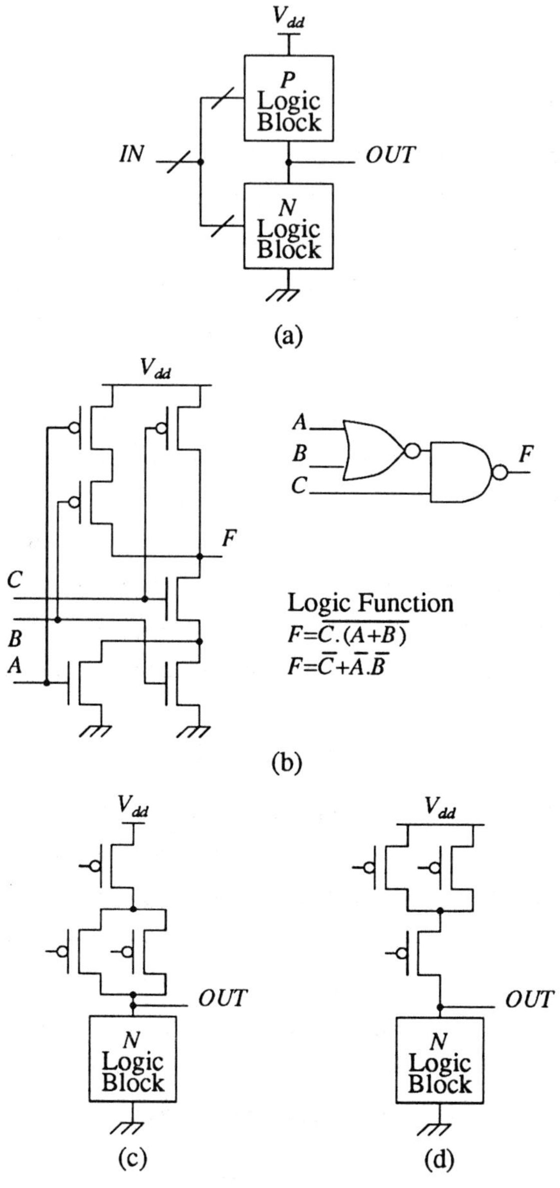

Figure 5.19 CMOS complex logic gate.

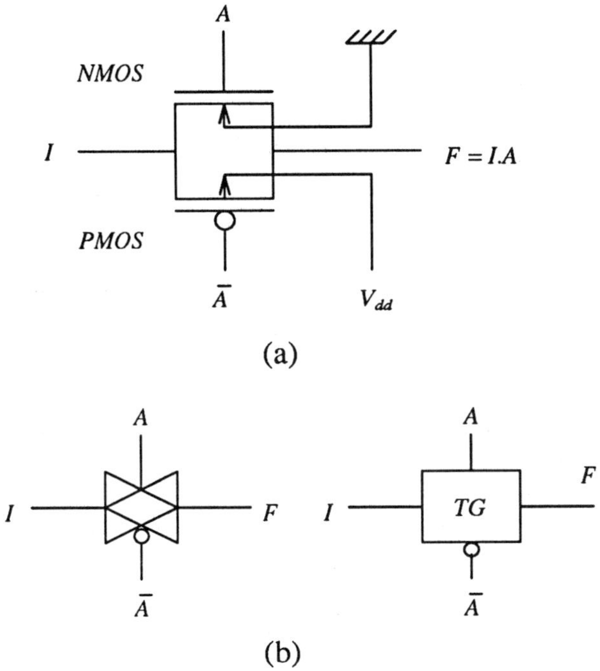

Figure 5.20 (a) CMOS transmission gate and (b) schematic symbols.

Fig. 5.21(a) shows these times for a reference case where the input logic level is $\approx(V_{dd}/2)$, and as a result $V_{TN} \approx |V_{TP}|$. In this case the two devices turn on at the same time. Fig. 5.21(b) shows these times for the case where the input logic level ≈ 0, and $|V_{TP}| > V_{TN}$ because of the back-gate bias effect. In this case the NMOS device turns on before the PMOS and the transmission gate delay time between the input and the output is almost the same as in case (a). Similarly, if the input logic level $\approx V_{dd}$ and as a result $V_{TN} > |V_{TP}|$, as shown in Fig. 5.21(c), the delay time of the gate will be unaffected. In conclusion, independent of the voltage level of the input variable of the CMOS transmission gate, the gate delay time is approximately the same. It is easy to see that this is *not* the case when single-channel type is used in designing transmission gates. A drawback of the CMOS TGs is that it consumes more area than a single-channel transmission gate (NMOS TGs or PMOS TGs). Thus, if the area is of prime concern, non-complementary NMOS TGs are used.

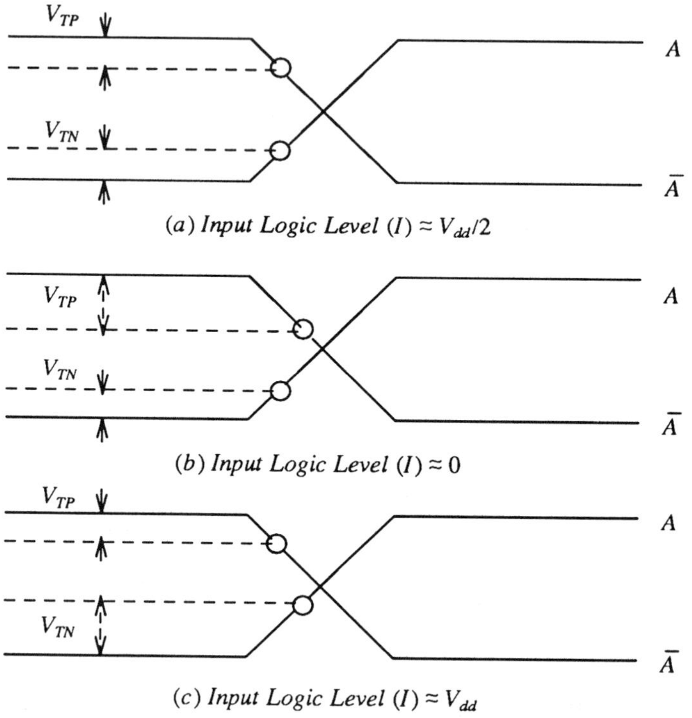

Figure 5.21 CMOS transmission gate waveforms.

5.4.2.4. CMOS TG Logic

The output of a CMOS TG can be used as a control input of another TG, and logic functions can be implemented accordingly offering CMOS TG logic. In such implementation the transistor count, hence the silicon area, is low compared to conventional CMOS implementation. This is highlighted in the implementation of such functions as multiplexing, demultiplexing and decoders. Fig. 5.22 shows an example where the control logic inputs A and B demultiplex the four inputs x_1, x_2, x_3, and x_4 into one output, F. This form of logic is also used when the inputs and their logic complement are available. The implementation does not need V_{dd} or ground lines. However, the implementation suffers from a number of drawbacks; the driving capability of the circuit is limited and the delay increases with long TG chains. Moreover, the circuit does not provide a restoration of the logic levels, i.e. the logic gates are passive with no gain elements. However, the CMOS TG logic can be used in CMOS circuit design offering an extra degree of circuit design freedom.

It should be noted that dynamic storage of logic levels occurs at different nodes of the TG logic trees. This could be desirable and use is made of it, or could be undesirable. As shown in Fig. 5.22, wired logic is widely used. Output function F is obtained by logic wiring the different branches of the logic tree. As a result, undefined potentials could develop during the logic operation. Formal design techniques along with CAD systems for CMOS TG designs can be used to alleviate these problems.

5.4.3. CMOS Static Flip-Flops

The basic building blocks of CMOS flip-flops are the CMOS inverter, logic gates, and transmission gates. Fig. 5.24 shows a cross-coupled CMOS static flip-flop. In the storage mode where V_A is high, V_B is low, Q_1 and Q_4 are on while Q_2 and Q_3 are off. Similarly, when V_A is low, V_B is high, Q_1 and Q_4 are off while Q_2 and Q_3 are on. The standby power dissipation of the cell is very small. The state of the flip-flop is changed by using two CMOS transmission gates connected to V_A and V_B nodes as shown in Fig. 5.23.

Fig. 5.24 shows a D-type CMOS master-slave static flip-flop. Each of the master and the slave requires the presence of the clock to latch up and store the information. Thus they differ from the basic cross-coupled configuration of Fig. 5.23. The transmission gates are represented with ideal switches: $(TG)_1$, $(TG)_2$, $(TG)_3$, and $(TG)_4$. The transmission gate $(TG)_1$ connects the master section to the input data (D) when the clock input (C) is low, while $(TG)_3$ connects the slave section to the output of the master section (node 1) when the clock input (C) is high. The gates $(TG)_2$ and $(TG)_4$ are identical to $(TG)_3$ and $(TG)_1$ respectively, and are used to latch the master and the slave sections during C and $\overline{C}$ time slots respectively. Fig. 5.24(c) shows the flip-flop timing diagram.

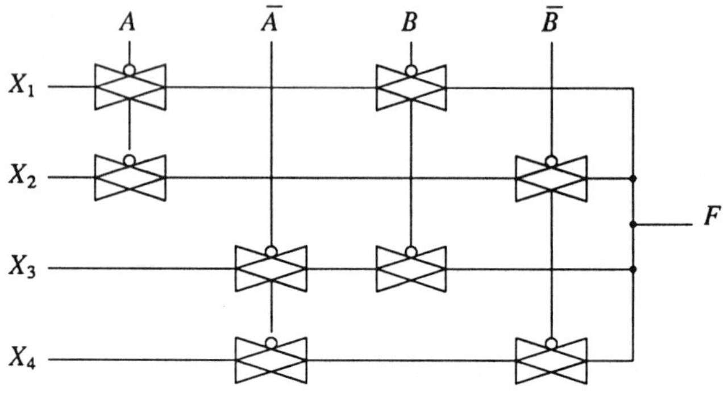

(a)

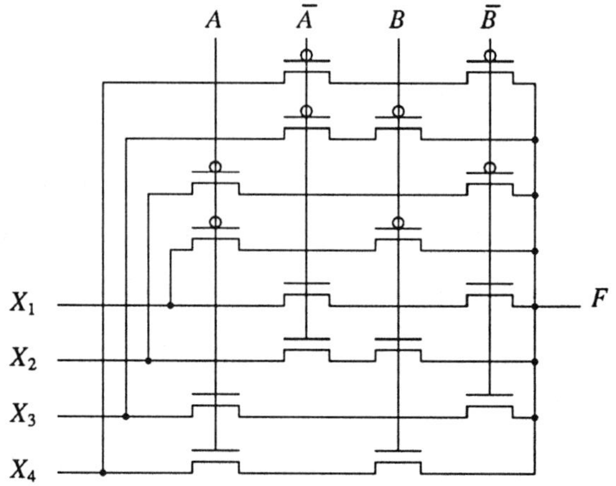

A	B	F
0	0	X_1
0	1	X_2
1	0	X_3
1	1	X_4

(b) (c)

Figure 5.22 CMOS TG Logic Example (a) Logic Diagram, (b) Circuit Schematic, (c) Logic Table.

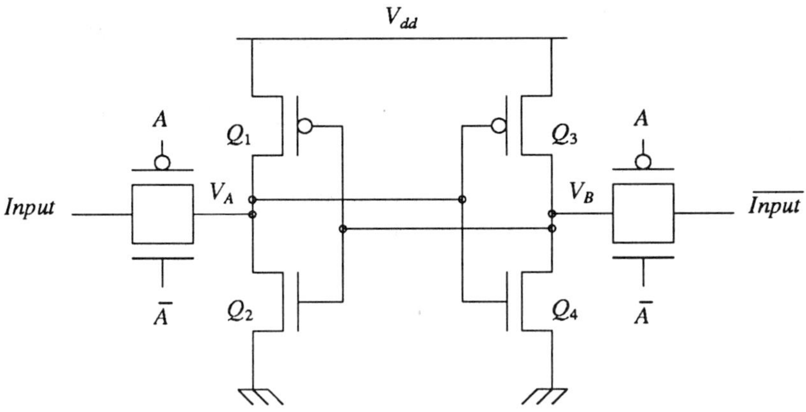

Figure 5.23 CMOS cross-coupled static flip-flop.

5.4.4. CMOS Dynamic Flip-Flops

The D-type static flip-flop can be modified to obtain the dynamic one shown in Fig. 5.25. Although the term dynamic flip-flop is commonly used to refer to Fig. 5.25, each stage is basically a dynamic shift-register stage. Each stage consumes less area than its corresponding stage of Fig. 5.24, because of the absence of the latching paths. The information is stored on the output node capacitances of the two inverters. As in dynamic circuits, the minimum clock frequency is limited by the leakage paths at these output nodes.

The D-type CMOS flip-flop is the simplest to realize and is advantageous to us in subsystem designs as a storage or shift register element. However, if other flip-flop types are required, e.g., JK-flip-flops, then the D-type flip-flop can be modified to realize these types.

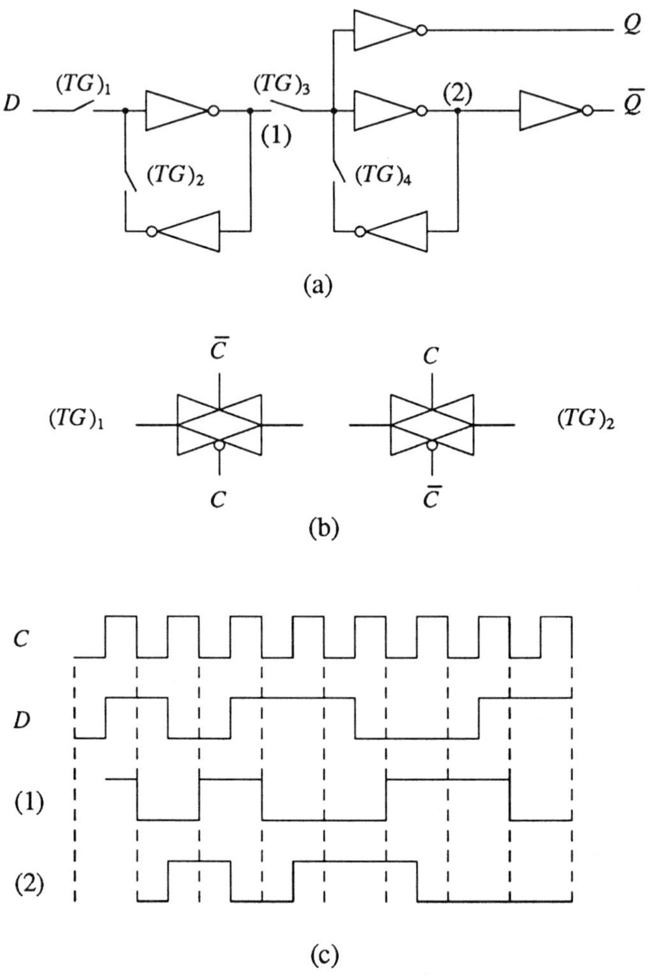

Figure 5.24 CMOS Master-Slave static D-type flip-flop (a) circuit diagram; (b) transmission gate types; (c) timing diagram.

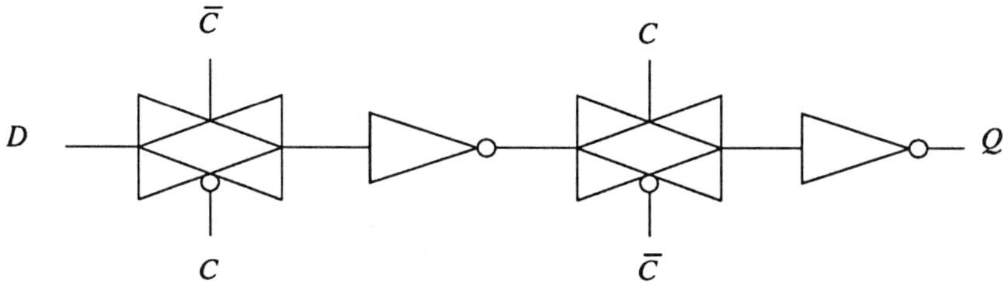

Figure 5.25 CMOS Master-Slave dynamic D-type flip-flop.

5.4.5. Pseudo-NMOS CMOS Logic

Static CMOS gates tend to become inefficient in terms of chip area because an n-input logic gate requires $2n$ transistors while the equivalent NMOS gate requires (n + 1) transistors. The gate area can be reduced if CMOS circuits are designed in a similar way to NMOS circuits; a PMOS device is used as a load to replace the depletion-type device in NMOS circuits. These types of circuits are referred to as pseudo-NMOS, and a NOR gate is shown in Fig. 5.26. The design of this class of circuits follows that of static E/D NMOS circuits. The advantage of using this circuit over NMOS is its compatibility with CMOS technologies. In addition, the back gate bias effect does not affect the load transistor. Although this circuit form offers a reduction in total transistor count and ease of layout, because each logic input is connected to one transistor rather than two as is the case in CMOS, it has the disadvantages of non-zero static power dissipation, a higher voltage for V_0 and ratioed operation which tends to increase the NMOS area and hence the input capacitance.

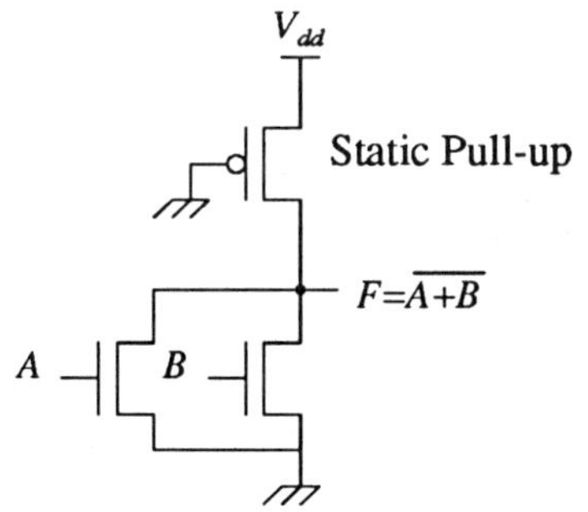

Figure 5.26 Pseudo-NMOS CMOS logic.

5.4.6. Tristate CMOS Logic

Another special CMOS circuit is one which is used to deliver a tristate output. Fig. 5.27(a) shows a tristate CMOS inverter, referred to as a clocked inverter. It is equivalent to a static inverter and a transmission gate, except that the clock ϕ controls both the NMOS and PMOS transistors of the transmission gate. When ϕ is high, the circuit acts as a normal inverter and when ϕ is low, the output goes into a high output independence state. The inverter of Fig. 5.27(a) does not provide a high driving capability to output capacitances because of the series connection of the transistors. Fig. 5.27(b) shows another implementation of the tristate CMOS inverter which avoids this problem.

5.4.7. Dynamic CMOS Circuits

To reduce the area taken by CMOS circuits, the dynamic mode of operation can be used. Both PMOS and NMOS devices are used to an advantage and the principles of dynamic logic outlined in Section 5.4. are applied. One such circuit is shown in Fig. 5.28, and is referred to as a domino CMOS circuit[10]. The domino gate shown in Fig. 5.28 consists of a dynamic CMOS circuit followed by a static CMOS buffer. The dynamic circuit consists of a PMOS precharge transistor Q_1, an evaluation NMOS transistor Q_2, a storage capacitor C, and an N logic block which is a series-parallel combination of NMOS devices activated by the inputs and implementing the

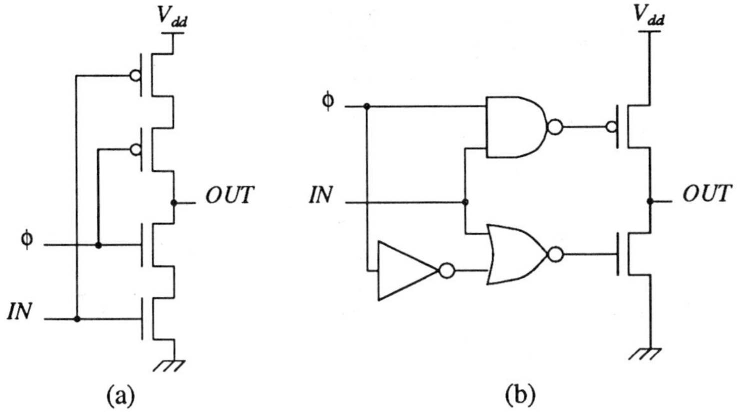

Figure 5.27 Tristate CMOS Inverters.

required logic.

During the precharge time T_1, the storage capacitance is charged through the PMOS device Q_1, the inputs have no effect during T_1, and the output of the buffer is low; hence the input transistors of the next stage are off. Also during T_1, the next stage is evaluating the logic state since it operates from $\bar{\phi}$.

During the evaluation time T_2, Q_2 is on, and depending on the logic performed by the N logic block, the voltage at node (a) is either discharged through the logic block and Q_2, or the voltage will be maintained. The output voltage at the OUT node will be the inversion of the voltage at node (a).

The circuit has the following features. Domino gate operation requires a single clock, ϕ to operate one stage and $\bar{\phi}$ to operate the loading stage. Since the dynamic circuits are buffered by the CMOS static inverters, the dynamic nodes [as node (a)] are isolated from the loading capacitances. Thus, the gate is suitable for a high fan-out operation. Because the gate is efficient in area for high fan-in (it requires $(n + 4)$ transistors compared to $2n$ for CMOS static gate), it is suitable for realizing complex logic functions. Because the output of the gate is always low during precharge time, the output node can only be high if switching occurs; hence the gate operation is

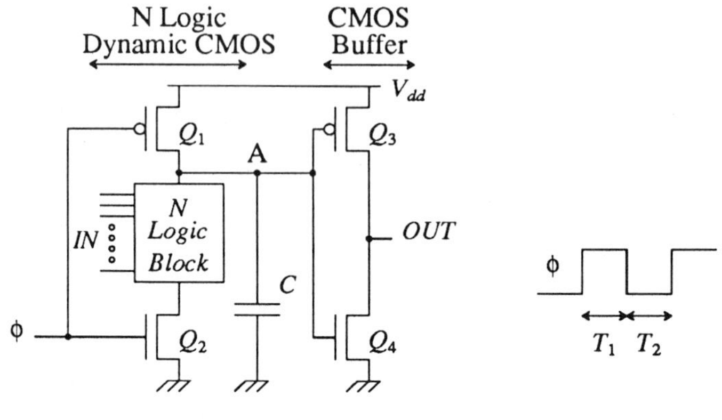

(a)

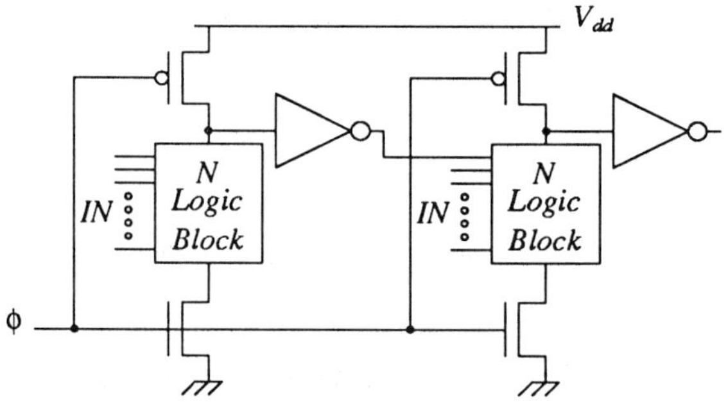

(b)

Figure 5.28 (a) Domino CMOS dynamic circuit with N logic block; (b) Two stage domino logic with N logic block.

glitch free if the inputs to the gates are set up during the precharge time.

One limitation of the gate is that it implements noninverting logic functions. However, this is not a serious limitation and can be overcome, if the need arises, by using CMOS static inverters. The designer can mix both static and dynamic CMOS logic circuits in a given design to optimize the overall performance.

In some applications, it is desirable to have a dynamic logic operation at low frequency. This can be obtained in a domino circuit by the addition of a low-current (low W/L) pull-up PMOS transistor *Q* as shown in Fig. 5.29(a). This transistor is chosen small enough so there is no significant impact on the pull-down operation during the evaluation time. Adding Q will allow a dc path during evaluation time and the gate power dissipation will increase. This is avoided in Fig. 5.29(b).

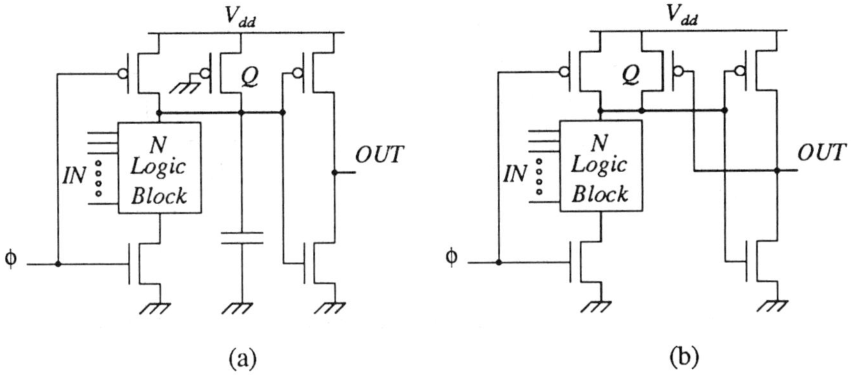

Figure 5.29 (a) The circuit of Fig. 5.28 with a pull-up PMOS transistor at the storage node; (b) The Pull-up PMOS transistor of (a) is controlled by the output node.

In Fig. 5.28 the domino gate uses a dynamic circuit with an N logic block. The complementary circuit is shown in Fig. 5.30 where the precharge transistor is an NMOS Q_2, the evaluation transistor is a PMOS Q_1 and the logic implementation uses PMOS transistors in the P logic block. The time slot T_1 is the precharge time, while T_2 is the evaluation time. This type of circuit can also be used in a domino dynamic arrangement separated by static CMOS buffers. Moreover, the two types of dynamic CMOS circuits, one N logic and the other with P logic, can directly drive each other without the use of buffers[11].

The circuit design of CMOS dynamic circuits follows that of NMOS dynamic circuits, and overall performance optimization of these circuits, including the buffering stages, can be performed[3].

Fig. 5.31 shows a two-stage pseudo 2ϕ CMOS shift register using two non-overlapping clocks ϕ_1, ϕ_2 similar to the NMOS circuit of Fig. 5.9. The circuit is similar to the CMOS circuit of Fig. 5.25 except that two clocks ϕ_1, and ϕ_2 are used

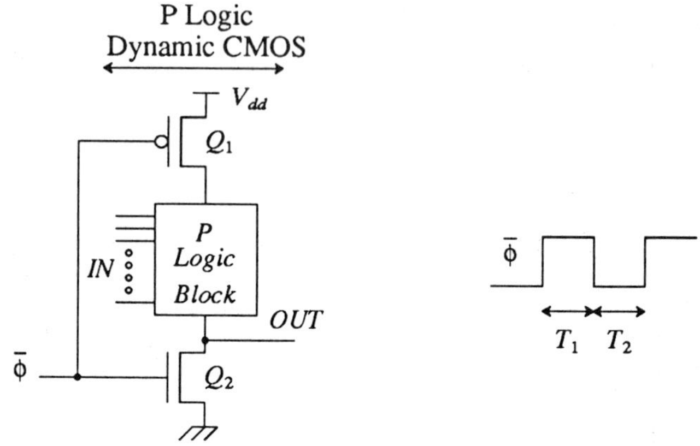

Figure 5.30 Domino CMOS dynamic circuit with P logic block.

instead of ϕ and $\bar{\phi}$. The circuit can operate with more tolerance on the clock waveforms compared to the case where only a single clock is used. The concept can be extended to implement combinational logic by replacing the two inverters with one or more levels of logic gates.

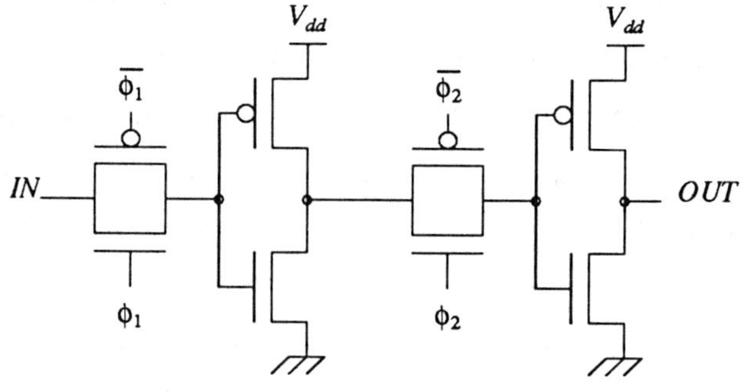

Figure 5.31 A two-stage Pseudo 2 ϕ CMOS register.

5.4.7.1. NORA (NO RACE) Logic

NORA (NO RACE) Logic is constructed by cascoding dynamic domino N and P logic blocks as shown in Fig. 5.32 followed by a C^2 MOS (clocked CMOS) output latch to avoid signal races[12]. The $\bar{\phi}$ section of Fig. 5.32 is created by interchanging ϕ and $\bar{\phi}$. A NORA pipeline is shown in Fig. 5.33 and it consists of alternating ϕ and $\bar{\phi}$ sections.

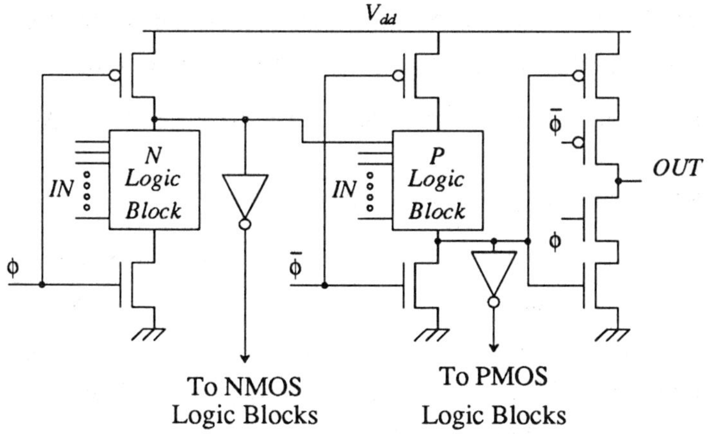

Figure 5.32 NORA circuits with latch; ϕ section.

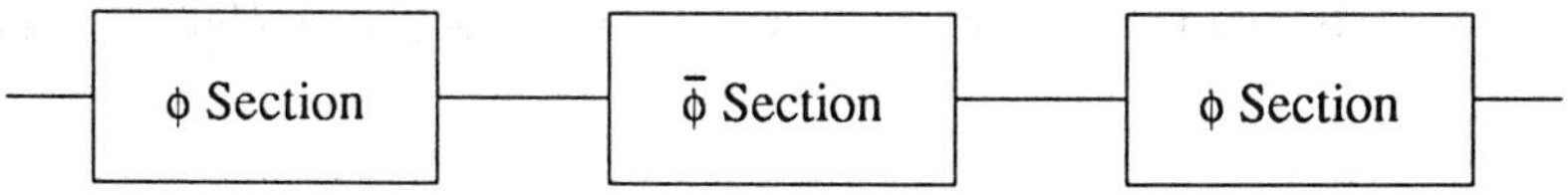

Figure 5.33 NORA Pipeline logic.

5.4.7.2. Zipper CMOS Logic

Zipper CMOS Logic is another dynamic CMOS system that overcomes charge sharing and soft node leakage errors by using additional clocking signals. Fig. 5.34 shows the 4 clocks ϕ, $\bar{\phi}$, ψ_1, ψ_2 generated from the masterclock. ϕ and $\bar{\phi}$ voltage swings are between V_{dd} and ground. ψ_1, swings between ground and $V_{dd}-|V_{TP}|$ and ψ_2 between V_{TN} and V_{dd}. Thus, ψ_1 in the N logic block circuits will set the PMOS transistors at the edge of conduction when ϕ is high and similarly ψ_2 will set the NMOS transistors in the P logic block circuits at the edge of conduction when $\bar{\phi}$ is low. The weakly conducting devices prevent leakage caused errors.

Fig. 5.35 shows the circuit that generates the clocks for Zipper CMOS.

5.4.8. CMOS Nonthreshold Logic (NTL)

The basic CMOS NTL logic gate, as shown in Fig. 5.36 consists of two fixed biased transistors P and N, capacitors C_S and an N logic block consisting of series/parallel combinations of NMOS transistors; although it is usually parallel devices offering a NOR gate. The principle of operation is based on the fact that P and N act as a potential divider allowing a small change at the input to change the resistive path in the N logic block and hence the output voltage. A dc path exists between V_{dd} and ground. The transistor N is the negative feedback path resulting in a nonthreshold-like dc transfer characteristic. The shunting capacitor C_s is used to reduce the negative feedback effect and increase the speed of operation. The circuit operates at lower logic swing than that of conventional CMOS and offers a higher speed than CMOS at the expense of higher power dissipation. However, the delay power product is nearly the same as conventional CMOS at high frequency of operation[14].

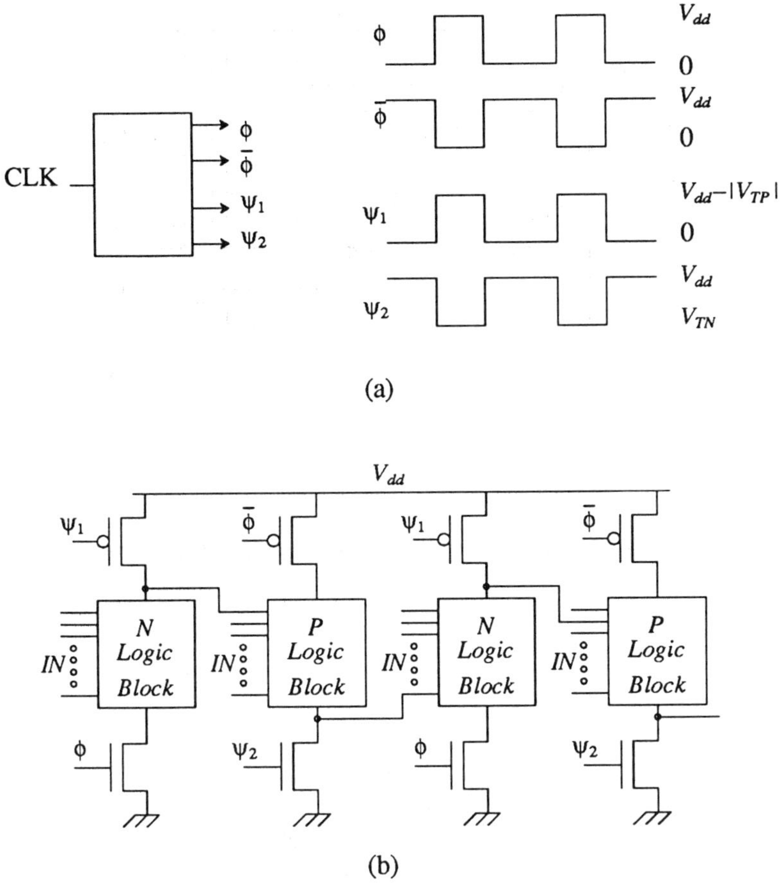

Figure 5.34 Zipper CMOS dynamic logic, (a) clocking signals (b) circuit structure.

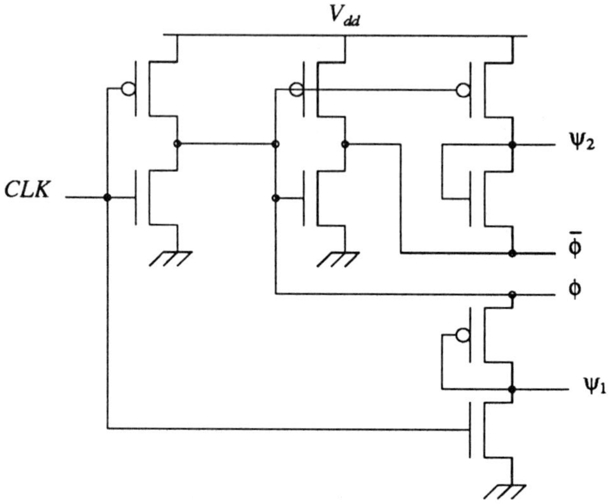

Figure 5.35 Clock generator for Zipper CMOS logic.

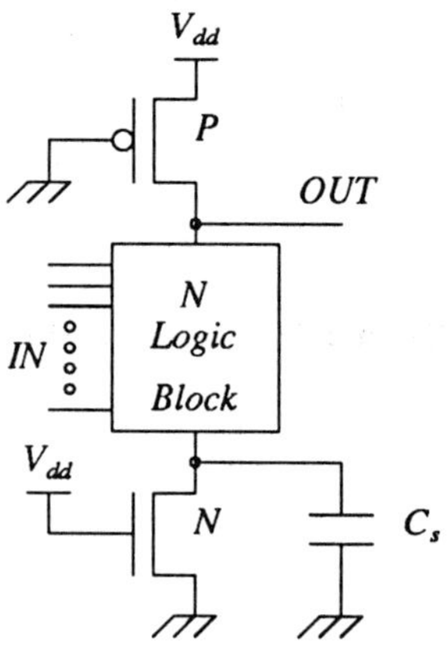

Figure 5.36 CMOS NTL logic gate.

5.4.9. Cascode Voltage Switch Logic (CVSL)

The cascode voltage switch logic (CVSL) is based on CMOS circuits that use positive feedback at the output to speed up the switching operation. The circuit as shown in Fig. 5.37 consists of two N logic blocks; one representing the logic tree and the other its complement. The two have cross coupled PMOS loads, providing the output Q and its complement $\overline{Q}$. Fig. 5.37(b) shows an example for function F where:

$$F = [(AB + C(D + E)]' \text{, where ' signifies the complement}$$

$$F' = [(A' + B')(D'.E' + C')]'$$

The CVSL is costly in chip area as both logic inputs and their complements are required. However, the output and its complement are provided.

5.5. Chapter Summary

This chapter deals with the NMOS inverter in detail as a basic circuit structure for understanding MOS circuits. It presents a wide variety of NMOS circuit configurations including NMOS current mode and dynamic NMOS circuits. It then deals with the CMOS inverter. The DC and transient analysis of the CMOS inverter is reviewed. Basic CMOS cells, such as complex gates and Flip-Flops, are introduced. The last part of the chapter presents different CMOS logic styles such as dynamic CMOS, nonthreshold CMOS and cascode voltage switch CMOS.

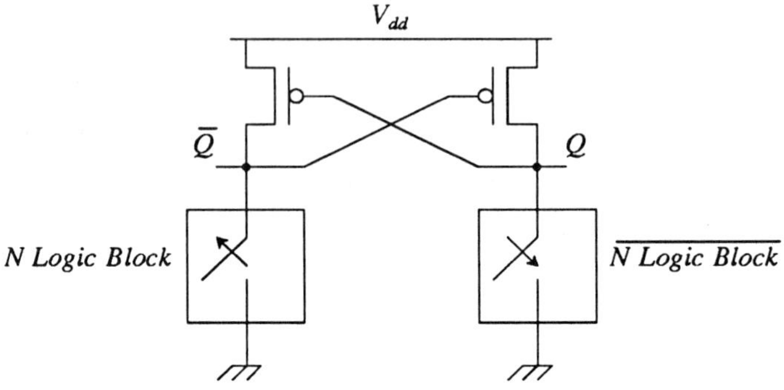

(a)

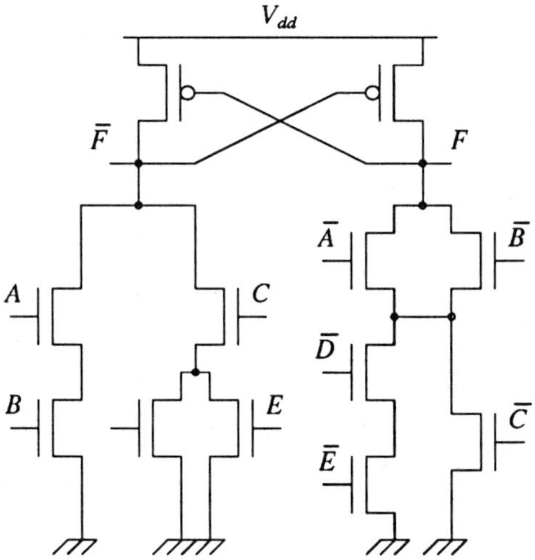

(b)

Figure 5.37 Cascode Voltage Switch Logic (CVSL), (a) Circuit Structure, (b) an Example.

References

1. M.I. Elmasry, *Digital MOS Integrated Circuits II*, IEEE Press (1992).

2. M.I. Elmasry, "Digital MOS Integrated Circuits: A Tutorial," in *Digital VLSI Systems*, ed. M.I. Elmasry, IEEE Press (1985). Also in M.I. Elmasry, ed., "Digital MOS Integrated Circuits", IEEE Press Book, 1981

3. J.P. Uyemura, *Fundamentals of MOS Digital Integrated Circuits*, Addison-Wesley (1988).

4. R.H. Crawford, *MOSFET in Circuit Design*, McGraw Hill (1967).

5. M.I. Elmasry, "Nanosecond NMOS VLSI Current Mode Logic," *IEEE Trans. on Electron Devices*, pp. 781-784 (1982).

6. Y.P. Tsividis, *Operation and Modeling of the MOS Transistor*, McGraw Hill (1987).

7. S.M. Sze, *Physics of Semiconductor Devices, 2nd Edition*, Wiley Interscience (1981).

8. T.E. Dillinger, *VLSI Engineering*, Prentice Hall (1988).

9. J.R. Burns, "Switching Response of Complementary Symmetry MOS Transistor Logic Circuits," *RCA Review*, pp. 627-661 (1964).

10. R.H. Krambeck et al, "High Speed Compact Circuits with CMOS," *IEEE Journal of Solid-State Circuits*, **SC-17** pp. 614-619 (1982).

11. V. Friedman and S. Liu, "Dynamic Logic CMOS Circuits," *IEEE Journal of Solid-State Circuits*, **SC-19** pp. 263-266 (1984).

12. N.F. Goncalves and H.J. DeMan, "NORA, A Racefree Dynamic CMOS Technique for Pipeline Logic Structures," *IEEE Journal of Solid-State Circuits*, **SC-18** pp. 261-266. (1983).

13. C.M. Lee and E.W. Szeto, "Zipper CMOS," *IEEE Circuits and Dev. Mag.*, pp. 10-17 (1986).

14. J.S. Wong et al, "CMOS Nonthreshold Logic (NTL) and Cascode Nonthreshold Logic (CNTL) for High-speed Applications," *IEEE Journal of Solid-State Circuits*, **24, No. 3** pp. 779-786 (1989).

CHAPTER 6

BIPOLAR CML INTEGRATED CIRCUITS [1]

Most of the bipolar logic families such as T^2L and I^2L are saturated-mode logic circuits. The fact that BJTs are driven into saturation slows the circuits' operation. Some techniques have been used to prevent hard saturation such as Schottky clamping. The purpose of this chapter is to introduce the nonsaturating current-mode bipolar logic. In essence this is achieved by limiting current and voltage swings to ensure only active-mode BJT operation. The benefits are major speed improvements due to decreased stored minority charge and reduced voltage swing. The operation of nonsaturating logic is based on current switching. Therefore, it is most commonly known as *Current Mode Logic* (CML). This chapter deals with the analysis and design of two types of current mode logic families; the *Emitter-Coupled Logic* (ECL)and the *Emitter Function Logic* (EFL).

6.1. Emitter Coupled Logic (ECL)

In order to increase the frequency of operation of a bipolar logic family it is necessary to prevent the transistor from operating in the saturation region. One method to do this is to make sure that the operating V_{CE} is greater than the soft saturation voltage and that the base current is just enough to maintain the transistor operating in the active region. Because it is difficult to control the base current at a level corresponding to the active region of operation ($I_B = I_C/\beta$) due to the wide tolerance in β, it is more practical to control the emitter current (or the collector current for that matter) at a constant value that is determined by the bias circuit.

---

[1] This chapter is based in part on the chapter titled "Logic Families for High-Speed LSI: ECL and EFL,"[1].

This leads us to the basic emitter-coupled logic gate of Fig. 6.1. It consists of a fixed bias transistor Q_o whose base is connected to a reference voltage V_r and a number of input transistors connected in parallel (Q_1 to Q_m).

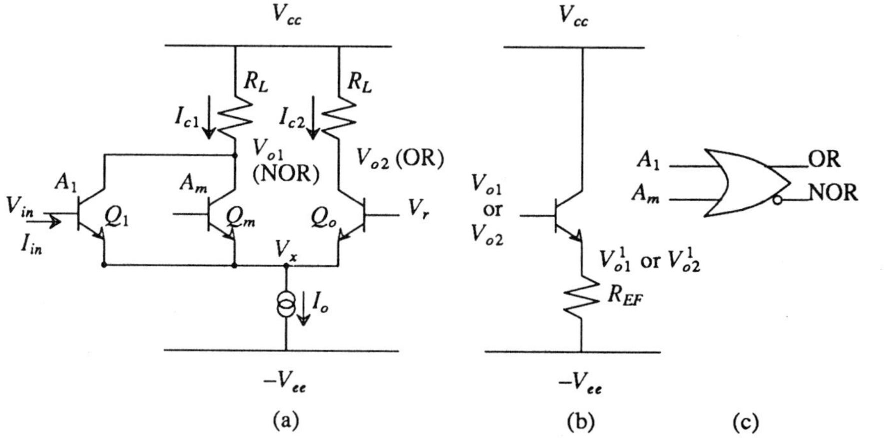

Figure 6.1 (a) ECL basic gate; (b) Emitter-follower stage; (c) Logic symbol.

The input transistors are able to steer the current I_o between Q_o and (Q_1 to Q_m). As the current is steered, the logic output voltage at V_{01} and V_{02} also will change giving a NOR and an OR logical output, respectively. Because of the fact the steering of the current [Fig. 6.2] changes the output voltage, the name current mode logic (CML) is also used. The current of Fig. 6.2 is normalized with respect to αI_o, which is the maximum collector current since I_o is the emitter current and $\alpha = I_C/I_E$. The abscissa represents the voltage excursion needed to steer current between Q_o and the input transistors. Since the collector currents are exponentially dependent on their base-emitter voltages, it only requires a change of mkT/q (m is about 2-3) in the input voltage with respect to V_r to switch the currents. This transition region is shown in Fig. 6.2. In the examples used in this chapter, V_r is set at an odd multiple of ½ V_{BEon}, where V_{BEon} is the base-emitter voltage required to turn a BJT on. This voltage is approximately equal to 0.7V. Because it is desirable to have symmetric noise margins at the logic high and low levels, a total logic swing of V_{BEon} is typical and it is higher than mkT/q. Referring to Fig. 6.1, the highest potential for V_{01} will be V_{cc} (typically ground) and the lowest potential will be $V_{cc}-V_{BEon}$, which is ½ V_{BEon} below V_r.

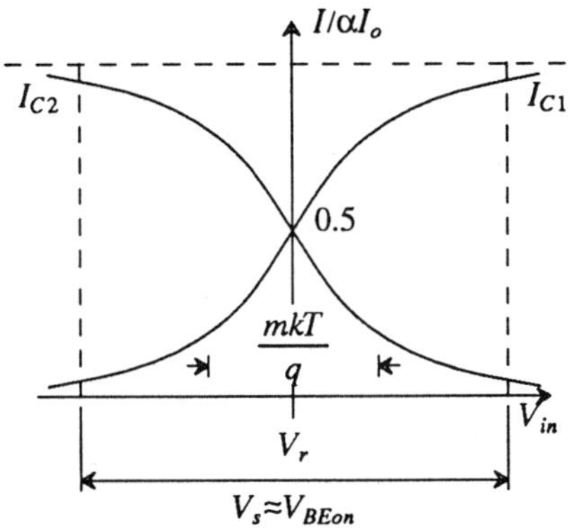

Figure 6.2 Normalized collector currents I_{C1} and I_{C2} vs V_{in}.

The threshold voltage of the ECL gate is V_r, which is the voltage level to which the input voltage is compared. If any of the input logic signals (A_i) is sufficiently higher than V_r, then Q_i is ON and Q_o is OFF. If all the input logic signals are sufficiently lower than V_r, then Q_o is ON and Q_i (i=1,...,m) are OFF. Allowing for a 0.7V drop across the current source to avoid saturation, V_r has to be 1.4V, i.e. two diode drops above ($-V_{ee}$). If Q_o is allowed to operate at V_{CE} of 350 mV, then its collector will be at 1.05V above ($-V_{ee}$). If the logic swing across R_L is a diode drop of 0.7V, then the required *minimum* V_{cc} is 1.75V above ($-V_{ee}$). However, in order to allow cascoding (stacking structures sharing the same current source) of ECL structures, V_{cc} is usually taken to be 5.2V above ($-V_{ee}$). This allows three ECL structures to be cascoded. If cascoding is not required, V_{cc} could be 1.75V above ($-V_{ee}$), assuming the current-source operating voltage and the logic swing are both 700 mV and the emitter-coupled pair operates with a minimum V_{CE} of 350 mV when they are on.

In ECL circuits the logic output is usually referred to as true ground potential, that is, V_{cc} is taken to be ground and the circuit will operate from a negative supply voltage. This provides a less noisy signal at the output. This can be explained as follows:

(1) If the "current source" is a resistor R_{EE}, $R_{EE} \gg R_L$, then noise at the V_{cc} node introduces a signal = $\Delta V_n R_{EE}/(R_{EE}+R_L)$, whereas noise introduced at the V_{ee} node creates a signal = $\Delta V_n R_L/(R_{EE}+R_L)$. The voltage divider ratio suggests that noise at the V_{cc} node is larger.

(2) If "current mirrors" are used as discussed next, changes at the V_{ee} terminal affect the bias level and hence the mirror current by the $\Delta V_n/R_{bias}$, whereas noise at the V_{cc} terminal appears directly at the output assuming the current sources have infinite output impedance.

In order to increase the driving capability of the basic ECL gate, an emitter-follower stage is added as shown in Fig. 6.1(b). The addition of emitter followers lowers the logic levels at the outputs by a V_{BEon}. Thus the reference V_r must also be lowered to a value of $1\frac{1}{2} V_{BEon}$. Fig. 6.3 shows the input current over the input logic swing about V_r. Normalized kT/q units ($V_t = kT/q$) are used, and it is observed that only a few V_t are required to go from the OFF to the ON states.

From Fig. 6.1(a) and Fig. 6.2 it is clear that if the input voltage changes by a diode drop V_{BEon}, the collector current will change by approximately I_o. The logic swing of the circuit V_s is equal to $I_o R_L$. Usually the output logic swing is limited to a diode drop and in many ECL circuits this is obtained by shunting the load resistance with a diode.

The current source is usually realized using a transistor working in the active region, as we shall discuss later.

In the circuit of Fig. 6.1(a) if we consider $-V_{ee}$ to be a reference voltage, the active current source needs a V_{CE} higher than the soft saturation voltage ($\approx$ 300 mV). Let this voltage be a diode drop of 700 mV. As a result, when any of the transistors (Q_o–Q_m) is ON, it is working in the active region ($V_{CE} \geq$ the soft saturation voltage of the transistor). Fig. 6.2 shows the relationship between the output currents I_{C1} and

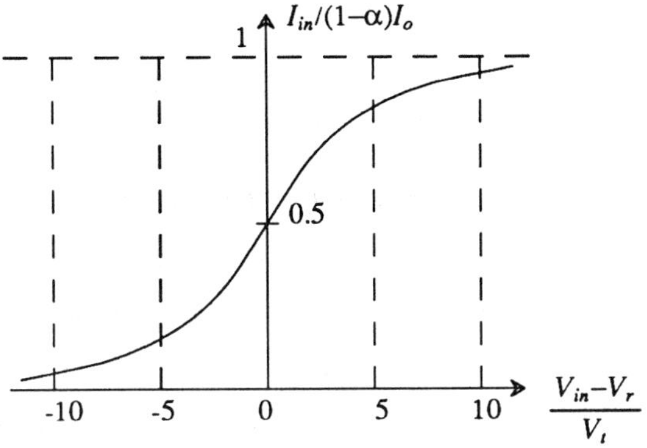

Figure 6.3 Normalized current I_{in} normalized $V_{in}-V_r$.

I_{C2} and the input voltage V_{in} assuming identical transistor characteristics and an ideal constant current source I_o. I_{C1} and I_{C2} is related to the input voltage by an exponential characteristic[2] :

$$I_{C1} = I_S \exp\left[\frac{V_{in} - V_x}{V_t}\right] \tag{6.1}$$

$$I_{C2} = I_S \exp\left[\frac{V_r - V_x}{V_t}\right] \tag{6.2}$$

where V_x is the common emitter node voltage shown in Fig. 6.1 and I_S is the bipolar transistor transport reference current. Hence, the ratio of the two collector currents is given by

$$\frac{I_{C1}}{I_{C2}} = \exp\left[\frac{V_{in} - V_r}{V_t}\right] \tag{6.3}$$

Using the equation $I_{C1}+I_{C2} = \alpha I_o$, then

$$\frac{I_{C2}}{\alpha I_o} = \frac{1}{1 + \exp[(V_{in}-V_r)/V_t]} \tag{6.4}$$

where α is the common-base short-circuit current gain. $V_t=kT/q$ is the thermal voltage. It follows that the input-current-input-voltage characteristic[3] [Fig. 6.3] is given by:

$$\frac{I_{in}}{(1-\alpha)I_o} = \frac{1}{1+\exp[(V_{in}-V_r)/V_t]} \tag{6.5}$$

Fig. 6.4(a) shows a complete realization of the classical two-input ECL OR/NOR gate with typical element values[4]. Transistors Q_{1A} and Q_{1B} provide the standard NOR gate and transistors Q_3 and Q_4 provide emitter-follower outputs from nodes 2 and 5. The 1.18 kΩ resistor from node 3 to V_{ee} sets the circuit operating current [for this circuit I_o = (5.2V - 2.5V_{BEon}) / 1.18-kΩ = 2.85 mA]. The voltages at nodes 3 and 4 are established at two and one V_{BEon} drops below the bias at node 8 which is at -0.44V, a value about ½ V_{BEon} below ground. The operation of the basic ECL gate dictates that I_o controls the maximum swing at nodes 2 and 5, hence V_{s2} = 290 Ω I_o = 0.825V. The output voltages at nodes 6 and 7 are shifted downwards by one V_{BEon} of Q_3 and Q_4. Hence the output levels of -0.7V and 1.525V shown in Fig. 6.4(b) are observed. One can see that the choice of the bias voltage at node 4 of V_r =-1.14V is exactly consistent with the midpoint output voltage swing. It is useful to define capabilities of ECL for more complex logic structures and the critical timing constraints of basic logic. Fig. 6.5 shows more complex ECL logic gates. The novel features are apparent as compared with Fig. 6.4. Diode clamps have been used as load devices [Fig. 6.5(b)] and a stack logic realization is used to generate two levels of functional complexity per bias current [Fig. 6.5(a)]. The diode clamp is useful in limiting the logic swing. The stacked circuit function is an efficient use of both power and device area. The one penalty is that two bias voltage levels are required although this is a minor internal problem as can be noted. This stacking feature is explained in Section 6.4. in association with EFL.

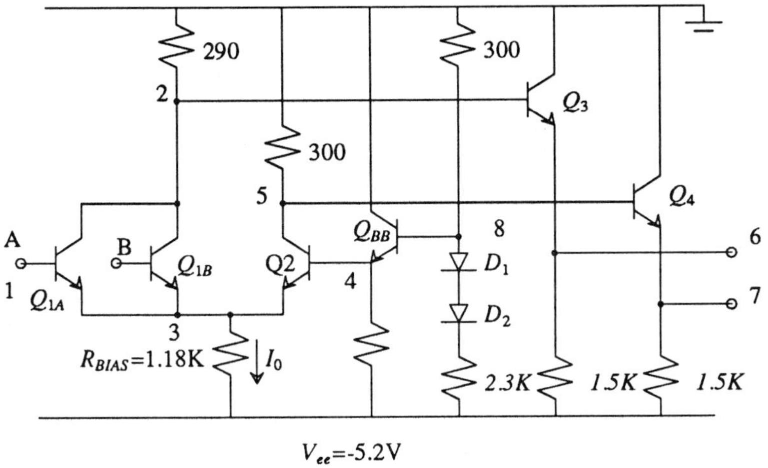

(a)

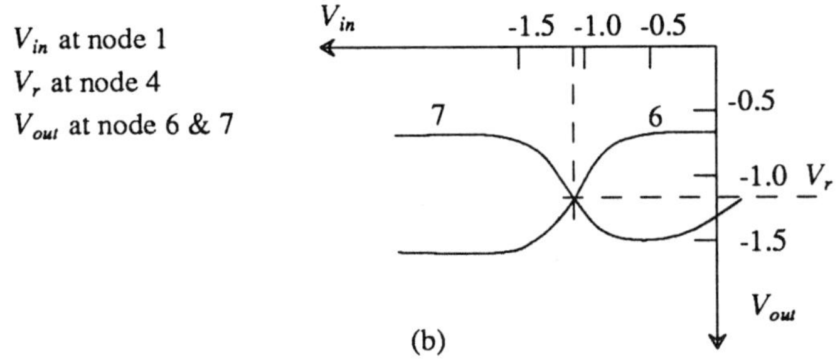

(b)

Figure 6.4 (a) An ECL NOR/OR gate; (b) V_{out} vs V_{in} for Fig. 6.4(a).

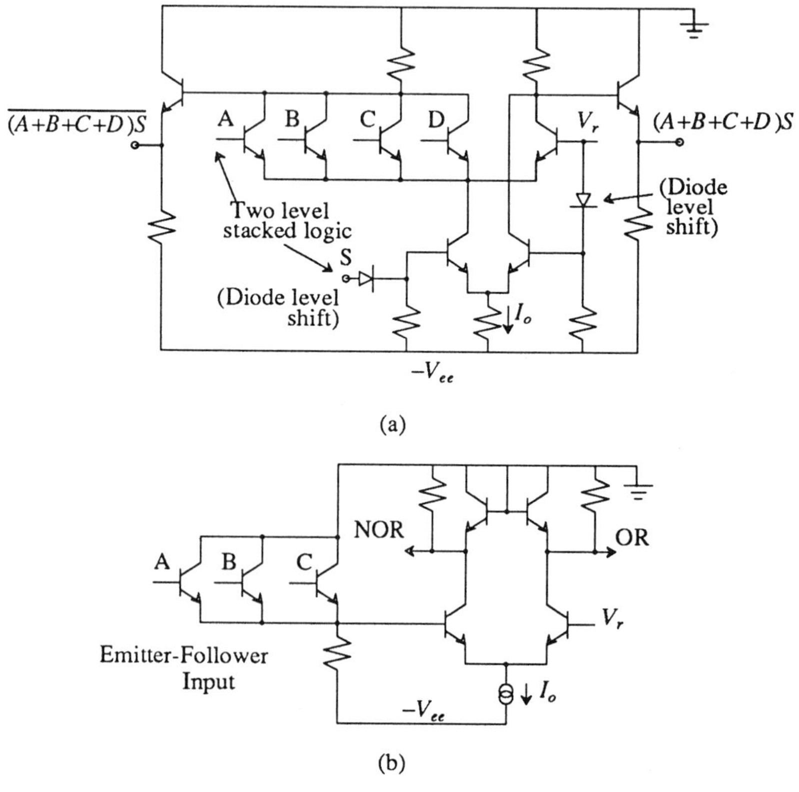

Figure 6.5 Modified ECL gates: (a) two-level stacked gates; (b) high-speed realization with diode clamps and emitter-follower inputs.

6.1.1. Transient Analysis of ECL and CML Gates

We will discuss the transient analysis of the CML gate first and subsequently add the analysis of the emitter follower[2].

---

[2] This section is based on the analysis of M. Ghannam[5].

6.1.1.1. Transient Analysis of the CML Gate

As the input of transistor Q_1 [Fig. 6.1(a)] increases, the collector current I_{C1} rises exponentially in time. If the natural final value of I_{C1} is equal to the maximum possible value (αI_o), it can be expressed as :

$$I_{C1} = \alpha I_o \left(1 - exp(-t/\tau_b)\right) \tag{6.6}$$

The time constant τ_b is a function of the parameters of the input circuit. An expression for τ_b will be derived later [Equation (6.19)]. Unfortunately, the final value towards which I_{C1} is rising is in general not known, however, it cannot exceed αI_o. Therefore, a factor D has to be added to Equation (6.6). Hence, I_{C1} is given by :

$$I_{C1} = D\alpha I_o \left(1 - exp(-t/\tau_b)\right) \tag{6.7}$$

As soon as the collector current of Q_1 reaches αI_o, it will cease to rise and will be clamped to that value. The base-emitter voltage of Q_1 has to adjust itself to I_{C1}. SPICE simulations have shown that the factor D is found to be about 1.5.

Fig. 6.6 shows the transient equivalent circuit of the CML gate of Fig. 6.1(a) (with only one input bipolar transistor Q_1). Using the charge control model, the base and collector currents of Q_1 can be written as follows :

$$I_{B1} = \frac{dq_{f1}}{dt} + \frac{q_{f1}}{\beta\tau_f} + \frac{dq_{E1}}{dt} + \frac{dq_{C1}}{dt} \tag{6.8.a}$$

$$I_{C1} = \frac{q_{f1}}{\tau_f} + \frac{dq_{C1}}{dt} \tag{6.8.b}$$

Where β is the dc current gain and τ_f is the forward transit time.q_{f1} is the total mobile base charge in the forward active region, q_{C1} and q_{E1} represent the charge in the collector-base and emitter-base depletion regions of Q_1 respectively. Similarly, the base and collector currents of Q_0 can be expressed as :

$$I_{B2} = \frac{dq_{f2}}{dt} + \frac{q_{f2}}{\beta\tau_f} + \frac{dq_{E2}}{dt} + \frac{dq_{C2}}{dt} \tag{6.9.a}$$

$$I_{C2} = \frac{q_{f2}}{\tau_f} + \frac{dq_{C2}}{dt} \tag{6.9.b}$$

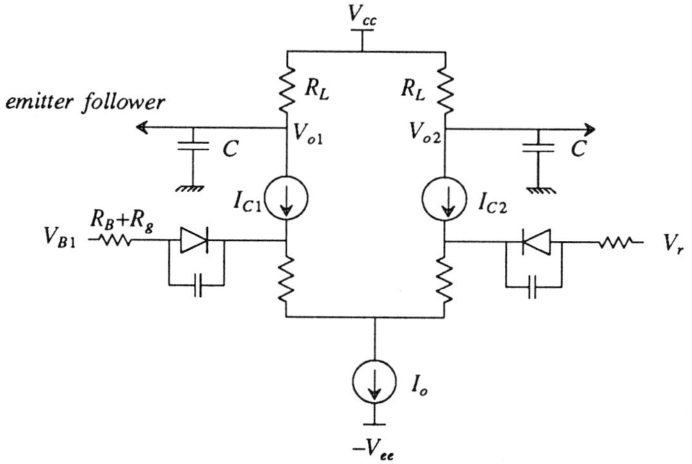

Figure 6.6 Transient equivalent circuit of the CML gate.

In this analysis, we will neglect the space-charge q_C and q_E because they are smaller than q_f. The reader can refer to [5] to study the effect of the space-charge on the delay of the CML gate.

From the equivalent circuit shown in Fig. 6.6, the following loop equation can be written :

$$V_{B1} = I_{B1}(R_g+R_{B1}) + V_{BE1} + (I_{B1}+I_{C1})R_{E1}$$

$$- (I_{B2}+I_{C2})R_{E2} - V_{BE2} - I_{B2}R_{B2} + V_r \tag{6.10}$$

Where R_g is the source resistance and R_B, R_E and R_C are the base, emitter and collector resistances respectively. The source resistance R_g is normally the output resistance of the previous stage, which is equal to $R_L = V_s/\alpha I_o$, where V_s is the swing of the input signal. Note that V_{B1} is equal to $V_r+V_s/2$.

The base-emitter voltage is in general given by :

$$V_{BE} = V_t \ln\left(\frac{I_C}{I_S}\right) \tag{6.11}$$

By applying the KCL at the common emitter node, we can write :

$$I_o = I_{B1} + I_{C1} + I_{B2} + I_{C2} \tag{6.12}$$

After neglecting q_E and q_C in Equations (6.8) and (6.9), the base currents can be expressed in terms of the collector currents. Hence, Equation (6.12) can be rewritten with I_{C1} and I_{C2} as the only variables. Equation (6.10) can also be written in terms of I_{C1} and I_{C2} by using Equation (6.11). Thus, I_{C2} can be expressed as a function of I_{C1}. Finally, Equation (6.10) can be rewritten as :

$$\tau_f R_t \frac{dI_{C1}}{dt} + \left(\frac{R_t}{\beta}+2R_E\right)I_{C1} - \left(R_E+\frac{R_B}{\beta+1}\right)I_o$$
$$-V_t \ln\left(\frac{I_o\beta}{I_{C1}(\beta+1)} - 1\right) - \frac{V_s}{2} = 0 \tag{6.13}$$

Where $R_t=R_g+2R_B+2R_E$. Introducing the normalized variable $i=I_{C1}/\alpha I_o$, the differential equation (6.13) becomes :

$$\tau_f \frac{di}{dt} = A + B\ln\left(\frac{1}{i} - 1\right) - Ci \tag{6.14}$$

where

$$A = \frac{V_s/2I_o+R_E+R_B/(1+\beta)}{\alpha R_t}$$

$$B = \frac{V_t}{\alpha I_o R_t}$$

$$C = \frac{1}{\beta} + \frac{2R_E}{R_t}$$

By separating variables and integrating both sides of Equation (6.14), we obtain :

$$\int_0^t \frac{1}{\tau_f}\, dt = \int_0^i \frac{di}{A + B\ln(1/i-1) - Ci} \tag{6.15}$$

The above integral does not have a closed form solution as it stands. However, it may be simplified by replacing $\ln(1/i-1)$ by $2.5-5i$. This approximation is very good within the required range, $0<i<1$. Hence, the integration in Equation (6.15) reduces to :

$$\int_0^t \frac{1}{\tau_f}\, dt = \int_0^i \frac{di}{A+2.5B - (5B+C)i} \tag{6.16}$$

Solving the above integration and substituting for i by $I_{C1}/\alpha I_o$ we get the following :

$$I_{C1} = \alpha I_o \frac{2A+5B}{10B+2C}\left(1 - \exp(-t(5B+C)/\tau_f)\right) \tag{6.17}$$

The comparison of Equations (6.17) and (6.7) yields the following expressions:

$$D = \frac{V_s + 5V_t + 2I_o[R_E + R_B/(1+\beta)]}{10V_t + 2I_o[\, 2R_E + (R_g+2R_B)/(1+\beta)\,]} \tag{6.18}$$

and

$$\tau_b = \frac{I_o\tau_f\beta(R_g + 2R_B + 2R_E)}{5V_t(1+\beta) + 2I_oR_E(1+\beta) + I_o(R_g + 2R_B)} \tag{6.19}$$

Note that R_g can be replaced by $V_s/\alpha I_o$ as explained earlier. Having derived the full current expression [Equation (6.7) with Equations (6.18) and (6.19)], we can determine the time required for I_{C1} to reach its maximum (αI_o) by substituting for I_{C1} by αI_o in Equation (6.7). This time is referred to as t_0 and is given by :

$$t_0 = \tau_b \ln\left(\frac{D}{D-1}\right) \tag{6.20}$$

So far we have considered the current transients. To complete the analysis, we have to study the transient response of the voltage at the collector node of Q_1 (the output voltage). The transient equivalent of the output circuit is shown in Fig. 6.7. The capacitance C is the summation of the collector-substrate capacitance of Q_1 (C_s), the Miller collector component of the collector-base capacitance of Q_1 and the collector-base capacitance (C_C) of the emitter follower in the case of an ECL gate. In the case of a CML gate, the collector-base capacitance of the emitter follower is replaced by the load capacitance C_L. The Miller collector capacitance is equal to $C_C(1-1/A_v)$, where A_v is the ratio (gain) of the collector voltage to the base voltage of Q_1. The gain A_v is much greater than unity, hence, the Miller capacitance is very close to C_C. Therefore,

$$\tau_c = R_L C = R_L(2C_C + C_S) \qquad \text{for } ECL \qquad (6.21.a)$$

$$\tau_c = R_L C = R_L(C_C + C_S + C_L) \qquad \text{for } CML \qquad (6.21.b)$$

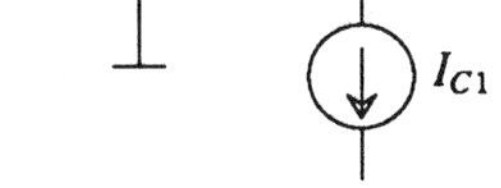

Figure 6.7 Transient equivalent of the collector circuit.

The transient behavior of the output node has to be analyzed for $t<t_0$, when I_{C1} is still rising [Equation (6.6)], and for $t>t_0$, when $I_{C1}=\alpha I_o$.

(i) $t < t_0$:

The nodal equation at the output node [Fig. 6.7] yields the following differential equation :

$$\frac{dV_{o1}}{dt} + \frac{V_{o1}}{\tau_c} = \frac{V_{cc}-I_{C1}(t)R_L}{\tau_c} \tag{6.22}$$

To solve Equation (6.22) we have to substitute first from Equation (6.7) into (6.22). Using Laplace transform and partial fraction, we get :

$$V_{o1}(s) = \frac{V_{cc}}{s} - D\alpha I_o R_L \left[\frac{1}{s} + \frac{1}{s+1/\tau_c}\left(\frac{\tau_c}{\tau_b-\tau_c}\right) - \frac{1}{s+1/\tau_b}\left(\frac{\tau_b}{\tau_b-\tau_c}\right)\right] \tag{6.23}$$

The output voltage in the time domain can be found by taking the inverse Laplace transform of (6.23).

$$V_{o1}(t) = V_{cc} - D\alpha I_o R_L \left[1 + \left(\frac{\tau_c}{\tau_b-\tau_c}\right)e^{-t/\tau_c} - \left(\frac{\tau_b}{\tau_b-\tau_c}\right)e^{-t/\tau_b}\right] \tag{6.24}$$

The output voltage might reach the 50% point ($V_{cc}-V_s/2$) within t_0. This can be determined by solving Equation (6.24) at $t=t_0$. If $V_{o1}<V_{cc}-V_s/2$, then the delay is less than t_0. In such case, the delay time can be found by substituting for V_{o1} by $V_{cc}-V_s/2$ in (6.15) and solving for t using iterative methods.

(ii) $t > t_0$:

In this interval the collector current of Q_1 is constant and is equal to αI_o. By substituting for $I_{C1}(t)$ in Equation (6.22) by αI_o and solving the resulting differential equation using Laplace transform, we obtain the following :

$$V_{o1}(t-t_0) = V_{cc} - V_s - \left(V_{o1}(t_0) + V_s - V_{cc}\right) e^{-(t-t_0)/\tau_c} \tag{6.25}$$

Hence, the 50% delay time is given by :

$$t_{cml} = t_0 + \tau_c \ln \left[2\frac{V_{o1}(t_0) + V_s - V_{cc}}{V_s}\right] \tag{6.26}$$

6.1.1.2. Transient Analysis of the ECL Gate

The delay of the ECL gate can be divided into two components, the delay of the CML and that of the emitter-follower stage. The delay of the CML stage has been already considered. The exact transient analysis of the emitter-follower stage is very complex[3]. However, for a first order analysis we can simplify the complexity of the problem. Let us consider the following two cases :

(i) Small Load Capacitance

In this case, the base-emitter voltage of the bipolar transistor of the emitter-follower remains almost constant. The output of the emitter follower stage tracks the output of the CML very closely. Hence, the recharging time of the emitter-base and diffusion capacitances of the bipolar transistor is negligible compared to the CML delay. Therefore, we can assume that the ECL speed is equal to that of the CML core.

(ii) Large Load Capacitance

When the emitter-follower is heavily loaded the output node is not capable of tracking the input node. This might cause the transistor to turn off during the transitions. For example, in the case where the base voltage of the bipolar transistor of the emitter-follower falls while the heavily loaded output node is unable to respond instantaneously, the base-emitter voltage decreases forcing the transistor to turn off. The output capacitance discharges through the emitter-follower resistance (R_{EF}) until the base-emitter voltage builds up to the level which allows the bipolar transistor to turn on again. Therefore, we can assume that the delay of the emitter-

---

[3] Some attempts have been made to derive the delay of the emitter-follower stage assuming a step input. This approach overestimates the delay of the emitter-follower.

follower stage is due mainly to the discharging of the output capacitance through R_{EF}. The 50% delay of the emitter-follower can be expressed as :

$$t_{ef} = 0.7 R_{EF} C_o \tag{6.27}$$

where the output capacitance, C_o, is the summation of the external load capacitance and the collector-substrate capacitance of the bipolar transistor.

The total delay of the ECL gate can be obtained by adding the delay components t_{cml} and t_{ef} from Equations (6.26) and (6.27).

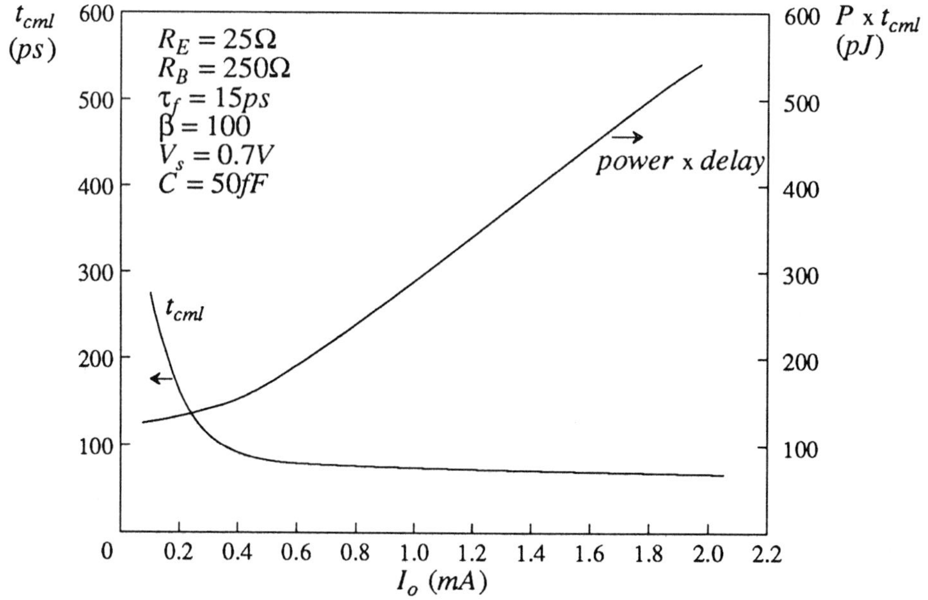

Figure 6.8 The delay and the delay power product as a function of the CML current.

Fig. 6.8 illustrates the delay as a function of the current I_o. The delay falls as the current increases, however, as I_o exceeds a certain level the delay ceases to improve. This can be best explained as follows. Initially, when I_o is too small R_L must be large (to maintain a fixed voltage swing), hence the output time constant (τ_c) becomes too large causing the second term of Equation (6.27) to dominate. As the current I_o increases τ_c is reduced causing the total delay to decrease. However, as I_o grows the contribution of t_0 becomes more significant (because the time required to switch I_o from one branch to the other increases). This means that by increasing I_o the first term of Equation (6.27) increases while the second term decreases. As a result, the delay remains almost constant and may even increase slightly at high values of I_o. To avoid high power dissipation one must select the current I_o that yields near minimum delay with the least possible power delay product. In the example shown in Fig. 6.8, it is obvious that I_o=0.4mA is a good compromise between speed and power.

The above transient analysis relates the device parameters, such as the forward transit time and the parasitic resistances and capacitances, to the circuit speed. The parameters that are of special importance to the delay are the base forward transit time, the base resistances and the parasitic capacitances represented by C [defined in Equation (6.21)]. Fig. 6.9 illustrates the effect of each of these parameters on the delay of a CML gate. The parameters are normalized with respect to the nominal values given in the insert of Fig. 6.9. It is interesting to note that the delay is very sensitive to τ_f, or equivalently f_T, for large values of τ_f. For small values of τ_f (< 25ps), the delay sensitivity to τ_f and R_B are similar. Therefore, if high f_T is achieved at the expense of the base resistance, the ECL speed might not improve[4]. This indicates that the cut-off frequency should not be used solely as a figure of merit for BJTs. It is more meaningful to use the ECL delay as a measure for the performance of bipolar transistors.

[4] If the cut-off frequency of the BJT is increased by reducing the base width, the intrinsic base resistance would increase.

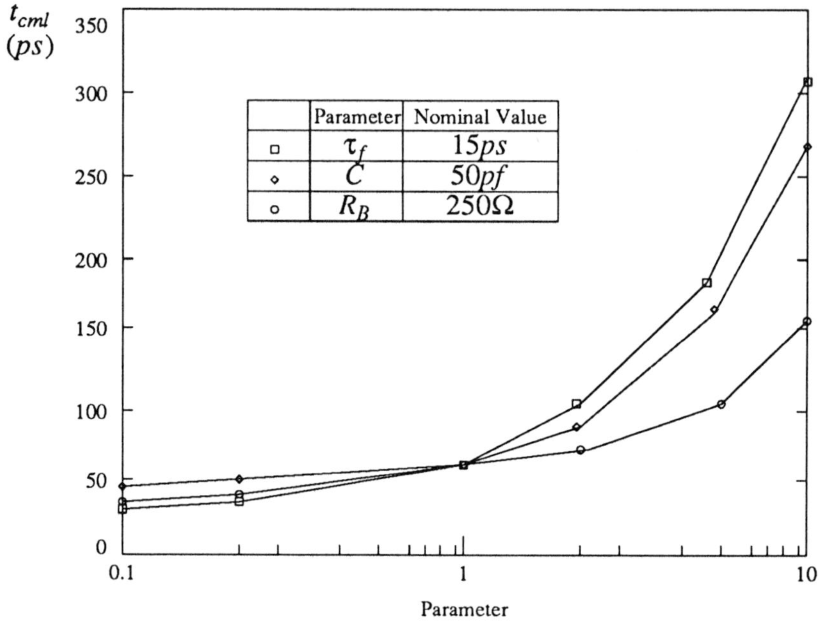

Figure 6.9 The sensitivity of the delay of the CML gate to the bipolar device parameters.

6.1.2. ECL Current Sources

Fig. 6.10 shows a simple biasing circuit to obtain n active current sources for ECL gates. The circuit consists of a biasing chain, a resistor R, and diode-connected transistor. For identical transistor characteristics

$$I_o = \frac{I_{ref}}{1 + (n+1)/\beta} \tag{6.28}$$

If $(n+1)/\beta \ll 1$ then $I_o \approx I_{ref} = (V_{cc} - V_{BEon})/R$.

The simple circuit shown in Fig. 6.10 can be modified to accommodate different requirements. For example, Fig. 6.11 shows a biasing circuit that provides different values of current sources. This may be required where different ECL gates are operating at different speeds.

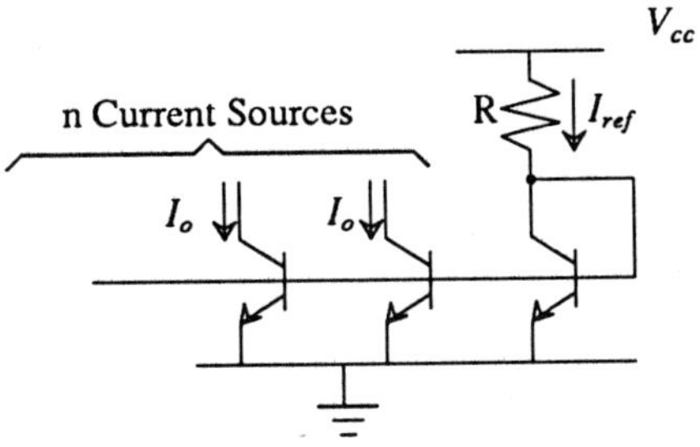

Figure 6.10 Biasing circuit to obtain I_o.

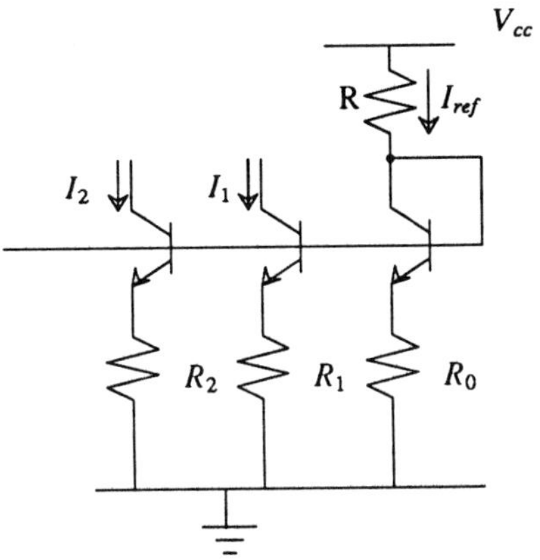

Figure 6.11 Biasing circuit to obtain different values of I_o.

In Fig. 6.11, we have

$$I_1 = I_{ref}\frac{R_0}{R_1}, \quad I_2 = I_{ref}\frac{R_o}{R_2}$$

$$I_{ref} \approx \frac{V_{cc} - V_{BEon}}{R+R_o} \tag{6.29}$$

It should be mentioned here that extensive work has been done on current sources for operational amplifiers[3], and it is applicable to the design of current sources for ECL.

6.1.3. ECL Reference Voltages

Fig. 6.12(a) shows a simple biasing circuit to provide a reference voltage V_r to ECL gates. It consists of a diode-resistor chain followed by an emitter-follower stage. The emitter-follower stage provides a low output resistance, hence V_r can feed more than one ECL gate. Neglecting the base current of the emitter follower transistor, we have,

$$V_r = V_{cc}\frac{R_2}{R_1+R_2} - V_{BEon}\frac{R_2 - R_1}{R_1+R_2} \tag{6.30}$$

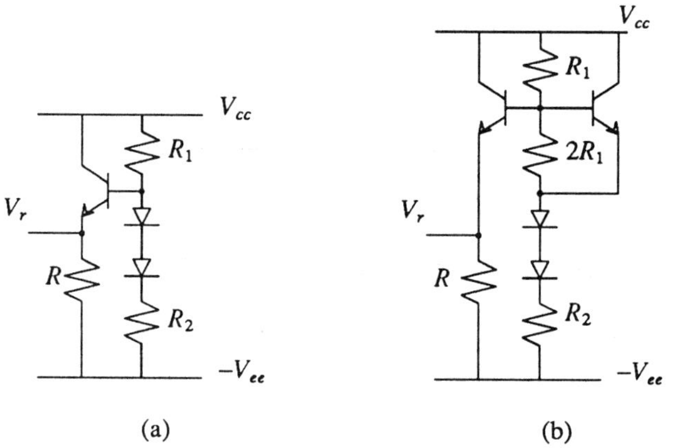

Figure 6.12 (a) and (b) Biasing circuits to obtain the reference voltage.

The purpose of the resistor R is to bias the emitter-follower stage in the active region. The circuit can be modified to meet different requirements[4].

As will be shown in Section 6.6, it is often desirable to set V_r at 1.5 V_{BEon}. The circuit shown in Fig. 6.12(a) can be designed to achieve the desired bias level by choosing R_1 and the bias current to set the reference voltage. A more exact method to set V_r is to replace R_1 with the circuit shown in Fig. 6.12(b). In this case, the resistive divider of R_1 and $2R_2$ gives 0.5 V_{BEon} across R_1, which is then applied at the base of the emitter follower. The output V_r is then $1.5V_{BEon}$.

6.2. Emitter Function Logic (EFL)

Emitter-function-logic (EFL) cells can be viewed as a repartitioned form of ECL[6-9]. The main feature of EFL is the increase in logic-function complexity at minimum increment in the device and circuit complexity. This is achieved without additional power dissipation in comparison to the single function current switch gate. This increase in functional complexity results in significant improvements in relative density, power per gate, and speed-power product. It should also be noted that all of these gains are realized only if this high functional complexity can be fully utilized. Fig. 6.13 shows similarities with ECL. Fig. 6.13(a) shows the conventional ECL gate with two collector loads. The time delay between the input and the OR output is less than that between the input and the NOR output. This is because of the Miller effect at the NOR output. At the NOR side the input and output voltages swing simultaneously in opposite directions. On the other hand, for the OR side the base is fixed so that only the collector-voltage swing affects speed. The time delay of the circuit would be that of the OR output if the NOR input is not used. Thus if only a single load, the one end connected to the fixed-bias transistor, is left in the circuit; two ECL gates connected in cascade will appear as shown in Fig. 6.13(b). Furthermore, if a simple repartition of the circuit is done, the input now is at the emitter of the fixed-biased transistor Q_{in} and the output(s) at the output transistor Q_{out}. This is shown in Fig. 6.13(c). The structure now realizes a noninverting gate, that is, the logical output is the same at the logical input, which is of limited use in logic design. However, if Q_{in} is made as a multiemitter transistor as shown in Fig.

6.13(d), the structure now realizes an AND gate, in that a_1, a_2, and a_3 must be high for the outputs to be high. The input pulldown can be implemented as (1) a resistor, (2) active current source, or (3) logically controlled switch current source. The logical capability of the structure is further enhanced if a two-level EFL cell, as shown in Fig. 6.13(e), is used as a building block for logic design. Fig. 6.13(e) shows Q_{in} with two emitters (the number of emitter inputs can be more than two as we shall see later), and one current source is used. This current source is controlled by a second level of current steering transistors Q_1 and Q_2, where Q_2 is a fixed bias transistor with reference voltage V_{r2}. As a result the output logic function F is given by:

$$F = A_1 a_1 + \overline{A}_1 a_2 \tag{6.31}$$

where A_1, a_1, and a_2 are in general wired-OR functions. That is, multiple logic outputs can be directly attached (*OR*ed) to a_1, for example.

Equation (6.31) can be explained more physically as follows. If A_1 is logical "1," that is, its voltage level is greater than V_{r2}, the current I_o is steered through Q_1 and, hence, the output logical level is controlled by a_1 and $F = a_1$. Alternatively if A_1 is logical "0," that is its voltage is less than V_{r2}, the current I_o is steered through Q_2 and, hence, the output logical level is controlled by a_2 and $F = a_2$.

In order to allow the outputs to be directly coupled either to an input emitter or input base(s), V_{r1} must be one V_{BEon} higher than V_{r2}. Hence the threshold voltage of the EFL gate is V_{r2}. It should be mentioned that each output emitter can fan-out to *only one* input emitter. The reason for this is that when the current I_o is steered through Q_1 or through Q_2 to the emitter input it will be further steered through Q_{in} or through Q_{out} *of the previous stage*. Allowing output emitters to fan out to only one emitter input prevents current sharing problem between inputs. On the other hand, Q_{out} can fan-out to any number of bases since these inputs have a negligible current input requirement.

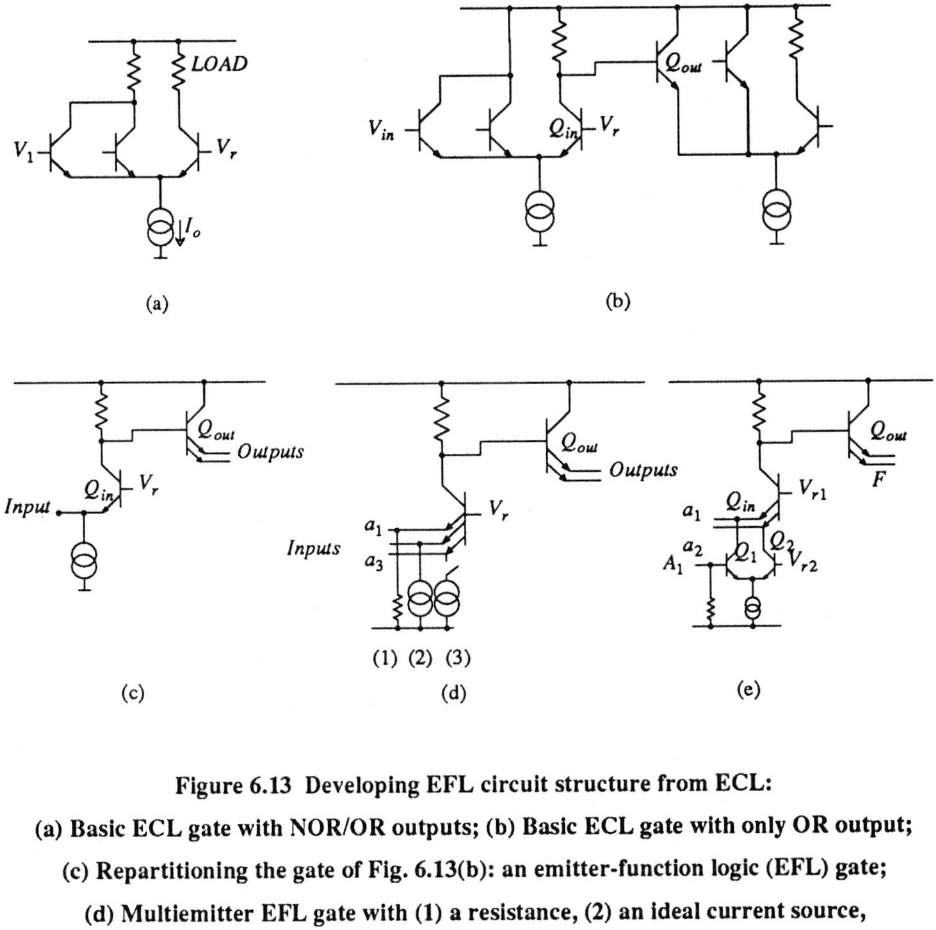

Figure 6.13 Developing EFL circuit structure from ECL:
(a) Basic ECL gate with NOR/OR outputs; (b) Basic ECL gate with only OR output; (c) Repartitioning the gate of Fig. 6.13(b): an emitter-function logic (EFL) gate; (d) Multiemitter EFL gate with (1) a resistance, (2) an ideal current source, or (3) a logically controlled switchable current source; (e) Two-level EFL cell, where a_i are emitter inputs and A_i are base inputs and the output $F=A_1a_1+\overline{A_1}a_2$.

Fig. 6.14(a) and 6.14(b) show an EFL cell with more than two emitter inputs. The cell has $(2n + 1)$ inputs, where $A_1, \cdots, A_n$ are the base inputs and are referred to as *control inputs* and $a_1, \ldots, a_{n+1}$ are the emitter inputs and are referred to as *direct inputs.* The output function F is logical "1," if both A_i and a_i are logical "1," for any i. If all A_i are "0" and a_{n+1} is "1," the output is also logical "1." Normally only one control input (A_i) is allowed to be "1" at a given time and when more than one "A_i" is "1," the output must be a DON'T CARE case. If F = "0" for the DON'T CARE cases, F can be expressed as

$$F = A_1 a_1 + A_2 a_2 + \cdots + A_n a_n + (\bar{A}_1 \bar{A}_2 \cdots \bar{A}_n) a_{n+1} \tag{6.29}$$

Each A_i and a_j can be a first order wired-OR function from the emitter outputs of previous stages.

In practice, the maximum wired fan-in numbers are limited because of chip-layout considerations or because of the required delay. A maximum number of four is feasible and it satisfies the requirements of most logic realizations.

In Fig. 6.14(a) and 6.14(b) only one control input is allowed to be "1" at a given time to prevent the current source I_o from being shared between Q_{out} of the previous stages. Multiple outputs will result in a nonspecific variation in delay. Although this restriction of allowing only one control input to be "1" at a given time presents a limitation on logic designs, in practice it does not present a serious drawback.

In EFL, the collectors of the lower transistors can be tied together as shown in Fig. 6.14(c) and 6.14(d) to form a wired-OR function of a_1 and f_1. Where f_1 is the NOR of $A_1, \cdots, A_m$:

$$F = (f_1 + a_1)(\bar{f}_1 + a_{n+1}) \tag{6.30}$$

where

$$f_1 = \overline{A_1 + A_2 + \cdots + A_m}$$

In this case no restriction on control inputs applies and any number of control inputs can be high, since all the control currents are summed at only one direct input.

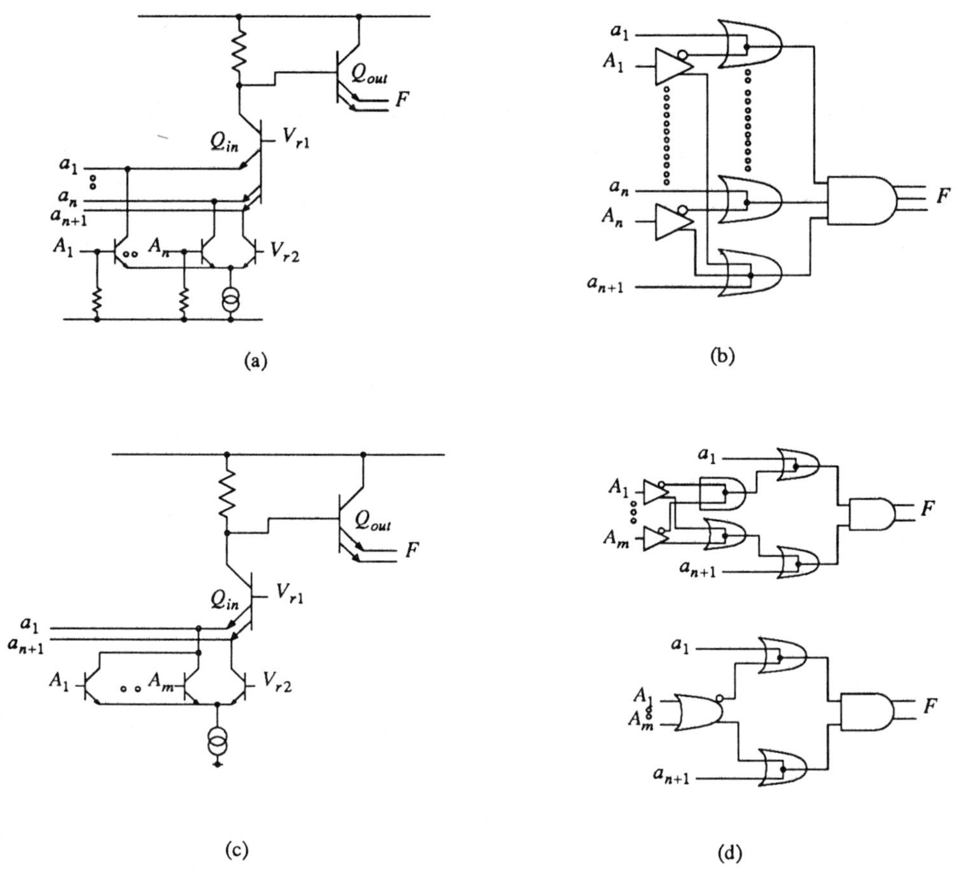

Figure 6.14 (a) Circuit diagram of a two-level EFL cell, with n emitters; (b) Logic gate equivalent; (c) Circuit diagram of a two-level EFL, with two emitters; (d) Logic gate equivalents.

6.2.1. EFL Cell Implementation

Having discussed the operation of an EFL cell it is useful to illustrate how the technology is used to realize the basic block shown in Fig. 6.14. Fig. 6.15 shows one possible symbolic layout of a four-input/four-output cell excluding the differential stages used to create $Q_1, Q_2, \cdots, Q_{n+1}$[7]. Two isolation regions are used. The load resistor R is created by extending the base diffusion in the "L" shape between the base contact of F_3 and F_4 to the pn^+ metal contact to the collector of Q_{out}. Fig. 6.15(c) shows the section view of the devices as viewed along the "L"-shaped section line shown in Fig. 6.15(a). In addition to the resistor and F_3/F_4 devices, a clamping-diode-connected transistor is shown. This diode appears in parallel with the resistor and guarantees a maximum logic swing of V_{BEon}. Note that the $+V_{cc}$ contact (usually ground potential) consists of ohmic contacts to *both* the p regions and the n^+ collector. A schematic circuit equivalent is shown in Fig. 6.15(d) and both the circuit element R as well as extrinsic device resistances are depicted. The input device(s) are nearly identical except for the resistor/diode combination. The design choice to use a distributed R is a "neat-trick" of IC technology and leads to higher packing density.

Unfortunately the superintegration of R also offers certain penalties - in particular the logic swing becomes sensitive to the control of sheet resistance and added parasitics must be considered during gate design. Moreover, sometimes well-isolated components are desirable, which can be controlled separately and even adjusted independently. The key advantage of removing R from the layout shown in Fig. 6.15(a) is that an alternate layout is possible that reduces R_C and R_B, which in turn increases the speed performance of the gate.

The next few subsections will consider examples, of logic, memory, and register design using EFL cells.

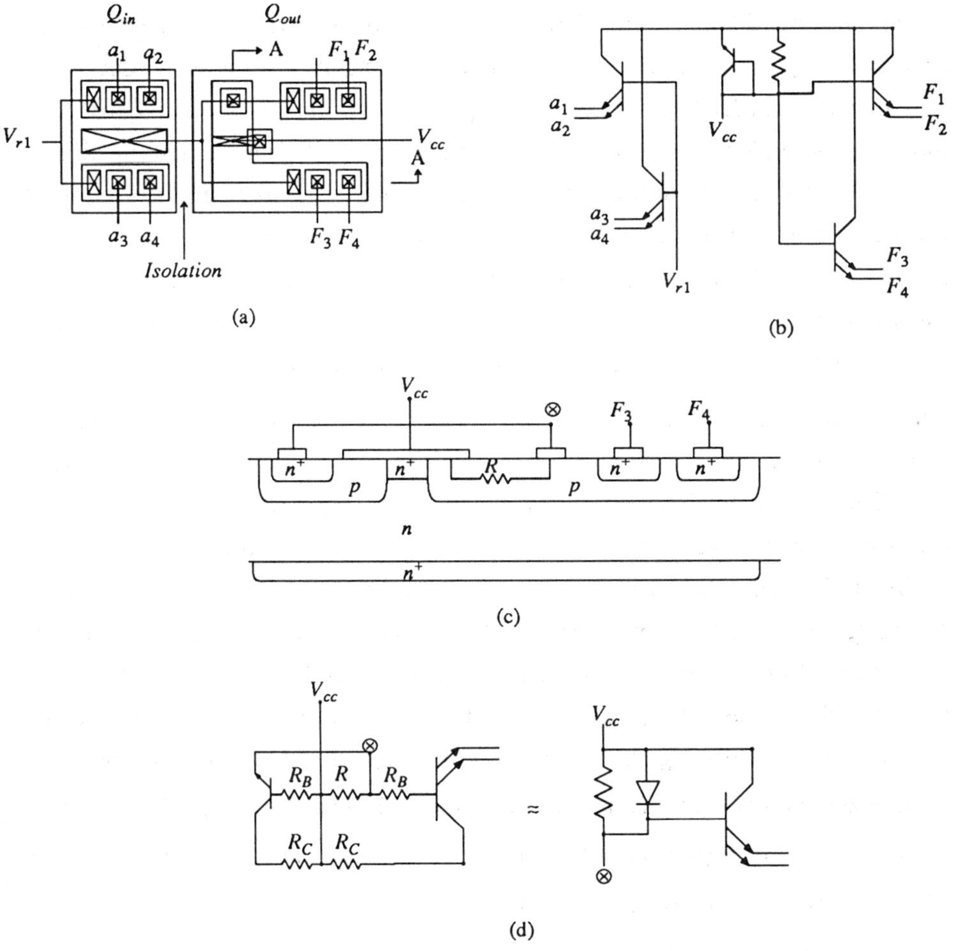

Figure 6.15 Basic Q_{in}, Q_{out} and the load for an EFL gate.

(a) Symbolic layout; (b) Circuit diagram corresponding to the layout.

(c) Cross section AA; (d) Equivalent circuit of cross section AA including parasitics.

6.2.2. EFL Logic Building Blocks: Combinatorial Logic Elements

Examples of simple combinatorial functions are given in Fig. 6.16. It should be realized that more than just single gate functions should be implemented in order to utilize fully the EFL cells. Fig. 6.16(a) shows the basic AND-OR implementation. Note that the power dissipation is a function of the number of the wired-OR's, and no extra power is added due to the AND function. Two-input AND and INHIBIT are shown in Fig. 6.16(b) and 6.16(c). The selector in Fig. 6.16(d) is the first function of sufficient complexity to utilize the EFL logic capability. This is, in fact, the most frequently used element in logic designs, as will be shown further in more complex designs. Fig. 6.16(e) through 6.16(g) show logic realizations that make more extensive use of the EFL gate structure. Fig. 6.16(e) is the single cell realization of the exclusive-OR function. Clearly this is a very efficient use of silicon area.

The exclusive NOR gate implementation in Fig. 6.16(f) uses one current source and requires two emitters for A, B inputs. Only the variables (not their complements) are used. The exclusive-OR realization shown in Fig. 6.16(g) has two current sources, and only direct inputs are used. The exclusive NOR gate shown in Fig. 6.16(h) is a modified EFL cell circuit, which shows the use of a multiemitter device at the lower switching level. Data multiplexing and demultiplexing are shown in Figs. 6.16(i) and 6.16(j). Note the presence of only one noninverting buffer in the data path in order to achieve the high speed. For the demultiplexer circuit the current sources shown as I^*_o are driving the inputs of following cells and should be properly associated with inputs for the next stage. It should be clear that $\overline{D}_o,...,\overline{D}_3$ may drive the $a_1, \cdots, a_n$ level inputs since current source I^*_o would make no logical sense if used on a base input on the $A_1, \cdots, A_n$ level.

A full-adder function implemented with four EFL cells (mostly selectors) is shown in Fig. 6.16(k).

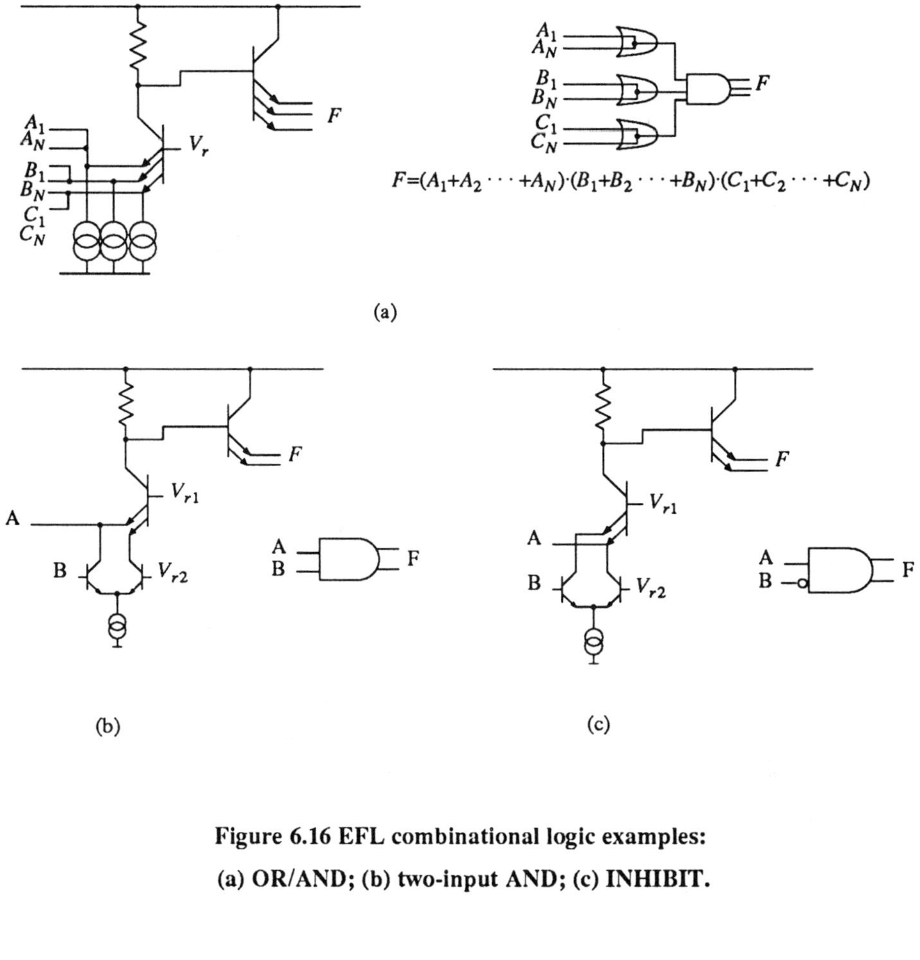

Figure 6.16 EFL combinational logic examples: (a) OR/AND; (b) two-input AND; (c) INHIBIT.

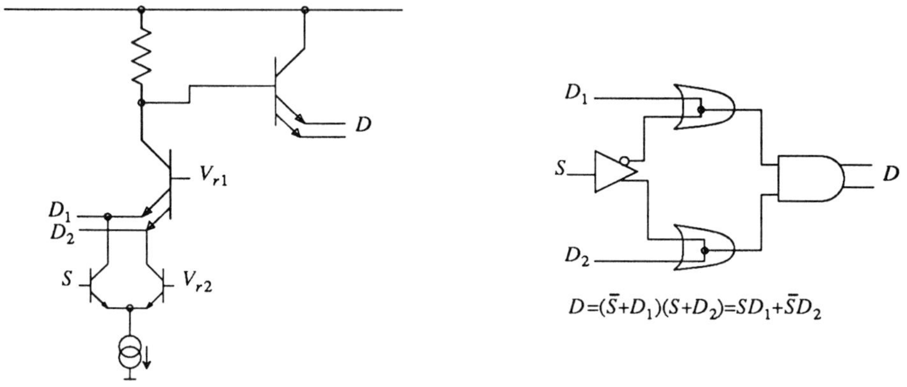

(d)

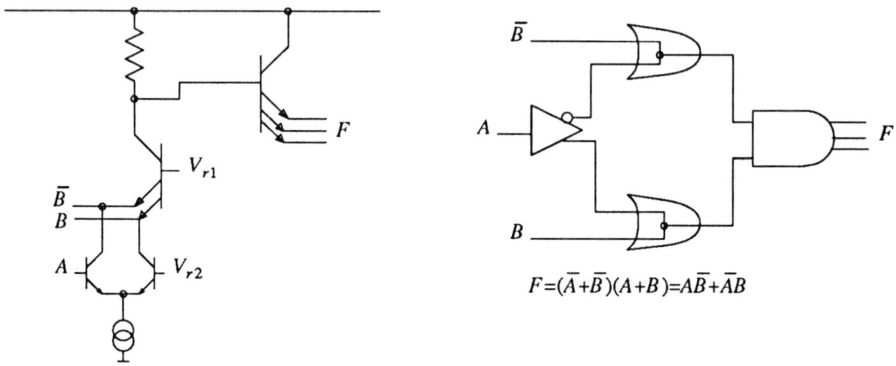

(e)

Figure 6.16 (Continued)

(d) selector; (e) exclusive-OR.

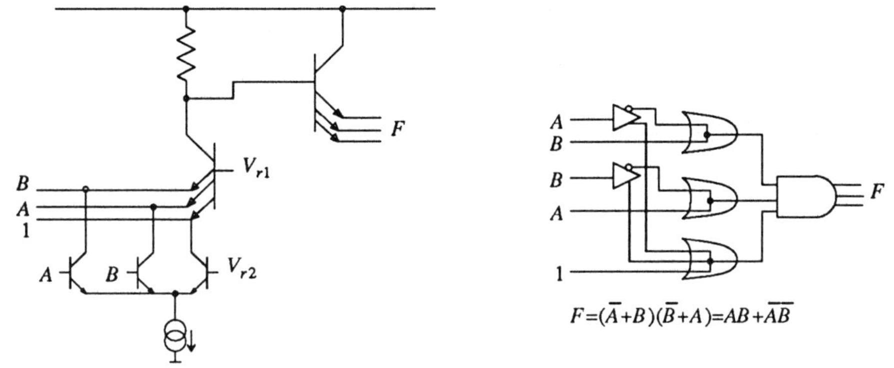

(f)

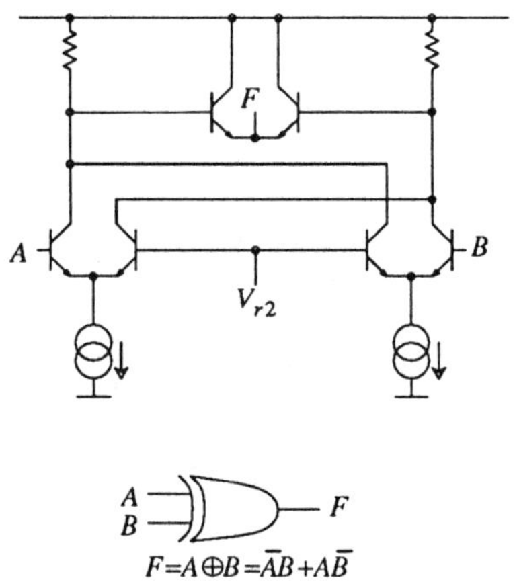

(g)

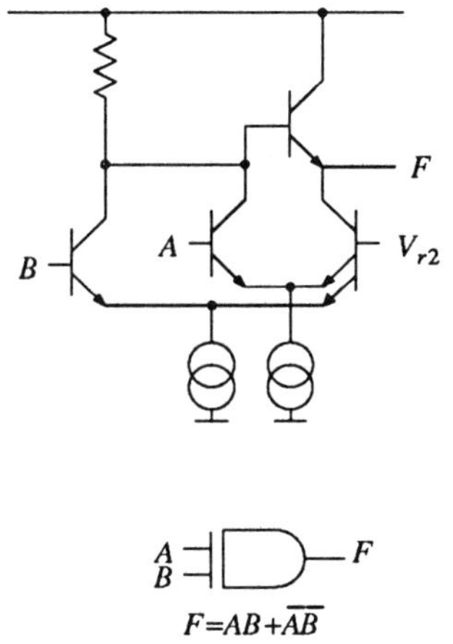

(h)

Figure 6.16 (Continued)

(f) exclusive-NOR; (g) exclusive-OR; (h) exclusive-NOR.

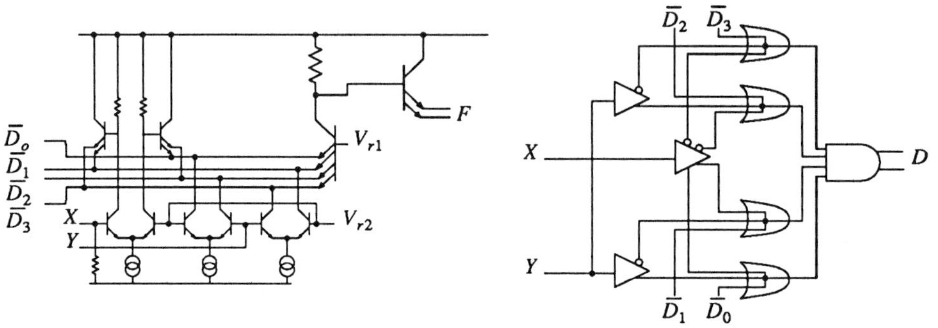

(i)

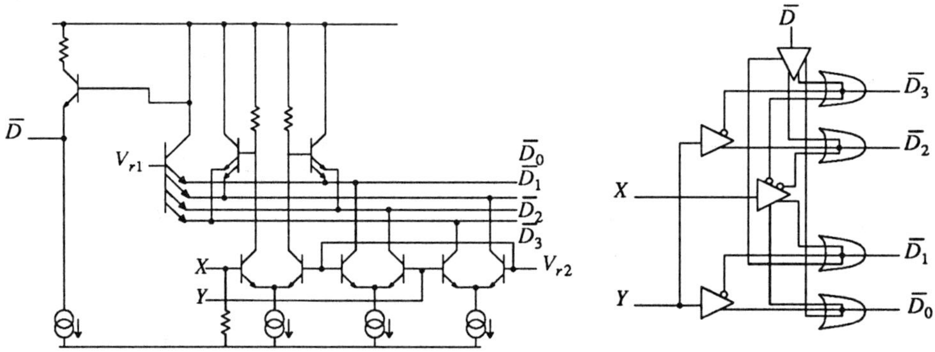

(j)

Figure 6.16 (Continued)

(i) 4:1 multiplexer; (j) 1:4 demultiplexer.

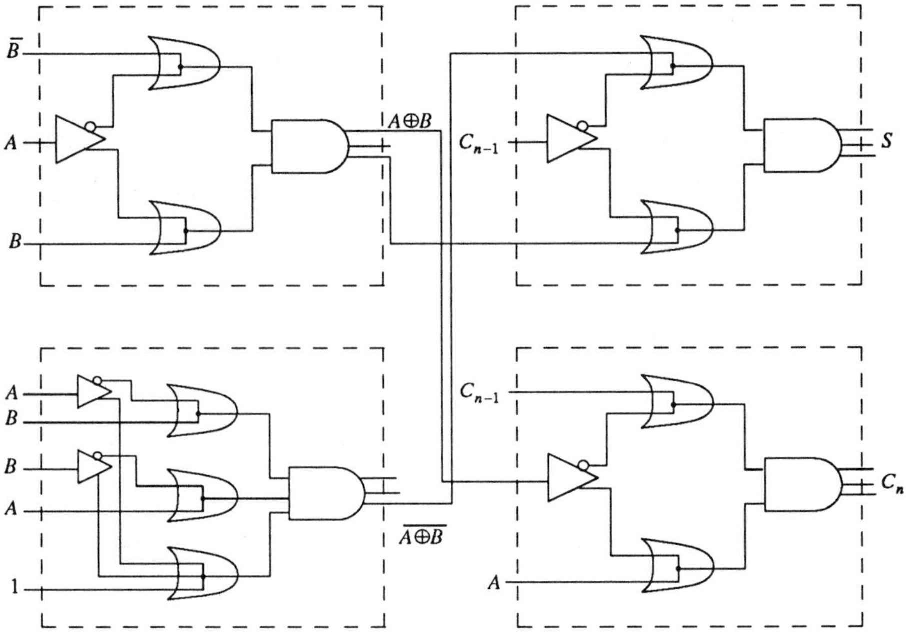

$S \equiv Sum$
$= C_{n-1}(\overline{A \oplus B}) + C_{n-1}(A \oplus B)$

$C_n \equiv Carry$
$= C_{n-1}(A \oplus B) + A(\overline{A \oplus B})$

(k)

Figure 6.16 (Continued)

(k) full-adder.

6.2.3. EFL Logic Building Blocks: Sequential Logic Elements

The ability to hold information is an essential function in digital circuits. Fig. 6.17(a) shows the simplest EFL realization of the RS latch constructed with a single bias reference. The operation of this circuit realization is as follows. One output and input are tied together with a bias resistor and provide the set (S) input. A second input is the reset (R). The set is normally low and the reset is high. For Q low the current through R_{EE} is provided by Q_{in}. To set Q high a positive pulse must be applied to S which turns Q_{in} off and forces Q_{out} to drive the current through R_{EE}. To reset the circuit a negative pulse is applied to R, which turns Q_{in} back on again, moving base current from Q_{out}, thus returning to the original Q-low condition with bias to R_{EE} provided by Q_{in}. This circuit provides the memory function.

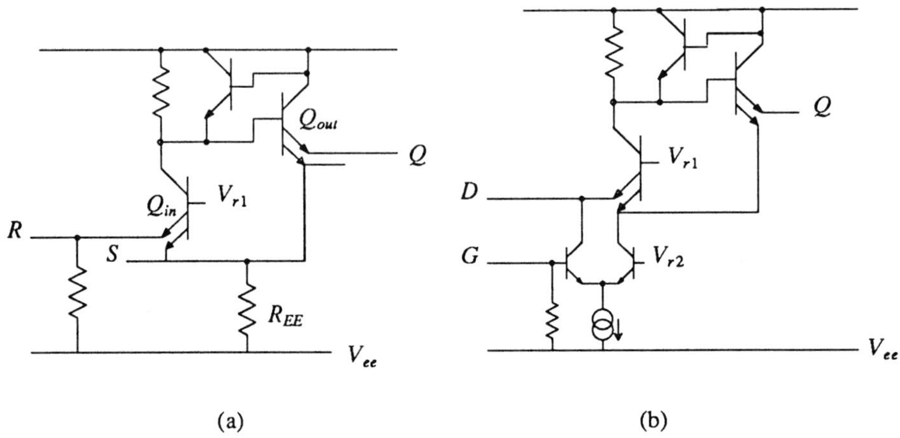

Figure 6.17 EFL latches: (a) simple RS and (b) gated latch.

The gated latch shown in Fig. 6.17(b) represents a circuit form more compatible with clocked logic and standard current-source biasing with two reference levels.

Starting from these basic EFL cells we can now move on to a more formal definition of several memory-element realizations. Two basic configurations are considered:

1. Configuration A, as shown in Fig. 6.18(a), where one of the output emitters, representing the output Q, is connected to the control input X. The cell has similar characteristics to the RS latch, as can be seen from the state table and state diagram shown in Fig. 6.18(b). This realization compares favorably with the conventional cross-coupled two-gate realization.

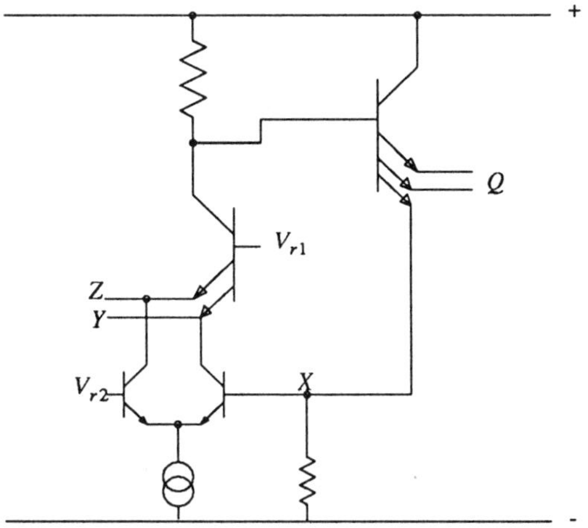

(a)

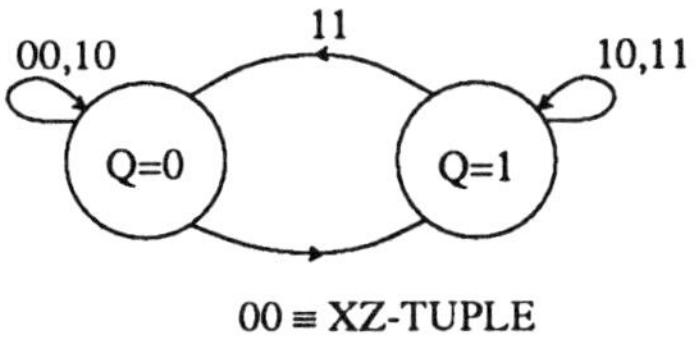

Y	Z	Q_t
0	0	0
1	1	1
1	0	Q_{t-1}
0	1	*ind.*

(b)

Figure 6.18 EFL latch, configuration A (analogous to RS-type latch): (a) circuit diagram and (b) state table and state diagram.

2. Configuration B, as shown in Fig. 6.19(a), where one of the output emitters, representing the output Q, is connected to the direct input Z (or Y). As shown from Fig. 6.19, the structure has the characteristic of a data latch, where the input X is the gating input (Fig. 6.17). This EFL realization consumes less power and less silicon area than the conventional realization of four gates.

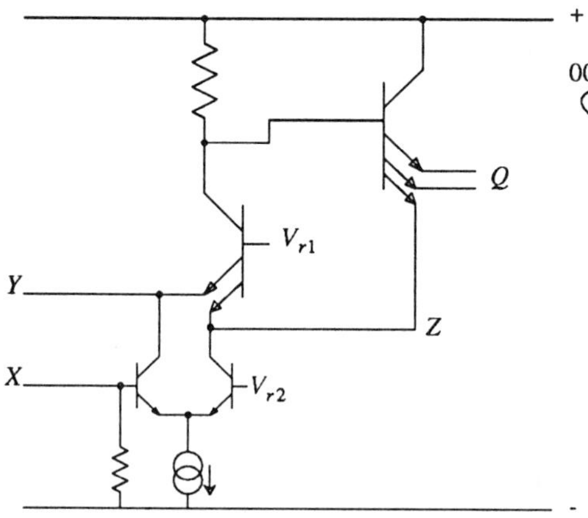

(a)

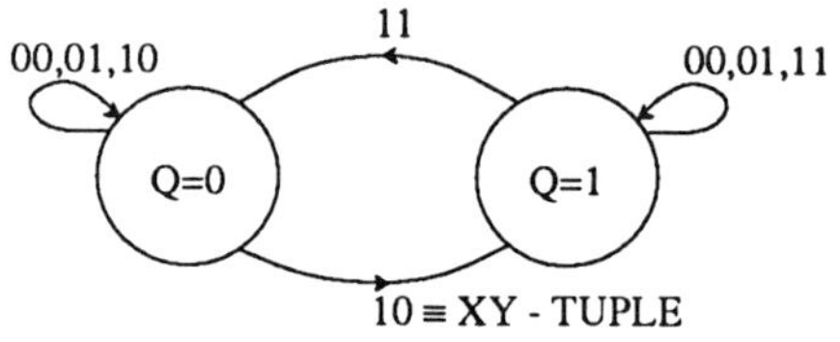

Y	Z	Q_t
0	0	Q_{t-1}
1	1	Q_{t-1}
1	0	Q
1	1	1

(b)

Figure 6.19 EFL latch, configuration B (analogous to D-type latch): (a) circuit diagram and (b) state table and state diagram.

The above two configurations can be used to design more-complex flip-flops. Two EFL cells are used to realize a gated RS flip-flop, as shown in Fig. 6.20, as compared with four gates used in conventional realizations. Fig. 6.21 shows a master-slave realization of an RS flip-flop. Two EFL cells are used to realize a master-slave D flip-flop, as shown in Fig. 6.22 as compared with 10 gates used in conventional realizations. The use of the same V_{r2} threshold for both master and slave latches is permitted only if minimum clock edge speed can be guaranteed. The use of two slightly offset thresholds, as in Fig. 6.22(b), guarantees hazard-free operation for even slow clock edges. The threshold difference of ± 100 mV will enable clock edge speed range from dc to subnanosecond. The JK master-slave circuit realization using

EFL cells is shown in Fig. 6.23.

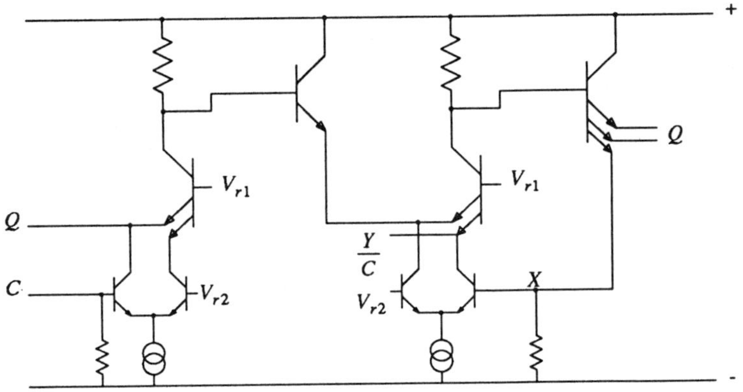

(a)

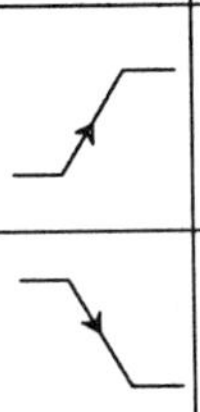

0	Y	Z	Q_t
	C	0	0
	1	1	1
	1	0	Q_{t-1}
	0	1	ind.
	φ	φ	Q_{t-1}

(b)

Figure 6.20 EFL gate RS flip-flop: (a) circuit diagram and (b) state table.

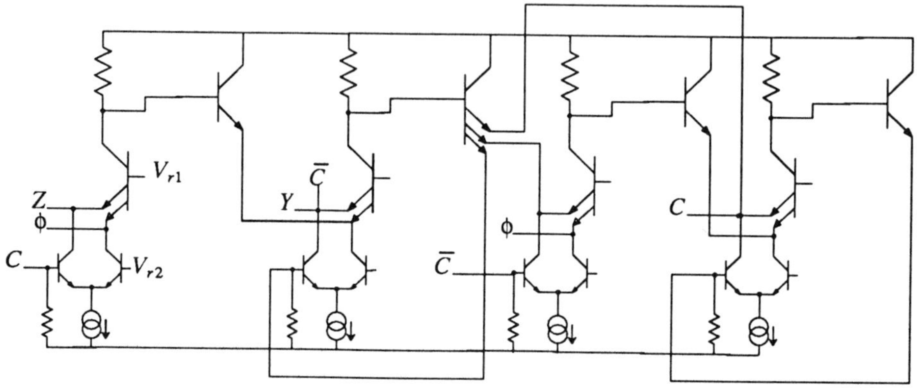

(a)

C	Y	Z	Q_t
rising edge	ϕ	ϕ	Q_{t-1}
falling edge	0 1 1 0	0 1 0 1	0 1 0 ind.

(b)

Figure 6.21 EFL master-slave RS-type flip-flop: (a) circuit diagram and (b) state table.

Although saving of silicon area and power consumption is possible for all commonly used flip-flop configurations, the most significant saving is for the master-slave D-type flip-flop. For this reason it is used in preference to the others, for the realization of functional blocks.

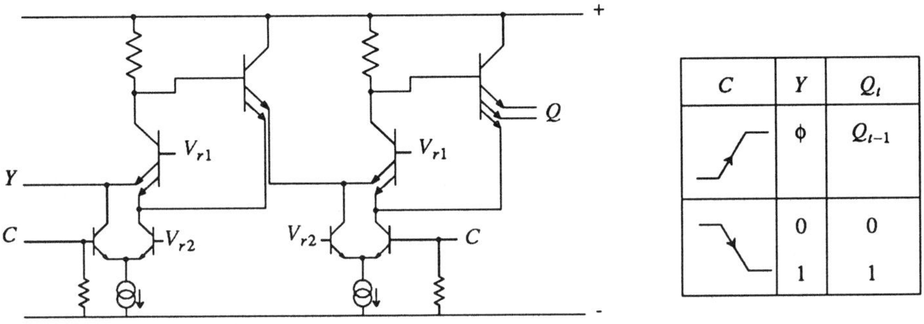

(a)

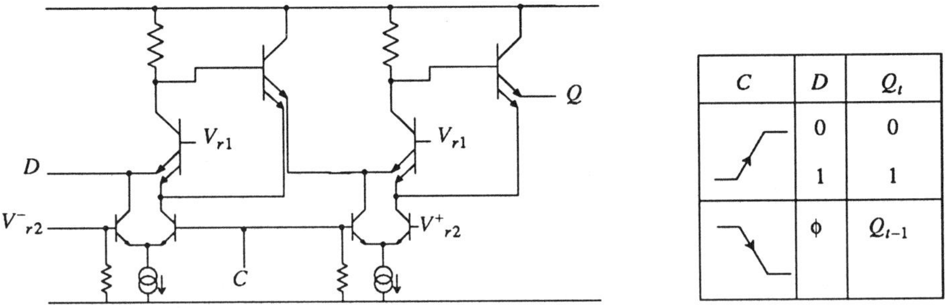

(b)

Figure 6.22 EFL master-slave D-type flip-flop:
(a) negative edge triggered; and (b) positive edge triggered.

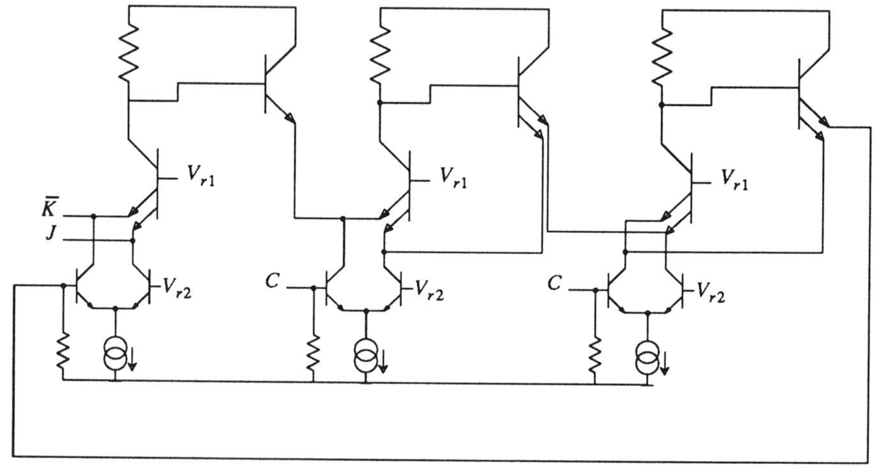

(a)

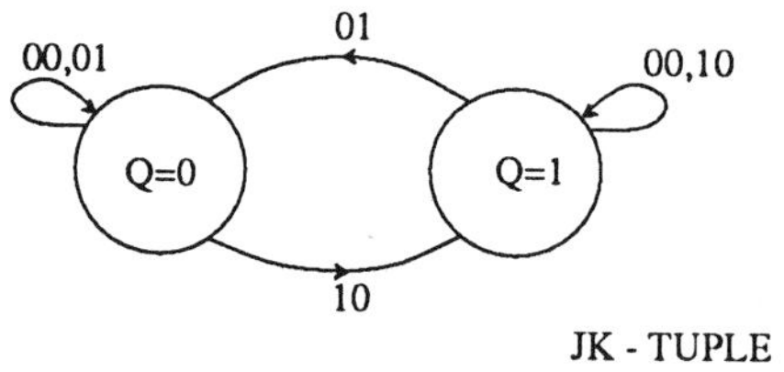

C	Y	Z	Q_t
↑	φ	φ	Q_{t-1}
↓	0	0	Q_{t-1}
	0	1	0
	1	0	1
	1	1	$\overline{Q}_{t-1}$

(b)

Figure 6.23 EFL master-slave JK flip-flop:
(a) Circuit diagram; and (b) state diagram and state table.

Table 6.1 compares the EFL to the NAND realization of commonly used logic and storage functions, a factor of 2-5 improvement in terms of gate to cell counts results. However, because the functions, where the improvement factor is 2, are not frequently used in a fully developed design, it is fair to assume an average improvement of 4.

Table 6.1 EFL Cells vs. NAND Realization of Logic Gates and Flip Flops

Function	Number of EFL Cells	Number of NAND Gates
Exclusive-OR	1	3
Equality	1	3
Full-adder	4	9
RS flip-flop	1	2
D-type latch	1	4
Gated *RS* flip-flop	2	4
Master-slave *D*-type	2	10
Master-slave JK	3	9

6.3. DC Specifications for a Practical EFL

There have been a variety of integrated realizations of EFL cells and some of these will be discussed. In the early EFL realization, Skokan[7] demonstrated that a complete flip-flop including 50-Ω line buffers could be fabricated in an area of 400 μm × 400 μm using conventional diffused transistors with f_T of 1.8GHz. This realization gave a master-slave D-type flip-flop, which operates at a clock speed of over 500MHz for a power dissipation of 16mW. Other development yielded 4mW circuit using 1-GHz transistors, which operated at a clock speed of 200MHz. Fig. 6.24 shows the voltage levels of EFL. This type of circuit is compatible in logic levels with several conventional on-chip emitter-coupled-logic (ECL) systems. The voltage levels in the system are dependent on the forward voltage of *pn* junctions and are shown for a junction voltage of 0.8V.

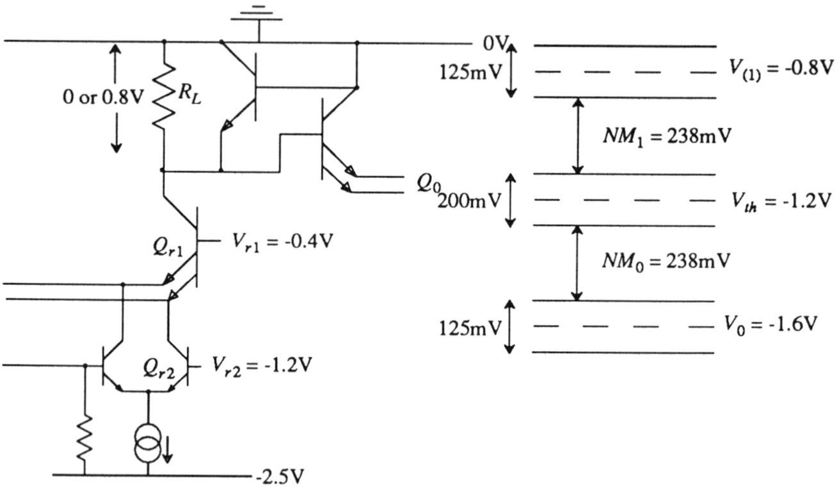

Figure 6.24 Logic levels and noise margins for an EFL cell.

In Fig. 6.24 the voltage at the collector of Q_{r1} can vary between 0 and -0.8V, while the base is biased at -½ V_{BEon} or -0.4V. Considering Q_{r2}, the collector potential can never fall below -1.2V because it is connected to an emitter of Q_{r1} and its emitter cannot be more positive than 2V while in conduction, so the minimum conducting value of V_{CE} is 0.8V. For the lower level transistors, with logic inputs high, the minimum value of V_{CE} is 0.4V. For the output transistor Q_o, V_{CE} varies between 0.8 and 1.6V. Since the worst-case soft saturation voltage (the V_{CE} at which f_τ for the transistor falls below its specified value) is 0.2V, there is a worst-case margin of 0.2V for V_{CE} of all the transistors. This allows the circuit to be operated above 100°C, where this margin reduces to 0.0V. A variety of realizations can be used for the current source I_o, and it is not a critical part of the circuit because there are no high-frequency requirements.

The logic level system shown in Fig. 6.24 provides a threshold tolerance of 200-mV, "0" and "1" level tolerances of 125-mV and noise margins of more than 200-mV.

6.4. Chapter Summary

This chapter deals with high speed bipolar current mode logic. We have reviewed the fundamentals of emitter coupled and emitter function logic circuits. The use of current mode logic for implementation of combinational and sequential digital building blocks was demonstrated.

References

1. M.I. Elmasry, *Digital Bipolar Integrated Circuits,* Wiley, New York (1983).
2. C.S. Meyer D.K. Lynn and D.J. Hamilton, *Analysis and Design of Integrated Circuits,* McGraw-Hill, New York (1968).
3. M.I. Elmasry and P.M. Thompson, "The Design of Load Structures for Current Mode Subnanosecond Logic," *IEEE Journal of Solid State Circuits,* pp. 72-75 (1975).
4. P.R. Gray and R.G. Meyer, *Analysis and Design of Analog Integrated Circuits,* Wiley, New York (1977).
5. M.Y. Ghannam, R.P. Mertens and R.J. Van Overstreten, "An Analytical Model for the Determination of the Transient Response of CML and ECL Gates," *IEEE Trans. on Electron Devices,* **Vol. 37** pp. 191-201 (Jan. 1990).
6. M.I. Elmasry and P.M. Thompson, "Logic Partition for Multiemitter Two-Level Structures," *IEEE Transactions on Circuit Theory,* pp. 354-359 (May 1974).
7. Z.E. Skokan, "Emitter Function Logic: Logic Family for LSI," *IEEE Journal of Solid State Circuits,* pp. 356-361 (1973).
8. M.I. Elmasry and P.M. Thompson, "Two-Level EFL Structures for Logic-in-Memory Computers," *IEEE Transactions on Computers,* pp. 250-258 (March 1975).

9. M.I. Elmasry, "Logic Design Using Emitter-Function-Logic," *IEEE Transactions on Computers*, pp. 952-956 (September 1976).

CHAPTER 7

BiCMOS DIGITAL INTEGRATED CIRCUITS

In this chapter, we introduce a variety of digital BiCMOS circuit structures such as the totem-pole BiCMOS gate and the merged MOS/bipolar current mode circuits. This chapter starts with a brief comparison between MOS and bipolar devices in terms of their current drive capabilities. The advantage of using bipolar devices for the totem-pole structure is demonstrated in Section 7.2. In that section, we study the DC and transient characteristics of the BiCMOS inverter. The transient behaviour is analyzed in detail to derive a delay model. This model is then used to understand the critical device and circuit parameters for the design of the BiCMOS gate. It also serves as the basis for a comparison between CMOS and BiCMOS for different loading conditions. In Section 7.3, we discuss the performance of conventional totem-pole circuits for low power supply voltage. Several full-swing BiCMOS structures, which are suitable for low voltage applications, are introduced. In Section 7.4, we also consider the use of BiCMOS for integrating CMOS with current mode logic in a single system. Some of the basic cells required for such applications are introduced.

7.1. Comparisons between Bipolar and MOS Transistors

MOSFETs have the following advantages over bipolar transistors; (i) high input resistance, (ii) self-isolation, which translates into lower silicon area consumption, and (iii) bidirectional operation, which allows the implementation of transmission gates. However, we should indicate that as the technology advances, the difference in the area between bipolar and MOS transistors is narrowing; especially when advanced process techniques such as trench isolation are used.

The bipolar transistor, however, is superior over MOSFETs in terms of current drive capability. This can be related to its superior transconductance[1]. The transconductance of a bipolar transistor is given by :

$$g_{m,bipolar} = \frac{I_s}{V_t} e^{V_{BE}/V_t} \tag{7.1}$$

The transconductance of the MOS transistor is given by :

$$g_{m,MOSFET} = \frac{\mu C_{ox}}{2} \frac{W}{L} (V_{GS} - V_T) \tag{7.2}$$

In contrast to the bipolar transistor, the transconductance of the MOSFET shows great technology dependence. More important to note is that the transconductance of the MOSFET is linearly dependent on the input voltage; whereas, this dependence is exponential for the bipolar transistor. Thus, the collector current of the bipolar transistor is more sensitive to the input voltage variations than the drain current. This feature makes the pull-up/pull-down (PUPD) bipolar buffers faster than their MOS counterparts. Fig. 7.1 shows a comparison between the saturation current of an NMOS transistor and the collector current of a bipolar transistor for different device areas; e.g. 90, 130, 170 and 210 μm^2. The drain current is limited by its saturation value. While the collector current is determined by the base-emitter voltage (note that Fig. 7.1 shows the external base-emitter voltages). The collector current, per unit area, is superior to the drain current, per unit area, provided that the base-emitter voltage is sufficiently large, which is most probably the case when the bipolar transistor operates in the high level injection regime. However, if we assume advanced isolation techniques, the bipolar area will be smaller and the current per unit area for the bipolar transistor will be further increased.

Comparing bipolar and MOS logic circuits depends on the circuit type. General statements such as "bipolar is faster than MOS" can be misleading. One must consider specific circuits in a comparison. In this section, we have restricted the comparison to PUPD structures, because of their importance in VLSI circuits. Our goal here is to show that bipolar PUPD circuits are faster than their MOS counterparts for the reasons mentioned above. To demonstrate that, we chose two simple circuits, an NMOS inverter and a bipolar RTL inverter [see Fig. 7.2]. Assume that both inverters are driving the same load capacitance. If the input and

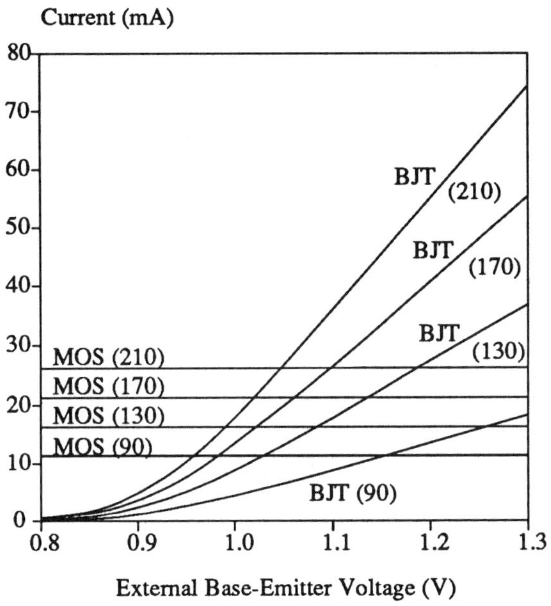

Figure 7.1 A comparison between the current drive of BJT and NMOS transistors for different areas.

output voltage swings, of each circuit, are equal, the fall time can be simply expressed as :

$$t_{fall} = \frac{C_L V_s}{I} = \frac{C_L}{g_m} \tag{7.3}$$

where I is either the collector or drain current. The above equation implies that the fall time is inversely proportional to the current or transconductance. At high enough base-emitter voltage, the collector current can be greater than the drain current as shown in Fig. 7.3(a). As a result, the fall time of the RTL circuit is smaller than that of the NMOS circuit as illustrated in Fig. 7.3(b).

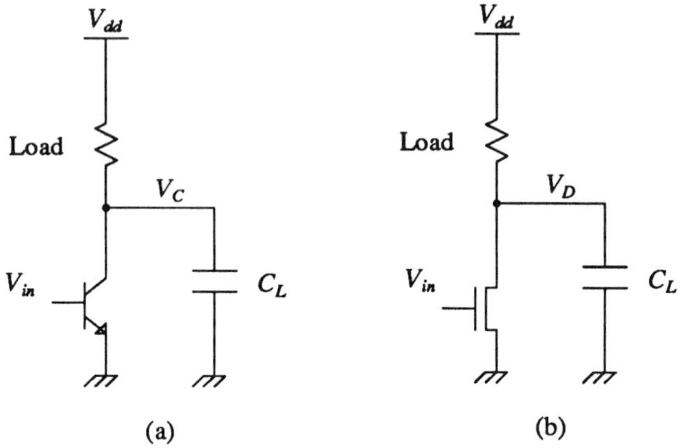

Figure 7.2 (a) Bipolar RTL inverter; (b) NMOS inverter.

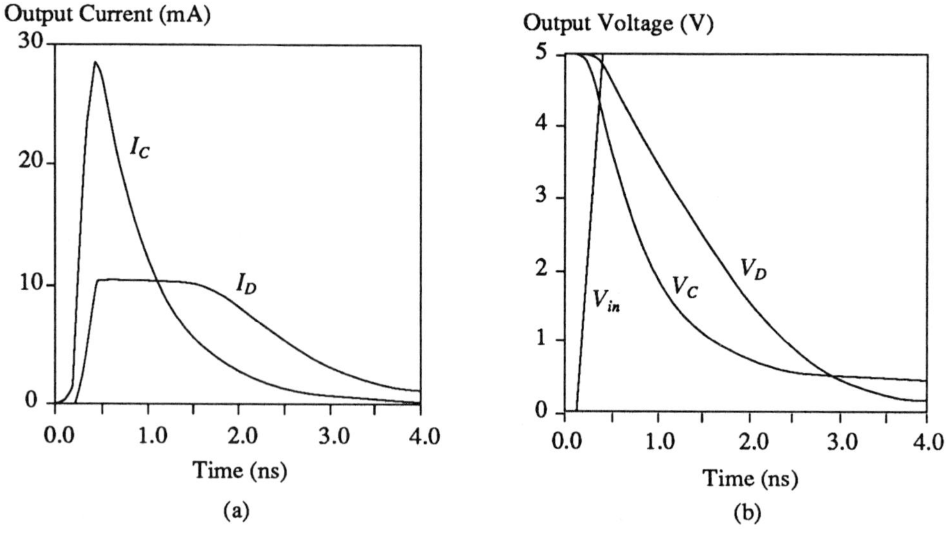

Figure 7.3 Current and voltage waveforms of the RTL and NMOS inverters.

7.2. BiCMOS Totem-Pole Inverter

The limited current drive of the CMOS inverter [Fig. 7.4] prohibits fast switching, especially if the load capacitance is large. To solve this problem, bipolar drivers can be used.

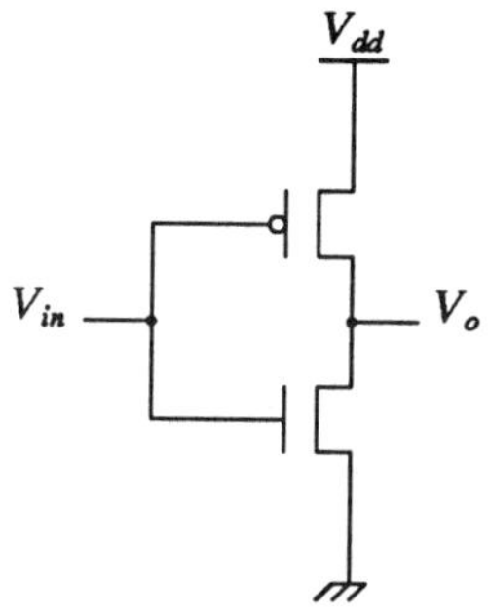

Figure 7.4 CMOS inverter.

The question now is: *What would be the best configuration of the bipolar driver ?*

One possible configuration would be a complementary NPN/PNP driver [Fig. 7.5][2]. However, it has limited application because it offers only noninverting logic. Moreover, its input impedance is too small.

Another solution was proposed in the late 1960's by Lin et al[3]. They proposed to increase the transconductance of the MOSFET by adding a bipolar transistor to its output as shown in Fig. 7.6. As a result, the transconductance of the composite MOS/BJT device is greater than that of the MOS transistor by a factor of β, the current gain of the bipolar transistor. The merged PMOS/BJT and NMOS/BJT switches can replace the P- and N- MOSFETs of the CMOS inverter to yield the BiCMOS inverter buffer illustrated in Fig. 7.7[3]. The operation of the BiCMOS inverter is similar to that of CMOS. When the input is LOW, the PMOS is ON and its drain current turns the transistor Q_1 ON. The collector current of Q_1 charges the output load capacitance. As the output voltage reaches $V_{dd}-V_{BEon}$, Q_1 gradually turns OFF. The HIGH DC level of the output is hence $V_{dd}-V_{BEon}$. If the input is HIGH,

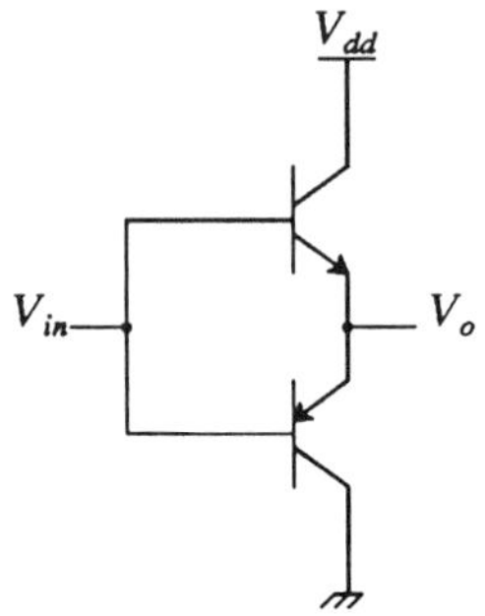

Figure 7.5 Complementary bipolar driver

transistor N is turned ON causing Q_2 to turn ON to discharge the output node. As a result, the output node is pulled down. As the output voltage reaches V_{BEon}, transistor Q_2 turns OFF and the DC LOW level stays at V_{BEon}. In Section 7.3, we will show that the HIGH and LOW levels may vary due to AC effects. The BiCMOS gate shown in Fig. 7.7 is apparently an efficient driver with zero static power dissipation. However, its speed is limited because of the long turn OFF time of the bipolar transistors. Therefore, bleeding elements must be added to speed the discharging of the base charge when the bipolar transistors are turning OFF. Fig. 7.8 shows one way of bleeding the base charge. The drawback of this circuit is that a large DC current would flow through P and Z_{d1} when P is turned ON.

To avoid the static power dissipation, Z_{d1} must be connected between the base of Q_1 and the output node. Fig. 7.9 shows the R-type BiCMOS inverter[4,5], in which resistors are used to bleed the base charge. Using MOS bleeding elements as shown in Fig. 7.10[6] offers an acceptable solution, because the path between the base of the bipolar and ground is created only when the bipolar transistor is to be turned OFF. For example, transistor N_{d1} turns ON only when the input is HIGH to discharge the base of Q_1. Transistor N_{d2} can be controlled by the output node or by the base of Q_1. However, using the base node results in faster operation, because the base of Q_1 is pulled up faster than the output node, especially if the output is heavily loaded, and

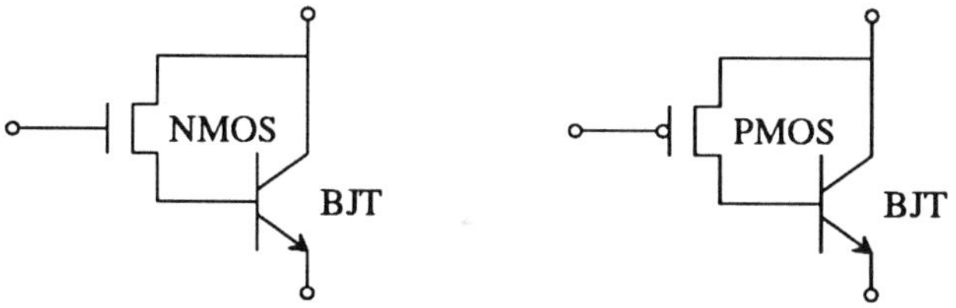

Figure 7.6 The merged PMOS/BJT and NMOS/BJT devices.

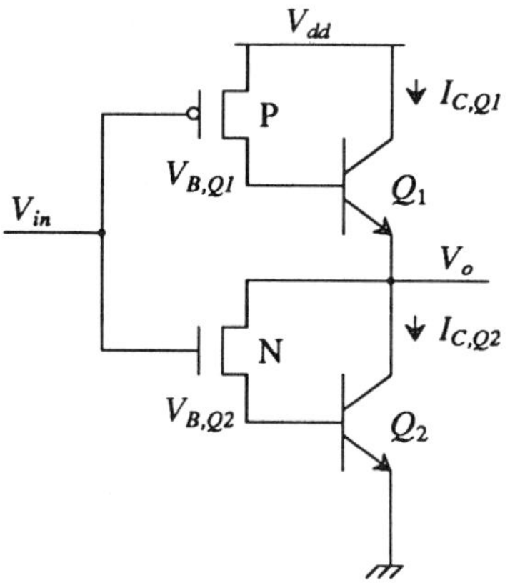

Figure 7.7 The early version of a BiCMOS inverter.

also because the voltage swing of the base node is larger.

The R-type BiCMOS circuit has two advantages, its voltage swing can be larger than that of the circuit shown in Fig. 7.10 and it occupies less area, especially for more complex gates. However, the drawback of the first circuit [Fig. 7.9] is that part of the current supplied by transistor P (N) is wasted through the bleeding resistors, which weakens the bipolar drive. The BiCMOS inverter of Fig. 7.10 is one of the earliest versions to be used in practice. We will, therefore, refer to it as the conventional BiCMOS circuit.

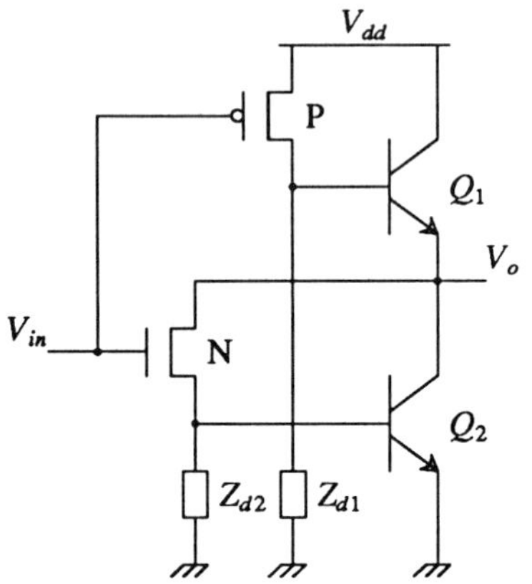

Figure 7.8 BiCMOS inverter with bleeding devices.

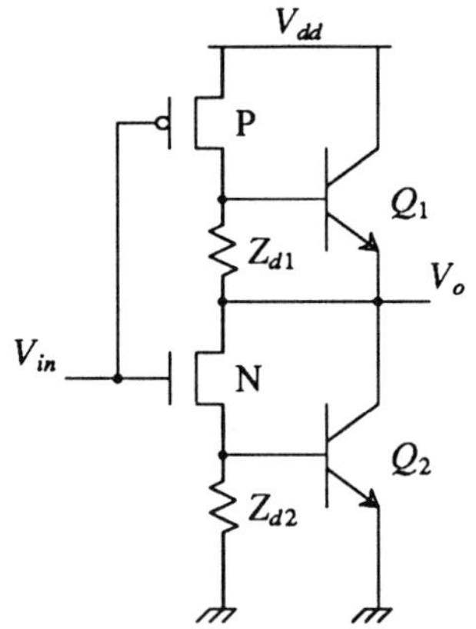

Figure 7.9 R-type BiCMOS inverter with resistive discharging.

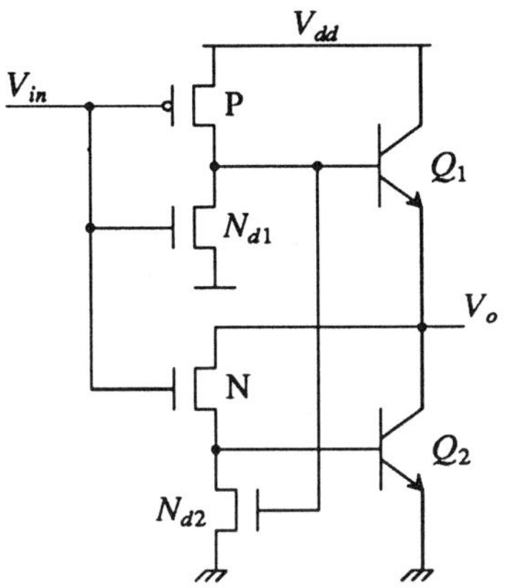

Figure 7.10 BiCMOS inverter with MOS discharge (conventional BiCMOS inverter).

7.2.1. DC Characteristics of the BiCMOS Inverter

In this section, we discuss the DC characteristics of the BiCMOS inverter shown in Fig. 7.7. Then we will comment briefly on the DC characteristics of the inverters shown in Fig. 7.9 and 7.10. Fig. 7.11 shows the DC transfer characteristics with the different regions of operation. Before explaining each region, we need to establish some concepts that are necessary for the discussion. In contrast to the concepts used in explaining the CMOS transfer characteristics[7], we do not use the expression "OFF" to describe the state of the MOSFETs when their gate-source voltage is smaller than the threshold voltage. Instead, we consider that the MOSFETs operate in the subthreshold region. This picture is more realistic and helps in understanding the operation of the BiCMOS circuit because the subthreshold currents of the P- and N-channel transistors, in the circuit shown in Fig. 7.7, may cause considerable base-emitter voltage drop. For example, a subthreshold current of $5pA$ results in a V_{BE} drop of $398mV$ if the forward saturation current of the bipolar transistor is $1\times10^{-18}A$. It is, therefore, not practical to assume zero drain current in the subthreshold region. Throughout the DC analysis we assume that the bipolar transistors do not saturate.

Therefore, the base currents are negligible compared to the collector currents. Hence,

$$I_{C,Q1} = I_{C,Q2} \tag{7.4}$$

As a result,

$$V_{BE,Q1} = V_{BE,Q2} \tag{7.5}$$

where $I_{C,Q1}$ and $I_{C,Q2}$ represent the collector currents of Q_1 and Q_2, and $V_{BE,Q1}$ and $V_{BE,Q2}$ are the base-emitter voltages of Q_1 and Q_2 respectively. The output voltage is therefore given by :

$$V_o = V_{B,Q1} - V_{BE,Q1} = V_{B,Q1} - V_{B,Q2} \tag{7.6}$$

where $V_{B,Q1}$ and $V_{B,Q2}$ are the voltages of the bases of Q_1 and Q_2 respectively.

The circuit operation can be divided into seven regions :

Region (A) : $V_{in} < V_{BE,Q2} + V_{TN}$

Both bipolar transistors are OFF and the transistor P operates in the linear region with zero drain-source voltage. Due to the subthreshold current of transistor N, the base-emitter voltage of Q_2 is around $0.4V$. As a result $V_{BE,Q1}$=$0.4V$. Hence, the output voltage is therefore $4.6V$ (assuming that V_{dd}=$5V$). As the input voltage increases the subthreshold current through N increases causing $V_{B,Q2}$ to rise and V_o to fall [see Equation (7.6)].

Region (B) : $V_{BE,Q2}+V_{TN} < V_{in} < V_{BEon}+V_{TN}$

Transistor N enters the saturation region, while both Q_1 and Q_2 are OFF. As the input voltage increases, the drain current of N increases until the base-emitter voltage of Q_2 is equal to V_{BEon}. At this point, Q_2 turns ON. As $V_{BE,Q1}$=$V_{BE,Q2}$, Q_1 also turns ON.

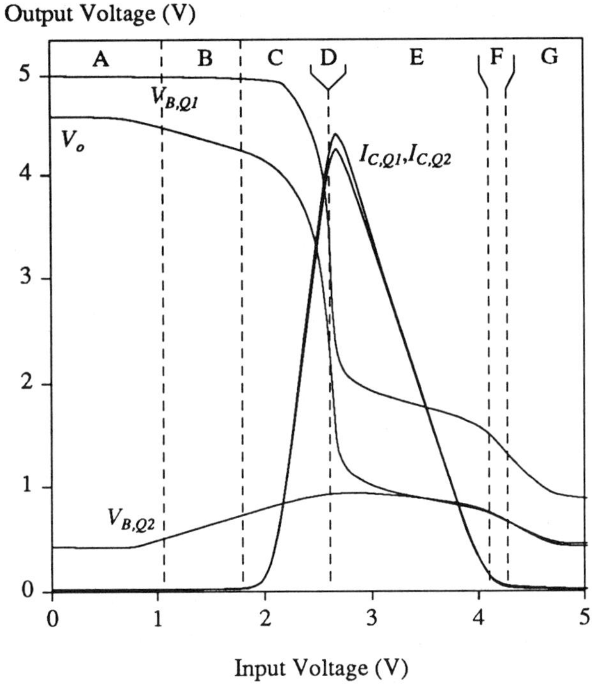

Figure 7.11 The DC transfer characteristics of the BiCMOS inverter.

Region (C) : $V_{BEon}+V_{TN} < V_{in} < V_{inv}$ [1]

In this region, both bipolar transistors are ON as seen from their collector current waveform in Fig. 7.11. Within this region, the output voltage falls sharply.

Region (D) : $V_{in} = V_{inv}$

When the drain-source voltage of *P* is equal to V_{DSsat}, transistor *P* enters the saturation region. In region (D), both bipolar transistors are ON and

---

[1] V_{inv} is defined as the input voltage at which the gain of the inverter is maximum.

both MOSFETs operate in the saturation region. As the collector currents of both transistors are equal [Equation (7.4)], the drain currents of transistors N and P are also equal. Therefore,

$$I_{D,P} = I_{D,N} \tag{7.7}$$

Hence,

$$W_P K_P C_{ox,P}(V_{dd} - V_{in} - |V_{TP}|) = W_N K_N C_{ox,N}(V_{in} - V_{BEon} - V_{TN}) \tag{7.8}$$

If the gate oxide thickness of the NMOS and the PMOS are equal, the input voltage can be expressed as :

$$V_{in} = \frac{V_{dd} - |V_{TP}| + \dfrac{K_N W_N}{K_P W_P}(V_{BEon} + V_{TN})}{1 + \dfrac{K_N W_N}{K_P W_P}} = V_{inv} \tag{7.9}$$

This is the only point where both MOS transistors are saturated. Therefore, region (D) corresponds to one point. At this point, the output voltage drops sharply towards its LOW level and is therefore called the inversion point.

Region (E) : $V_{inv} < V_{in} < V_{dd} - |V_{TP}|$

As the input voltage increases, the gate-source voltage and the drain current of P are decreasing. The drain-source voltage of transistor N is approaching zero. In this region, the output voltage is approximately equal to $V_{BE,Q2}$.

Region (F) : $V_{dd} - |V_{TP}| < V_{in} < V_{dd} - V_{BEon}$

This region is very narrow because $|V_{TP}|$ and V_{BEon} are fairly close. In this region, P operates in the subthreshold regime. Its drain current drops sharply.

Region (G) : $V_{in} > V_{dd} - V_{BEon}$

The base-emitter voltage of both transistors is decreasing as the subthreshold current of the MOS devices is decreasing. The final level of the output voltage is determined by the subthreshold current of the P transistor when V_{in}=5V and its drain-source voltage is zero. If $|V_{TP}|<V_{BEon}$, the order of regions (F) and (G) may be reversed.

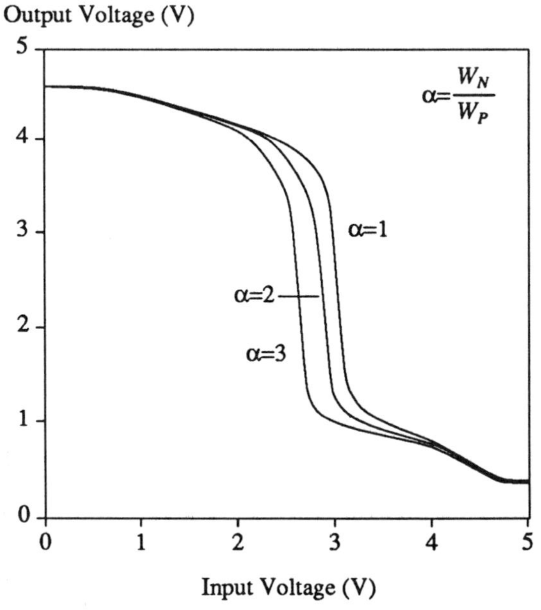

Figure 7.12 The DC transfer characteristic for different W_N/W_P.

From Equation (7.9) one can calculate the ratio W_N/W_P required to obtain symmetrical noise margins or in other words, $V_{inv}=V_{dd}/2$. For a typical 0.8μm BiCMOS technology, this ratio is found to be greater than 1, which means that transistor N is to be wider than transistor P. This can be explained by the fact that the gate-source voltage of transistor P is larger than that of N by V_{BEon}. Fig. 7.12 shows the transfer characteristics for different values of W_N/W_P. As the ratio decreases, the inversion voltage increases. However, the requirement of symmetrical rise and fall times can not be satisfied if $W_N/W_P>1$. Therefore, there is a tradeoff between the symmetry of the noise margins and the symmetry of the fall and rise times.

The DC characteristics of the conventional and R-type BiCMOS inverters look different than that shown in Fig. 7.11 because of the existence of the bleeding elements. The bleeding resistor Z_{d2} [Fig. 7.9] and the bleeding transistor N_{d2} [Fig. 7.10] pull the output node to ground under steady state conditions. This is shown in Fig. 7.13(a) and (b) when the input voltage is greater than $V_{dd}-|V_{TP}|$. Similarly, when the input is below $V_{dd}-V_{TN}$, the resistor Z_{d1} [Fig. 7.9] pulls the output node to V_{dd} [see Fig. 7.13(a)].

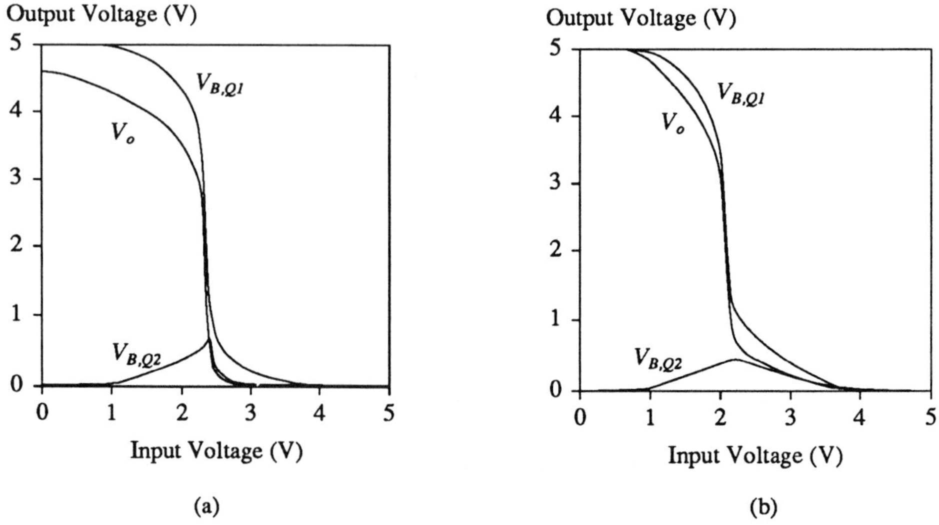

Figure 7.13 The DC transfer characteristics of:
(a) the conventional BiCMOS inverter; (b) the R-type BiCMOS inverter.

7.2.2. Transient Analysis of the Totem-pole BiCMOS Inverter

In this section, we study the transient behaviour of the conventional BiCMOS inverter of Fig. 7.10. The treatment presented here can be easily modified to fit other BiCMOS PUPD circuits as well. The purpose of this analysis is two-fold; it serves to understand the transient behaviour of the circuit and also to derive an analytical expression to calculate the delay time between the mid point of the input swing and the mid point of the swing of the output signal. For analysis simplicity we assume a step input. We first analyse the pull-up transient and then discuss the differences in the case of the pull-down transient.

(a) Transient Behaviour of the BiCMOS Inverter

We need to study the transient behaviour of the BiCMOS inverter qualitatively to appreciate the importance of each component in the circuit. If the input falls to its LOW level, transistor P turns ON and operates initially in the saturation region. Its drain current charges the base-emitter capacitance of Q_1 until $V_{BE,Q1}=V_{BEon}$ when Q_1 turns ON. The emitter current of Q_1 increases in a relatively short period of time to reach its peak. The output node is pulled-up following the base of Q_1 as illustrated in Fig. 7.14. As the base voltage of Q_1 exceeds V_{TN}, N_{d2} turns ON to discharge the base charge of Q_2. However, the capacitive coupling between the output node and the base of Q_2 tends to pull up the base node of Q_2. This delays the turning OFF of Q_2 especially if the width of N_{d2} is too small. Therefore, part of the emitter current of Q_1 is used to support the collector current of Q_2, the displacement current through the base-collector capacitance of Q_2 and to charge the collector-substrate capacitance of Q_2. We therefore see that actual charging current of the load capacitance is smaller than the current sourced by Q_1 [see Fig. 7.14(b)]. It is, therefore, important that the base-collector and collector-substrate capacitances are small. This means that the area of the bipolar transistor has to be as small as possible. Further discussion on the choice of the area of the emitter area of the bipolar transistors can be found in Section 7.3.4.1. As the base voltage of Q_1 reaches $V_{dd}-V_{DSsat}$, transistor P enters the triode region and its drain current drops gradually. Consequently, the collector current of Q_1 starts falling. As the output voltage approaches the $V_{dd}-V_{BEon}$ limit, transistor Q_1 is expected to gradually turn OFF. However, due to the capacitive coupling between the output node and the base of Q_1, the output voltage exceeds $V_{dd}-V_{BEon}$ as shown in Fig. 7.14(a). This phenomena is explained in detail in Section 7.2. The simulated waveforms during the falling transition are also illustrated in Fig. 7.14.

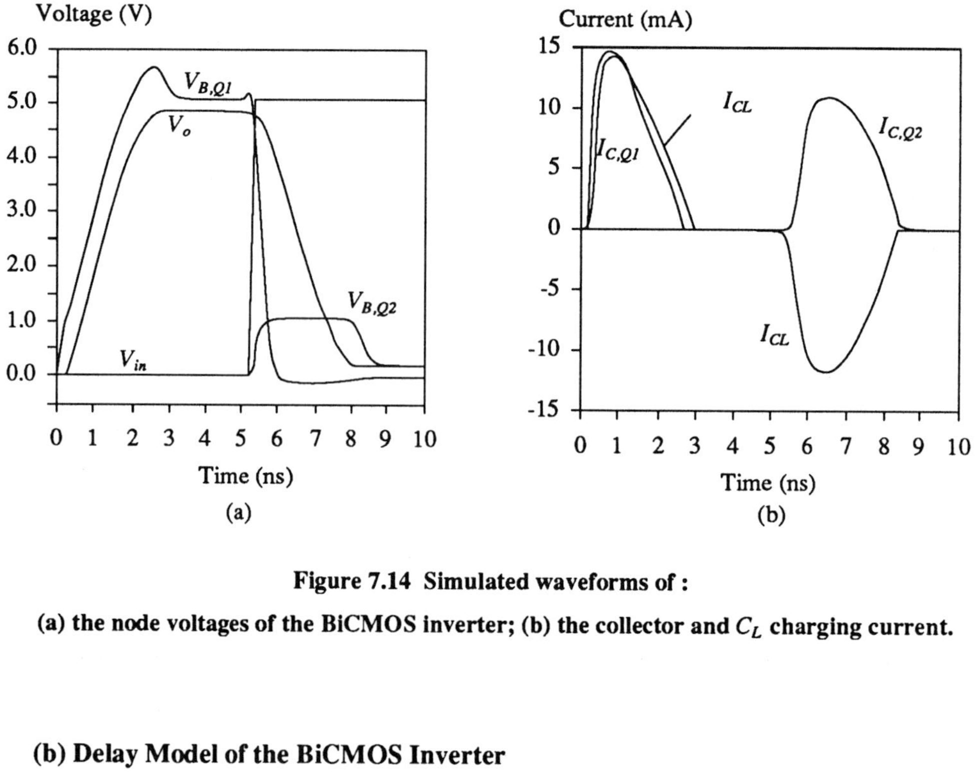

Figure 7.14 Simulated waveforms of :
(a) the node voltages of the BiCMOS inverter; (b) the collector and C_L charging current.

(b) Delay Model of the BiCMOS Inverter

The delay model of the BiCMOS inverter must account for the following :

(1) The parasitic capacitances.

(2) The parasitic resistors of bipolar transistors and especially the collector resistance.

(3) The effect of the high currents on the current gain of bipolar transistors.

(4) The effect of the high currents on the forward transit time of bipolar transistors.

Several models for the delay of the conventional BiCMOS inverter have been proposed[6, 8-14]

In this section, we present a delay model for the BiCMOS inverter, in which we account for most of the device parasitics and high level injection effects on the current gain and forward transit time. The model yields a closed form analytical delay expression, which can be used to understand the relations between the delay and the different device and circuit parameters.

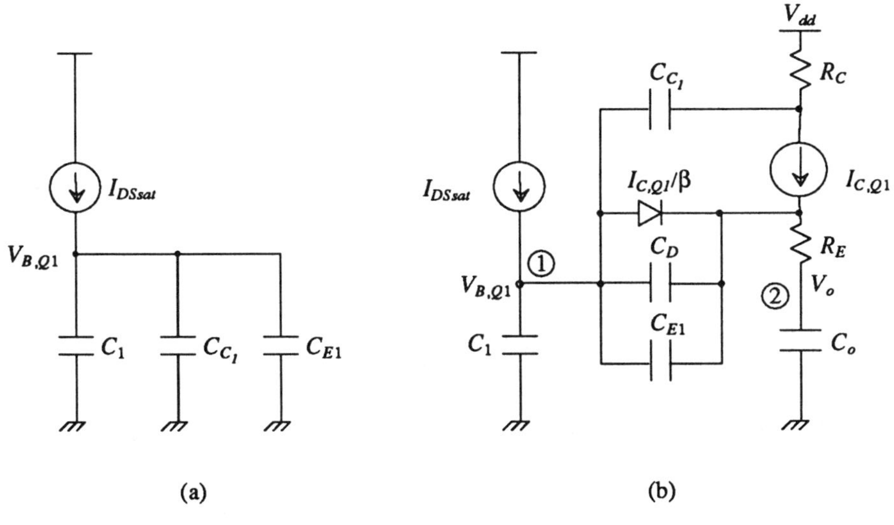

Figure 7.15 The transient equivalent circuit of the pull-up section of the BiCMOS inverter when: (a) Q_1 is OFF; (b) Q_1 turns ON.

Fig. 7.15 illustrates the transient equivalent circuit of the pull-up section of the inverter shown in Fig. 7.10. The drain-source voltage of the P transistor is small ($\approx 2V$). It will therefore operate in the saturation region while the output voltage is less than $V_{dd}/2$. As we are interested in deriving the 50% rise time, we can model the P transistor by a constant current source whose current is equal to the saturation current of the device. As shown in Section 4.1.3.[Equation (4.98)], the saturation current can be expressed as:

$$I_{DSsat} = K_P\, C_{ox}\, v_{sat}\, W_P\, (\,|V_{GS}| - |V_{TP}|\,) \tag{7.10}$$

where the gate-source voltage is equal to $V_{in,low} - V_{dd}$, where $V_{in,low}$ is the low level of the input. The capacitance C_1 accounts for all the parasitic capacitances at the drains of P and of N_{d1}. Therefore, it is given by :

$$C_1 = C_{d,P} + C_{d,Nd1} + C_{ox,Nd2} \tag{7.11}$$

where $C_{d,P}$ and $C_{d,Nd1}$ are the drain-substrate depletion capacitances of the transistors P and N_{d1}, and $C_{ox,Nd2}$ is the gate capacitance of N_{d2}. The base-emitter capacitance C_{E1} and the diffusion capacitance (C_D) of Q_1 are also included in the model. Recall from Section 4.2.3. that the diffusion capacitance is given by:

$$C_D = \frac{dQ_b}{dV_{BE}} \tag{7.12}$$

where Q_b is the base charge of Q_1 and is given by :

$$Q_b = \tau_f\, I_{C,Q1} \tag{7.13}$$

Finally, the capacitance of the output node is modeled by C_o, which includes the load capacitance, the collector-substrate and the average base-collector capacitances of Q_2.

The base resistance is not included in the equivalent circuit, because the drop across it is typically small. The collector resistance has to be accounted for in the model to determine whether the bipolar transistor enters the saturation region or not; as explained in Section 7.3.2.3.

In this analysis, we will use the simple expression of the current gain given in Equation (3.128), hence, the gain of Q_1 is given by:

$$\beta(I_{C,Q1}) = \frac{\beta_o}{1 + \frac{I_{C,Q1}}{I_K}} \tag{7.14}$$

The expression of the forward transit time, which was given by Equation (3.142) is presented here for convenience.

$$\tau_f = \tau_{fo}\left(1 + XTF\left(\frac{I_C}{I_C + ITF}\right)^2 e^{V_{BC}/1.44VTF}\right) \tag{7.15}$$

where VTF is a fitting parameter to model the change of τ_f as a function of V_{BC} (or V_{CE}), ITF models the change due to I_C and XTF controls the increase of τ_f. It is difficult to use the complex function of the forward transit time [Equation (7.15)] in the derivation of a closed form analytical delay model. However, we can still accommodate the high current effect on the forward transit time by using the average of the forward transit time over the operating range of currents and voltages. The average of the transit time, denoted $\tau_{f,av}$, is given by :

$$\tau_{f,av} = \tau_{fo}\left(1 + \frac{1.44\, XTF\ VTF\ G(x)}{v_1 - v_2}\left(e^{v_1/1.44VTF} - e^{v_2/1.44VTF}\right)\right) \tag{7.16}$$

where

$$v_1 = V_{dd}/2 - V_{BEon}$$

$$v_2 = V_{dd}$$

and

$$G(x) = \frac{x+2}{x+1} - \frac{2}{x}\ln(x+1)$$

where

$$x=\frac{I_{C,peak}}{ITF}$$

The delay time can be divided into two intervals.

(i) The first component, t_{Q1}, is defined as the time required to turn Q_1 ON. During t_{Q1}, $I_{C,Q1}$ is practically zero and the output voltage is hardly changing. The transient equivalent circuit, which can be used to model the circuit during t_{Q1}, is shown in Fig. 7.15(a). Writing the current equation at the base node of Q_1, yields the following differential equation :

$$I_{DSsat}=(C_1+C_{E1}+C_{C1})\frac{dV_{B,Q1}}{dt} \tag{7.17}$$

Solving the above differential equation under the condition that, $V_{B,Q1}=V_{BEon}$ when $t=t_{Q1}$, we get :

$$t_{Q1}=(C_1+C_{E1}+C_{C1})\frac{V_{BEon}}{I_{DSsat}} \tag{7.18}$$

In the derivation of Equation (7.18), it was assumed that the initial output voltage was $0V$ as shown in Fig. 7.14(a), however, the actual value is V_{BEon}. The initial voltage of the base of Q_1 is $0V$, therefore, the base-emitter voltage changes from $-V_{BEon}$ to $+V_{BEon}$ during t_{Q1}. Hence, t_{Q1} in Equation (7.18) has to be multiplied by a factor of 2.

(ii) The second component, t_2, is the time required to charge the output node to the middle point of the swing.

Applying KCL at nodes (1) and (2) [see Fig. 7.15(b)] yields the following differential equations :

$$I_{DSsat}-C_1\frac{dV_{B,Q1}}{dt}-\frac{I_{C,Q1}}{\beta_o}(1+\frac{I_{C,Q1}}{I_K})-C_D\frac{dV_{BE}}{dt}$$

$$-C_{C1}\frac{d}{dt}(V_{BE}+I_{E,Q1}R_E+V_o-V_{dd}+I_{C,Q1}R_C)=0 \tag{7.19}$$

$$I_{E,Q1} = C_o \frac{dV_o}{dt} \tag{7.20}$$

For simplicity, we assume that $I_{E,Q1} \approx I_{C,Q1}$. When Q_1 is on, its base-emitter voltage is almost constant, hence $dV_{B,Q1}/dt \approx dV_o/dt$. For the same reason we can neglect the term $C_D dV_{BE}/dt$ in Equation (7.20). However, the same term cannot be neglected with respect to the other terms in Equation (7.19). Using Equations (7.12) and (7.13), the term $C_D dV_{BE}/dt$ in Equation (7.19) can be replaced by $\tau_{f,av} dI_{C,Q1}/dt$. In doing so we have assumed that the forward transit time is constant and equal to the average value $\tau_{f,av}$. By substituting from Equation (7.19) in Equation (7.20), we can rewrite Equation (7.20) as,

$$I_{DSsat} - \left(\frac{1}{\beta_o} + \frac{C_{C1} + C_1}{C_o}\right) I_{C,Q1} - \frac{I^2_{C,Q1}}{\beta_o I_K} = \left(\tau_{f,av} + C_{C1}(R_C + R_E)\right) \frac{dI_{C,Q1}}{dt} \tag{7.21}$$

Solving the above differential equation yields,

$$I_{C,Q1} = \frac{2a}{b+c} \frac{1 - exp(-t/\tau)}{1 - \alpha exp(-t/\tau)} \approx \frac{2a}{b+c} (1 - exp(-t/\tau)) \tag{7.22}$$

where

$$\alpha = \frac{b - c}{b + c} \tag{7.23}$$

and

$$a = \beta_o I_K I_{DSsat} \tag{7.24}$$

$$b = (1 + \beta_o \frac{C_{C1}+C_1}{C_o}) I_K \tag{7.25}$$

$$c = \sqrt{b^2+4a} \tag{7.26}$$

$$\tau = \frac{\beta_o I_K(\tau_{f,av}+C_{C1}(R_C+R_E))}{c} \tag{7.27}$$

The average base transit time is given by Equation (7.16) and $I_{C,peak}$ is given by $2a/(b+c)$. Substituting from Equation (7.22) in Equation (7.20) and integrating we obtain the expression of the output voltage :

$$V_o = \frac{2a\tau}{C_o}\left(\frac{1}{b+c}\ln(\frac{\alpha - exp(t/\tau)}{\alpha-1}) - \frac{1}{b-c}\ln(\frac{1-\alpha exp(-t/\tau)}{1-\alpha})\right) \tag{7.28}$$

This expression can be simplified to,

$$V_o = \frac{2a\tau}{C_o}\left(\frac{1}{b+c}\ln(\frac{\alpha - exp(t/\tau)}{\alpha-1}) - \frac{1}{b-c}\ln(\frac{1}{1-\alpha})\right) \tag{7.29}$$

The time, t_2, at which V_o reaches $V_{dd}/2$ is given by :

$$t_2 = \tau \ln\left(\alpha - (\alpha-1)\exp\left[\frac{\Delta V_o C_o}{2a\tau}(b+c) + \frac{1}{\alpha}\ln(\frac{1}{1-\alpha})\right]\right) \tag{7.30}$$

where ΔV_o is defined as the difference between $V_{dd}/2$ and the low level of the output voltage.

t_2 can be approximated by the summation, $t_{Cp}+t_{RC}+t_{Tf}+t_{Co}$. Hence, the total delay t_d can be written as:

$$t_d = t_{Q1} + t_{Cp} + t_{RC} + t_{Tf} + t_{Co} \tag{7.31}$$

where

$$t_{Q1} = (C_I + C_{EI} + C_{CI})\frac{V_{BEon}}{I_{DSsat}} \tag{7.32}$$

$$t_{Cp} = (C_{CI} + C_I)\frac{\Delta V_o}{2I_{DSsat}} \tag{7.33}$$

$$t_{RC} = C_{CI}(R_C + R_E)\frac{2\beta_o I_K}{c-b}\ln(\frac{2c}{b+c}) \tag{7.34}$$

$$t_{Tf} = \tau_{f,av}\frac{2\beta_o I_K}{c-b}\ln(\frac{2c}{b+c}) \tag{7.35}$$

$$t_{Co} = C_o \Delta V_o \frac{I_K + c}{2a} \tag{7.36}$$

The term t_{Cp} is the delay component associated with the parasitic capacitances, t_{RC} is the delay component associated with the parasitic RC; t_{Tf} is the delay component associated with the forward transit time and t_{C_o} is a function of the load capacitance.

For an advanced state-of-the-art BiCMOS process, τ_{fo}=12ps, β_o=70, I_K=6mA, I_{DSsat}=2mA, C_1=25fF, C_{CI}=25fF, C_{EI}=36fF, C_{SI}=75fF, R_E=3Ω and R_C=33Ω, we have calculated all the delay components. Fig. 7.16 illustrates the values of the different delay components as a function of the load capacitance for the above process. t_{Co} is the only strong function of the load capacitance. It therefore dominates over the other delay components as the load capacitance increases (C_L>3pF). The contribution of t_{Tf} can not be neglected for smaller capacitances, and becomes more important as the forward transit time increases. However, as technology advances τ_{fo} falls below 10ps, which means that the delay is mainly determined by t_{Co} in the case of large C_L.

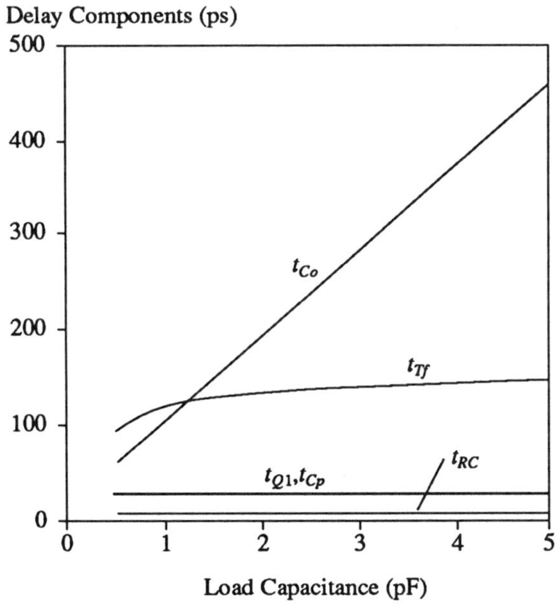

Figure 7.16 The delay components as a function of the load capacitance.

Fig. 7.17 demonstrates the reasonable agreement between the delay times calculated from the model [Equation (7.31)] and those extracted from the HSPICE simulations. It is interesting to see the effect of using the average of the forward transit time in comparison to using the low current forward transit time (τ_{fo}). As illustrated by Fig. 7.17, the thick solid curves, which correspond to $\tau_{f,av}$, agree better with HSPICE than the thin curves, which have been generated using τ_{fo}.

The delay expression in Equation (7.31) is general and is valid for all levels of injection. The condition for high level is:

$$C_o > \frac{\beta_o(C_{CI} + C_I)}{(\beta_o I_{DSsat}/I_K)^{1/2} - 1} \qquad (7.37)$$

It can be proven that if the above condition is not satisfied, the delay expression of Equation (7.31) reduces to the form that can be derived under low level injection condition (i.e. the current gain is β_o). In practice, the bipolar driver of the BiCMOS circuit operates under high level injection conditions. Under such conditions, Equation (7.37) implies that:

$$b < c \qquad (7.38)$$

and

$$c \approx 2\sqrt{a} \qquad (7.39)$$

The delay components t_{RC}, t_{Tf} and t_{Co} can be simplified to :

$$t_{RC} = 0.7C_{CI}(R_C + R_E)\left(\frac{\beta_o I_K}{I_{DSsat}}\right)^{1/2} \qquad (7.40)$$

$$t_{Tf} = 0.7\tau_{f,av}\left(\frac{\beta_o I_K}{I_{DSsat}}\right)^{1/2} \qquad (7.41)$$

$$t_{Co} = \frac{C_o \Delta V_o}{I_{DSsat}}\left(\frac{I_{DSsat}}{\beta_o I_K}\right)^{1/2} \qquad (7.42)$$

Equations (7.40) to (7.42) show that increasing the current gain of the bipolar transistors tends to increase t_{RC} and t_{Tf} and decrease t_{Co}. The effect of β_o is insignificant for the case of small capacitances. For large load capacitance, where

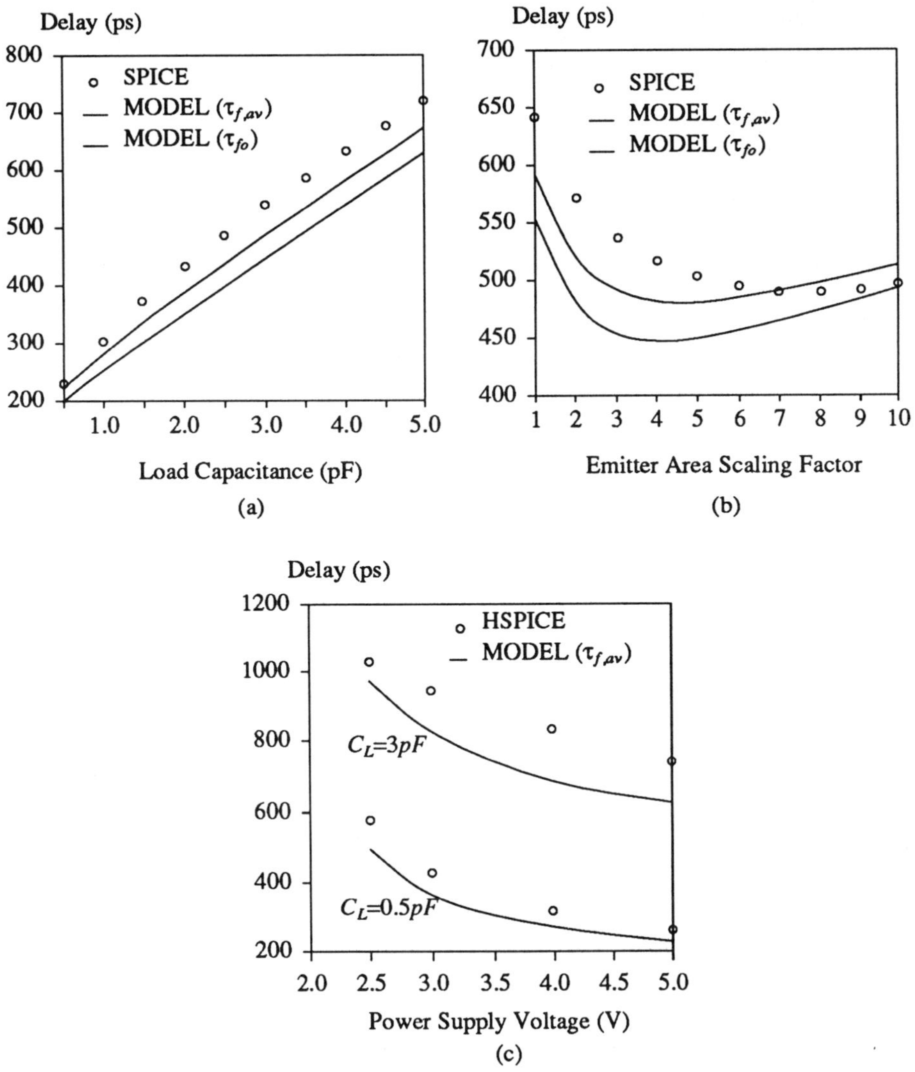

Figure 7.17 A comparison between the analytical model and HSPICE simulations.

t_{Co} is the dominant delay component, the dependency of the delay on the current gain is weak because t_{Co} is proportional to $\sqrt{\beta_o}$. This means that increasing β_o does not offer significant delay improvement. The same conclusion can be drawn for I_K. It is worth noting that the total delay is not sensitive to the parasitic resistance and capacitance for the typical values of β_o, I_K and I_{DSsat}. The fact that t_{Tf} is linearly proportional to τ_{fo} indicates that the delay is sensitive to the forward transist time.

The expression of the pull-down time is similar to that of the pull-up time except for the value of the drain current of transistor N [Fig. 7.10]. For the same reasons explained before, transistor N can be simply modeled by a current source whose current is given by :

$$I_{DSsat} = K_N \, C_{ox} \, v_{sat} \, W_N \, (V_{GS} - V_{TN}) \tag{7.43}$$

where the factor K_N is not equal to K^P (the saturation current constant of the PMOS transistor). The gate-source voltage, in Equation (7.43), is also less than that of the transistor P by V_{BEon}. To realize symmetrical rise and fall times the currents of transistors P and N have to be equal. Assuming reduced input swing, the gate-source voltage of P is $V_{dd}-V_{BEon}$ and that of the N is $V_{dd}-2V_{BEon}$. Let us take the ratio K_N/K_P to be 2.5 (which is the case for a typical submicron BiCMOS process) and assume that $V_{TN}=|V_{TP}|\approx 1V$. By equating (7.10) to (7.43) we find that $W_P/W_N=1.7$. Accurate circuit simulation show that this ratio is 1.75.

(c) Saturation Model

The bipolar transistors of the BiCMOS inverter should idealy operate in the forward active region. However, they may saturate if the collector resistance is large[2]. This might occur before or after the output has reached the 50% point depending on the value of the collector resistance. The delay model given by Equation (7.31) was derived with assumption that the BJT operates in the forward active region. In this section, we will consider the case when the BJT saturates before the output has reached the 50% point.

---

[2] This may be the case for a CMOS based BiCMOS process, where the collector resistance might be in the $k\Omega$ range.

Transistor Q_1 saturates if the following condition is satisfied.

$$V_{dd} - V_{CEsat} - V_o - I_{C,Q1}(R_C + R_E) - I_{DSsat}R_E = 0 \tag{7.44}$$

The time, t_s, at which Q_1 enters the saturation region, can be obtained by solving Equations (7.22), (7.28) and (7.44) numerically. Beyond t_s, the circuit can be regarded as an RC network as shown in Fig. 7.18.

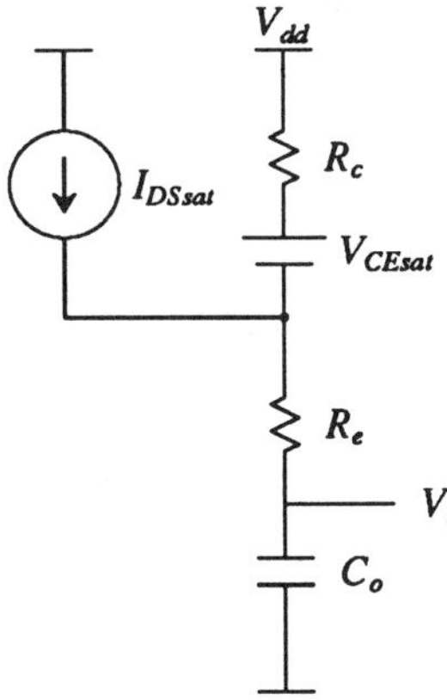

Figure 7.18 The transient equivalent circuit of the pull-up section of the BiCMOS inverter when Q_1 operates in the saturation region.

Applying the KCL at the output node yields the following :

$$I_{C,Q1} = \frac{V_{dd} - V_{CEsat} - V_o - I_{DSsat}R_E}{R_C + R_E} = C_o \frac{dV_o}{dt} - I_{DSsat} \tag{7.45}$$

Solving the above differential equation, we get the time required to charge the output node to $V_{dd}/2$ while Q_1 is saturated. This time is given by :

$$t_{sat} = C_o(R_C + R_E) \ln \frac{V_{dd} + I_{DSsat}R_E - V_{CEsat} - V_o(t_s)}{V_{dd}/2 + I_{DSsat}R_E - V_{CEsat}} \tag{7.46}$$

The total delay in such a case is given by the following summation :

$$t_d = t_{Q1} + t_s + t_{sat} \tag{7.47}$$

7.2.3. Delay Dependence on the Device Parameters

Fig. 7.19(a) and (b) illustrate the effect of varying the parameters of the bipolar transistors (Q_1 and Q_2) on the delay time. The data used for Fig. 7.19 were obtained from circuit simulations (HSPICE). Each parameter P is normalized to some typical value P_o. The typical values of the parameters are: β_o=100, I_K=2mA, τ_{fo}=12ps, C_C=10fF, C_E=12fF, C_S=25fF, R_B=450Ω, R_C=100Ω, R_E=10Ω.

From Fig. 7.19, we can make the following observations: (i) the effect of the parasitic resistances and capacitances is insignificant especially as the load capacitance is increased, (ii) the variation of the knee current and the current gain have similar impact on the delay and (iii) the delay is most sensitive to forward transit time variations. Similar conclusions were drawn from the discussion on the delay model in Section 7.2.2.(a).

7.2.4. BiCMOS Circuit Design

The circuit performance is affected by both the device and the circuit parameters. In this section, we discuss the effect of the circuit components to highlight the important design issues of each of them. The circuit components can be classified into; (i) the bipolar drivers (Q_1 and Q_2), (ii) the input MOSFETs (P and N) and (iii) the discharging MOSFETs (N_{d1} and N_{d2}) [see Fig. 7.10].

(a) Bipolar Drivers

The choice of emitter area is the main issue in designing the bipolar driver. The sizing of the emitter area depends on the level of the collector current. If the collector current is much higher than the knee current, the current gain and the forward transit time are significantly degraded. Therefore, it is necessary to increase the emitter area to ease the high current effects. Because the delay time is approximately proportional to $\sqrt{I_K}$ as shown by Equation (7.42), it follows that :

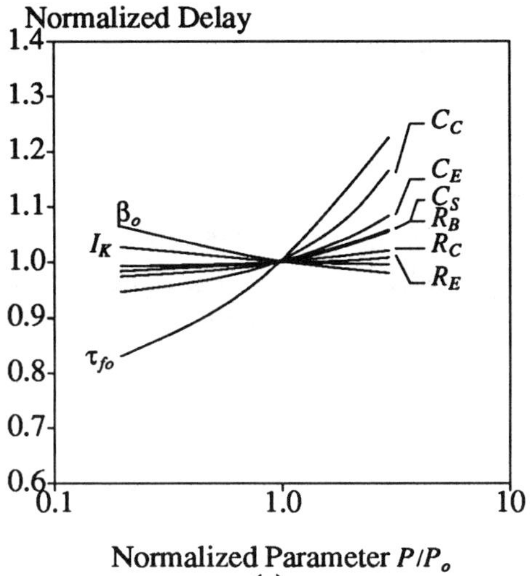

(a)

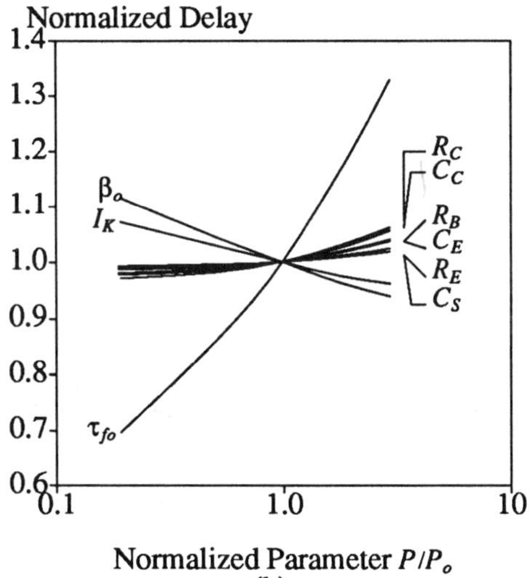

(b)

Figure 7.19 The dependence of the delay on the device parameters for :

(a) C_L=0.5pF; (b) C_L=3pF.

$$t_d \propto 1/\sqrt{A_E} \tag{7.48}$$

This means that the initial increase of the emitter area causes a substantial delay improvement especially when the load capacitance is large. However, the delay starts to saturate as the area increases, as shown in Fig. 7.20, because of the increasing parasitic capacitances of the bipolar transistor. If the load capacitance is small, the parasitic capacitance dominate and the delay degrades as the emitter area is sized up as shown in Fig. 7.20 for the case of C_L=0.2pF.

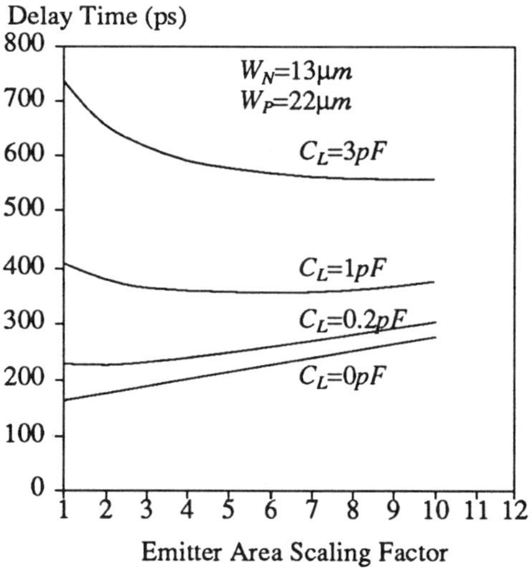

Figure 7.20 The delay dependence on the emitter area.

(b) Input MOSFET Devices

Finding the proper sizing of the input MOSFETs (P and N) is not trivial. The delay of the BiCMOS inverter is monotonically decreasing as the size of the input devices are scaled up as demonstrated by Fig. 7.21. As the delay is approximately proportional to $\sqrt{I_{DSsat}}$ [Equations (7.40) to (7.42)], it follows that :

$$t_d \propto 1/\sqrt{W} \tag{7.49}$$

where W is the width of either P or N. If the load capacitance is small, the MOS parasitic capacitances dominate as the MOS transistors are sized up which cancels the improvement achieved by increasing the drain current. In the case of larger load capacitance, increasing drain current causes the high current effects to be more influential, and therefore, the delay improves at a slower rate as the widths of transistors N and P are increased. So far we have considered the delay of the BiCMOS inverter only and neglected the delay of the stage that drives the BiCMOS inverter. The delay of the driver stage is increased as the input MOSFETs of the BiCMOS inverter are sized up.

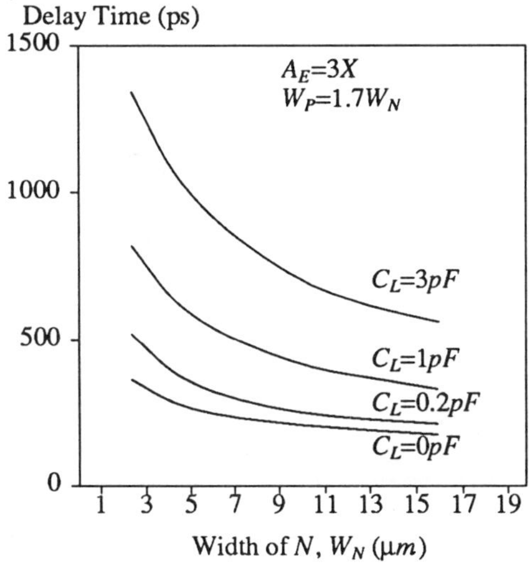

Figure 7.21 The dependence of the delay of the BiCMOS inverter on the size of the input MOS transistors.

It is more meaningful to consider the summation of the delay of the driver ($t_{d,driver}$) and the delay of the BiCMOS inverter ($t_{d,BiCMOS}$) rather than the delay of the inverter alone. Fig. 7.22(a) and 7.22(b) show the delay of the driver stage ($t_{d,driver}$), the delay of a BiCMOS inverter ($t_{d,BiCMOS}$) and the summation of both (t_d) for $0.5pF$ and $3pF$ load capacitances. All three delay times have been simulated for different sizes of the input MOSFETs of the BiCMOS inverter. The driver stage is a minimum size CMOS inverter. The delay of the BiCMOS inverter flattens while that of the driver increases as the input MOSFETs of the BiCMOS inverter are sized up. As a result, the total delay is increasing. However, the minimum of the total delay is close to the point where the delay of the BiCMOS inverter starts to saturate. For example, for the C_L=$3pF$ case, the minimum of the total delay occurs when W_N=$10\mu m$, while the delay of the BiCMOS inverter flattens at W_N=$12\mu m$.

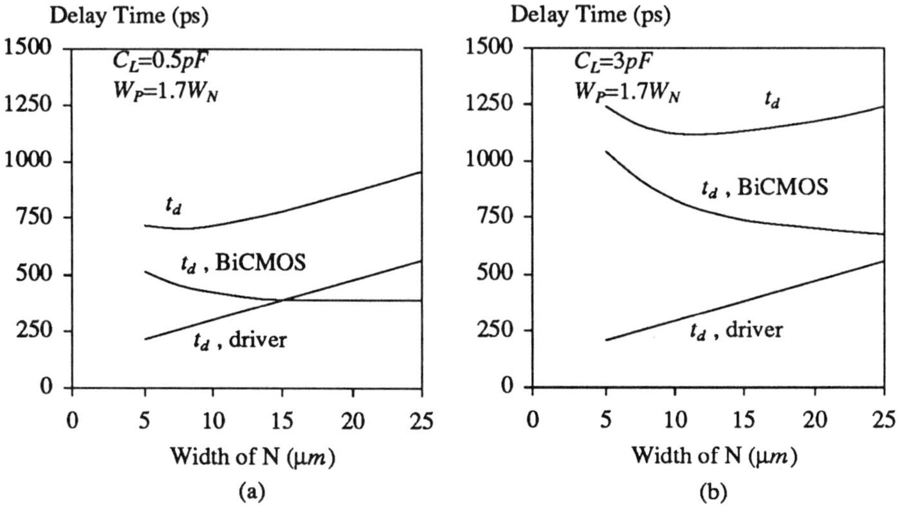

Figure 7.22 The total delay of the driver and the BiCMOS inverter as a function of the size of the input MOS transistors of the BiCMOS inverter.

(c) Bleeding Devices

In the case of the conventional BiCMOS inverter shown in Fig. 7.10, the bleeding MOSFETs (N_{d1} and N_{d2}) must be sized to allow fast discharge of the base charge to reduce the turn OFF time of the bipolar transistors. For typical applications, it is enough to use near minimum size MOSFETs for bleeding purposes.

The bleeding resistors of the R-type BiCMOS circuit must be large enough to prevent leaking the drain current of P and N through the resistors. If we assume that only 10% of the drain current is allowed through the resistor and if the drain current is $2mA$, then the resistor must be equal to $V_{BEon}/0.1I_{DSsat} = 4k\Omega$. The optimum value of the resistor must be determined by more accurate simulations. Fig. 7.23 illustrates the dependence of the delay of the R-type BiCMOS inverter on the bleeding resistors. Increasing the resistors means better drive for the bipolar driver and at the same time more resistance to the discharge of the base charge during turn OFF; thus, imposing two conflicting effects on the delay. The first effect dominates when the resistance is less than $1k\Omega$. Therefore we see a substantial delay improvement as the resistance increases up to $1k\Omega$. Beyond that, both effects cancel each other and the delay remains unchanged as seen from the simulation results illustrated in Fig. 7.23.

7.2.5. Comparison of the Delays of the CMOS and the BiCMOS Inverter

In this section, we compare the delay of a BiCMOS and a CMOS inverter whose input MOSFETs have similar sizing (i.e. same input capacitance). Let us consider two cases :

(i) Unloaded Inverters :

When C_L tends to go to zero, the delay of the BiCMOS inverter given in Equation (7.31) reduces to :

$$t_{d,BiCMOS} = (C_I + C_{CI} + C_{EI})\frac{V_{BEon}}{I_{DSsat}} + (C_{CI} + C_I)\frac{\Delta V_{o,BiCMOS}}{I_{DSsat}}$$

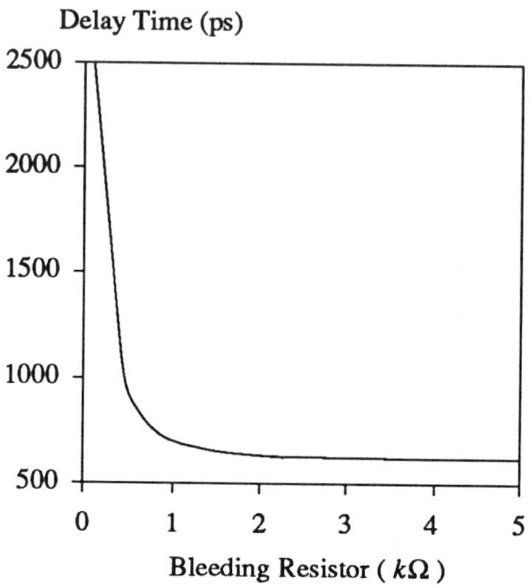

Figure 7.23 The dependence of the delay of the R-type BiCMOS inverter on the bleeding resistance.

$$+\frac{C_{CI}C_o}{C_{CI}+C_I}(R_C+R_E)$$

$$+\frac{C_o}{C_{CI}+C_I}\tau_{fo}$$

$$+C_o\frac{\Delta V_{o,BiCMOS}}{\beta_o I_{DSsat}} \qquad (7.50)$$

where C_o represents the summation of the parasitic capacitances of the bipolar transistors at the output node. $\Delta V_{o,BiCMOS}$ is the voltage swing of the BiCMOS inverter from V_{BEon} to $V_{dd}/2$. We use $\Delta V_{o,BiCMOS}$ instead of ΔV_o to distinguish it from the swing of the CMOS inverter. The voltage swing of the CMOS inverter to the 50% point can be expressed as :

$$\Delta V_{o,CMOS} = \Delta V_{o,BiCMOS} + V_{BEon} \qquad (7.51)$$

Equation (7.50) can hence be rewritten in the following form :

$$t_{d,BiCMOS} = (C_{CI} + C_I)\frac{\Delta V_{o,CMOS}}{I_{DSsat}} + C_{EI}\frac{V_{BEon}}{I_{DSsat}} + \frac{C_{CI}C_o}{C_{CI}+C_I}(R_C + R_E) + \frac{C_o}{C_{CI}+C_I}\tau_{fo} + C_o\frac{\Delta V_{o,BiCMOS}}{\beta_o I_{DSsat}} \tag{7.52}$$

The delay of an unloaded CMOS inverter can be approximated by :

$$t_{d,CMOS} = C_I\frac{\Delta V_{o,CMOS}}{I_{DSsat}} \tag{7.53}$$

The difference between the delays of the CMOS and BiCMOS inverters can be expressed as :

$$t_{d,BiCMOS} - t_{d,CMOS} = C_{CI}\frac{\Delta V_{o,CMOS}}{I_{DSsat}} + C_{EI}\frac{V_{BEon}}{I_{DSsat}} + \frac{C_{CI}C_o}{C_{CI}+C_I}(R_C + R_E) + \frac{C_o}{C_{CI}+C_I}\tau_{fo} + C_o\frac{\Delta V_{o,BiCMOS}}{\beta_o I_{DSsat}} \tag{7.54}$$

Equation (7.54) implies that the delay of unloaded BiCMOS is greater than that of the unloaded CMOS [see Fig. 7.24]. The difference between the delays of the CMOS and BiCMOS inverters is mainly a function of the parasitic resistances and capacitances of the bipolar transistor and its forward transit time. When the loading increases, t_{Co} dominates causing the BiCMOS circuit to outperform CMOS. The crossover capacitance, denoted C_x in Fig. 7.24, is typically in the order of $50fF$ to $100fF$. The crossover capacitance must be reduced to increase the load range over which BiCMOS outperforms CMOS. To do that, the unloaded delay of BiCMOS must be reduced. The only way to achieve this is by optimizing the bipolar device to minimize the parasitic capacitances and especially the collector-base and collector-substrate junction capacitances. However, as the supply voltage is scaled down, all the terms of Equation (7.54) become comparable. Therefore, the forward transit time and the collector and emitter resistances must also be improved.

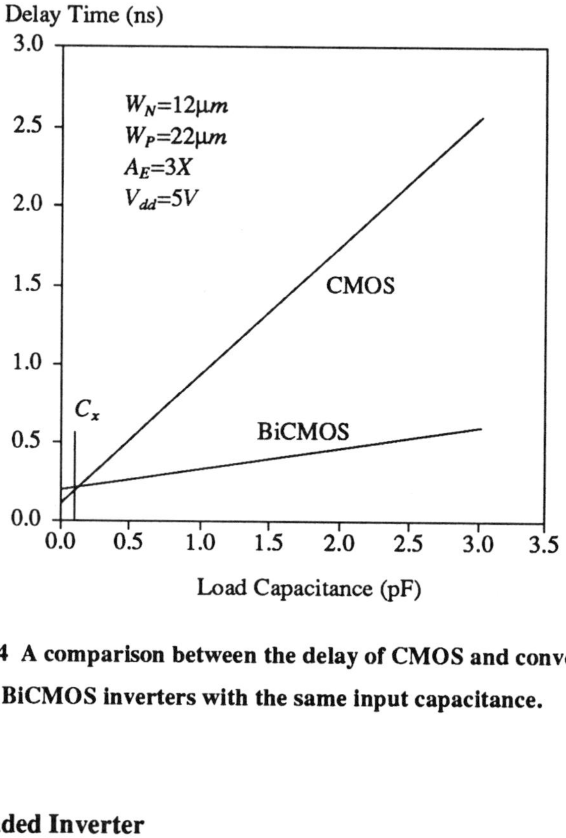

Figure 7.24 A comparison between the delay of CMOS and conventional BiCMOS inverters with the same input capacitance.

(ii) Heavily Loaded Inverter

The delay component t_{Co} [Equation (7.42)] dominates over all other components as the loading increases. We can, therefore, assume that the overall delay is equal to t_{Co}.

$$t_{d,BiCMOS} = \frac{C_o \Delta V_{o,BiCMOS}}{I_{DSsat}} \left(\frac{I_{DSsat}}{\beta_o I_K} \right)^{1/2} \tag{7.55}$$

The delay of the CMOS inverter, which has similar input capacitance to that of the BiCMOS inverter, can be approximated by :

$$t_{d,CMOS} = \frac{C_o \Delta V_{o,CMOS}}{I_{DSsat}} \tag{7.56}$$

The ratio between the CMOS and BiCMOS delays is given by :

$$\frac{t_{d,CMOS}}{t_{d,BiCMOS}} = \frac{\Delta V_{o,CMOS}}{\Delta V_{o,BiCMOS}} \left(\frac{\beta_o I_K}{I_{DSsat}} \right)^{1/2} \tag{7.57}$$

The ratio $\Delta V_{o,CMOS}/\Delta V_{o,BiCMOS}$ is slightly greater than unity. The square root of $\beta_o I_K/I_{DSsat}$ is the crucial term, which determines the actual improvement gained by adding bipolar drivers to the CMOS gate. It is important to note that the actual gain in speed is less than the expected β_o. This is due to the high level injection effects on the current gain of the bipolar transistor.

7.2.6. Layout of the BiCMOS Inverter

The layout of the conventional BiCMOS inverter [Fig. 7.10] is shown in Plate III. The layers of the used process are introduced in Section 2.7. The design rules can be found in Table 2.2. All the metal lines in this layout use first metal level (metal 1) to allow the second metal level to run vertically over the cell. The second metal is typically used to connect the cells to each other. Two gate fingers are used for the input MOSFETs (*P* and *N*), and four emitter strips were used for the NPNs of the bipolar driver. This way the inverter cell can fit within the required pitch of the rail lines. Substrate contacts have to be used for both the PMOS and the NMOS transistors to reduce the body effects. The pull-up NPN and the *P* transistors can both be merged in the same N-well, because the collector of Q_1 and the substrate of *P* share the same voltage (V_{dd}). This saves the isolation area that would be required if two separate wells are used. Using interdigitated bipolar structure has the advantage of reducing the base resistance. Note also that the collector deep diffusion runs in the horizontal direction to reduce the collector current path in the buried layer. An important issue in laying out a circuit, and especially a driver, is the reliability. To avoid electromigration failure, the width of the metal line and the number of contacts must be selected carefully based on the operating currents and the maximum allowable current densities in either the metal lines or the contacts.

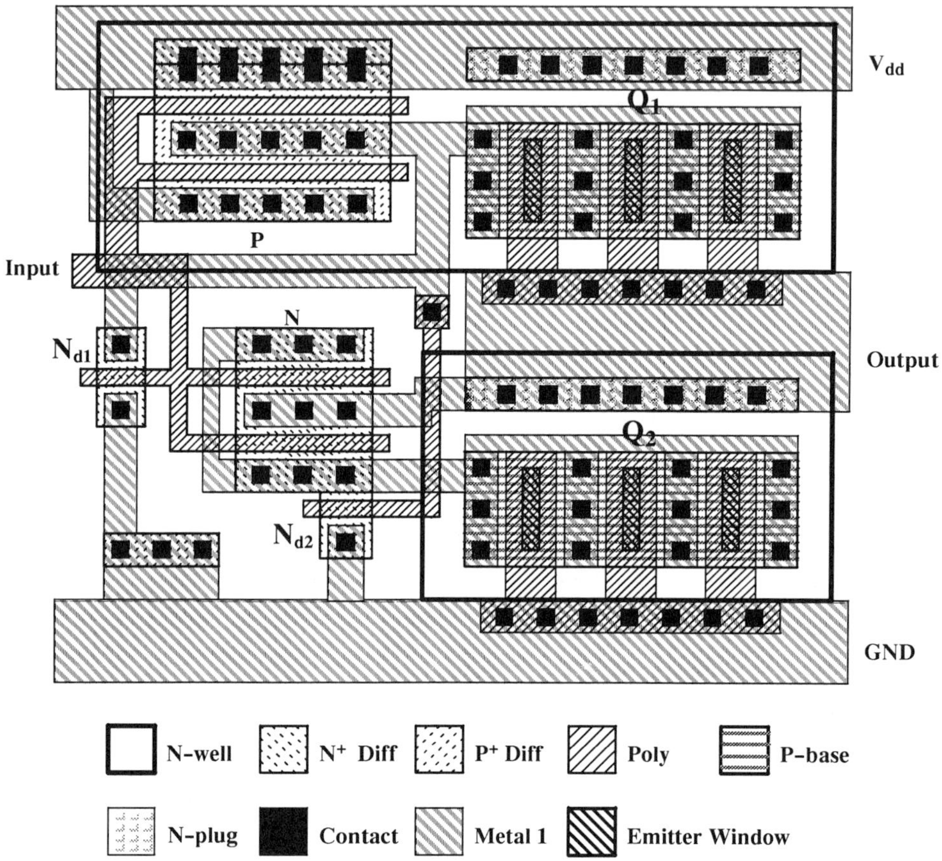

Plate III The layout of a BiCMOS inverter.

7.2.7. BiCMOS Gates

Two-input NOR and NAND gates are shown in Fig. 7.25(a) and 7.25(b). The logic function is implemented using the PMOS and NMOS blocks. The bipolar transistors Q_1 and Q_2 are used as current drivers. The discharging MOS transistors can be replaced by bleeding resistors as in the R-type BiCMOS inverter. Using bleeding resistors reduces the transistor count.

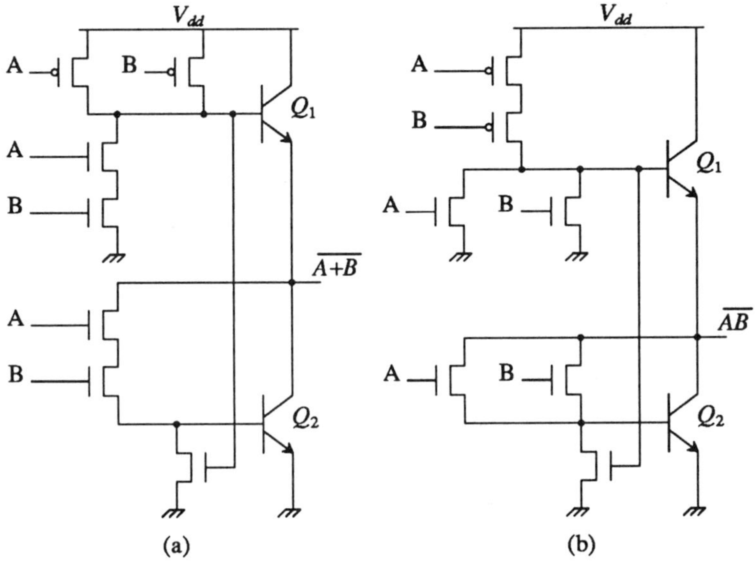

Figure 7.25 Circuit schematic of: (a) BiCMOS NOR gate; (b) BiCMOS NAND gate.

7.3. Full-Voltage-Swing BiCMOS Drivers

The reduced swing of the conventional BiCMOS driver degrades the speed of the driven gates and especially as the supply voltage is scaled down. The increase of the subthreshold currents in the MOS devices, which are driven by the reduced swing, leads to higher DC power consumption. Another disadvantage of the reduced swing is the noise margin reduction.

Full-swing operation can be achieved either by using conventional BiCMOS circuits with base-emitter or collector-emitter shunting techniques or by using other circuits such as the BiNMOS or the complementary BiCMOS. The use of complementary BiCMOS has been encouraged by the recent advances in bipolar technology, which led to high performance PNP transistors[15-17][3]. It is expected that the NPN and PNP transistors will exhibit close performance as the dimensions scale down and the base dopings increase[18].

7.3.1. The AC Effects on the Swing of the Conventional BiCMOS Gate

The static swing of the BiCMOS inverter shown in Fig. 7.7 is $V_{dd}-V_{BEon}$ to V_{BEon} as mentioned in Section 7.3.1. However, the actual swing of the output voltage during operation is found to be greater than $V_{dd}-2V_{BEon}$. This increase in the swing can be explained by studying the transient behaviour of the BiCMOS inverter as the output voltage exceeds $V_{dd}-V_{BEon}$ or falls below V_{BEon}.

When the output node is charged through Q_1 and while the output voltage is still less than $V_{dd}-V_{BEon}$, we can write the following continuity equation in the base :

$$I_b = \frac{q_b}{\beta_o \tau_{fo}} + \frac{dq_b}{dt} \tag{7.58}$$

The first term of the RHS of Equation (7.58) represents the recombination in the base and the second represents the rate of change of the base charge. Equation (7.58) neglects the effects of the space charge layers and does not account for the high

[3] The PNP transistors used throughout this chapter are assumed to be vertical transistors with a current gain of about 25 and an f_T of 2GHz.

current effects on the current gain and forward transit time. This simplified picture is, however, suitable for our purpose. As the output voltage approaches $V_{dd}-V_{BEon}$, and the base voltage of Q_1 is equal to V_{dd}, the base current falls to zero. Hence, Equation (7.58) reduces to :

$$\frac{q_b}{\beta_o \tau_{fo}} + \frac{dq_b}{dt} = 0 \tag{7.59}$$

Solving this differential equation yields the following expression for the base charge:

$$q_b = q_b^o \, e^{-(t-t_x)/\beta_o \tau_{fo}} \tag{7.60}$$

where q_b^o is the base charge at t_x, which is the time at which $V_o=V_{dd}-V_{BEon}$. The base charge is related to the collector current through the following expression :

$$I_{C,Q1} = \frac{q_b}{\tau_{fo}} \tag{7.61}$$

Hence, Equation (7.60) can be written in terms of $I_{C,Q1}$:

$$I_{C,Q1} = I_{C,Q1}^o \, e^{-(t-t_x)/\beta_o \tau_{fo}} \tag{7.62}$$

where $I_{C,Q1}^o$ is the collector current at t_x. As the collector and emitter currents of Q_1 are almost equal, the collector current is also given by :

$$I_{C,Q1} = C_o \frac{dV_o}{dt} \tag{7.63}$$

Substituting from Equation (7.63) into (7.62) and solving the resulting differential equation yields the following output voltage equation :

$$V_o = V_{H,RS} + \frac{\beta_o \tau_{fo} I_{C,Q1}^o}{C_o}(1 - e^{-(t-t_x)/\beta_o \tau_{fo}}) \tag{7.64}$$

where $V_{H,RS}$ is the output voltage at t_x and is equal to $V_{dd}-V_{BEon}$.

The above analysis shows that due to the base charge, Q_1 continues sourcing current beyond t_x. This causes the output voltage to increase as shown by Equation (7.64). Fig. 7.26 shows the waveforms of the output voltage and the base voltage and the emitter current of Q_1 for two cases of loading, $0.5pF$ and $5pF$. At the time when

$V_{B,Q1}=V_{dd}$, the output voltage reaches $V_{dd}-V_{BEon}$. Beyond that point, the emitter current continues to charge the load capacitance causing the output voltage to continue to rise. Due to the capacitive coupling, the base voltage of Q_1 is also pulled up beyond V_{dd}, causing a reverse drain current in P, which helps to bleed the base charge. This will eventually cause Q_1 to turn OFF. The final output voltage depends on the amount of charge stored in the base at t_x, which in turn depends on the collector current level. That is why the HIGH level of the output increases as the load capacitance increases [see Fig. 7.26(a)].

Fig. 7.26(b) shows the waveforms of the output voltage, the base voltage and the collector current of Q_2 during the pull-down transition for $0.5pF$ and $5pF$ load capacitance. The output voltage continues to fall below V_{BEon} until the charge stored in the base of Q_2 is removed. This could lead to the saturation of Q_2 causing the output to fall to V_{CEsat}. Fig. 7.26(b) shows that the V_o falls below the base voltage of Q_2. As a result, the drain current of N is reversed to aid discharging the base of Q_2 and speed its turn OFF. The LOW level of V_o is almost independent on the loading.

As the supply voltage is reduced, the collector current levels of the bipolar drivers are also reduced. This means that the amount of charge stored in the base of Q_1 and Q_2 is decreasing, and therefore we find that the current sourced by Q_1 (sunk by Q_2) after V_o exceeds $V_{dd}-V_{BEon}$ (falls below V_{BEon}) is also decreasing. Consequently, the relative loss in the actual logic swing is significantly increasing as the supply voltage is scaled down as shown by the simulation results [Fig. 7.27].

In short, we can say that the actual swing of the conventional BiCMOS circuit [Fig. 7.10] is degraded as the supply voltage is scaled down (< 3.5V).

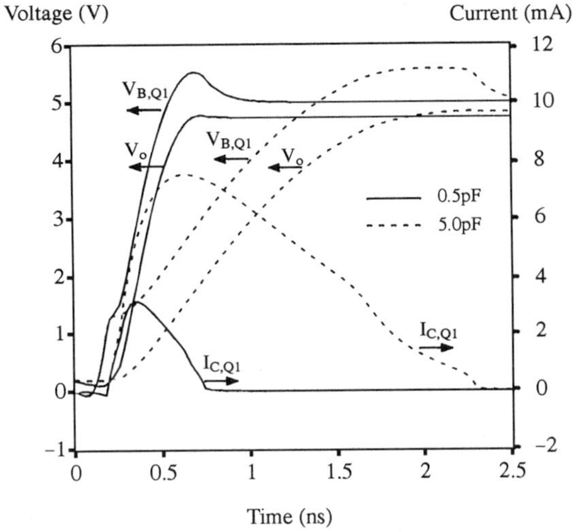

(a)

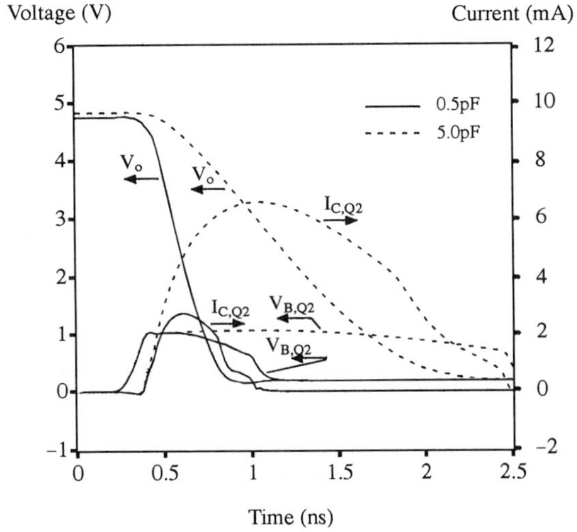

(b)

Figure 7.26 The waveforms of (a) voltages and (b) currents for $0.5pF$ and $5pF$ loading.

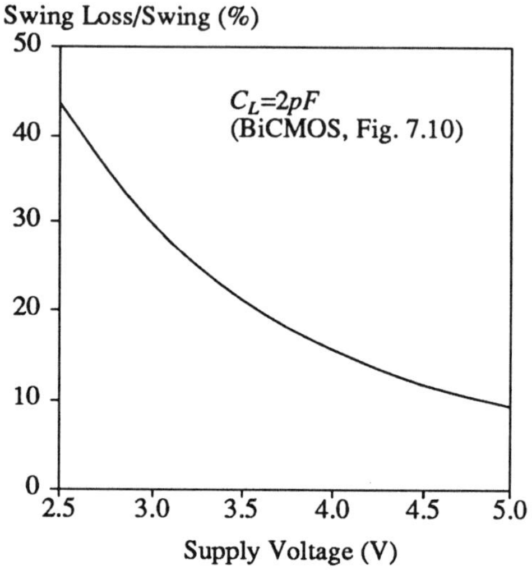

Figure 7.27 The ratio of the swing loss to the actual swing of the conventional BiCMOS inverter as the supply voltage is scaled down.

7.3.2. BiNMOS Circuits

The BiNMOS circuit shown in Fig. 7.28[13] combines the pull-up section of a conventional BiCMOS circuit [Fig. 7.10] and an NMOS transistor for pull-down. This circuit can also be viewed as a CMOS circuit in which the slow pull-up PMOS logic block is replaced by a fast BiCMOS pull-up block. The voltage swing of the BiNMOS circuit is larger than that of the conventional BiCMOS by about V_{BEon}. However, its fall time is limited by the transconductance of the N transistor. The poor sinking capability of the NMOS transistor may prolong the fall time especially in the case of heavy loading. Fig. 7.29 demonstrates the average delay of a BiNMOS inverter in comparison with that of a CMOS and a BiCMOS inverter as the supply voltage is scaled down. The delays of the different circuits are calculated from the frequency of a 5-stage ring oscillator simulated by HSPICE. All circuits have the same input capacitance. The BiNMOS oscillator operates at 2V supply voltage where the conventional BiCMOS fails. The BiNMOS circuit seems to be faster than both CMOS and BiCMOS when the load capacitance is small (e.g. 0.5pF as shown in Fig.

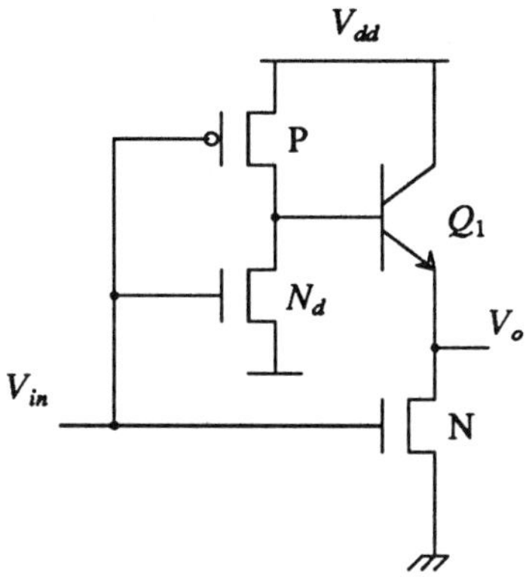

Figure 7.28 The circuit schematic of the BiNMOS inverter.

7.29(a)). In the case of heavy loading, the BiCMOS outperforms the BiNMOS circuit down to 3V supply voltage. In addition, the fall time is much longer than the rise if the BiNMOS would drive large load capacitance.

7.3.3. Full-Swing BiCMOS Circuits with Shunting Techniques

Fig. 7.30(a) to (c) show different types of BiCMOS circuits, in which the base-emitter shunting is used. The last circuit [Fig. 7.30(d)] uses collector-emitter shunting[19].

1- R-Type BiCMOS Circuit

The circuit shown in Fig. 7.30(a) was introduced in Section 7.3.1. As the output voltage exceeds $V_{H,RS}$, Q_1 ceases to source current to the load, and the load capacitance charges only through transistor *P* and the bleeding resistance (*R*). In the pull-down phase, the load capacitance is discharged only through *N* and the resistance *R* when V_o falls below V_{BEon} ($V_{L,RS}$). The DC transfer characteristics of the R-type BiCMOS shows a full output voltage swing [see Fig. 7.13(b)]. However, the actual swing depends on the load capacitance and the frequency of operation. As the capacitance of the load increases, the time required to charge

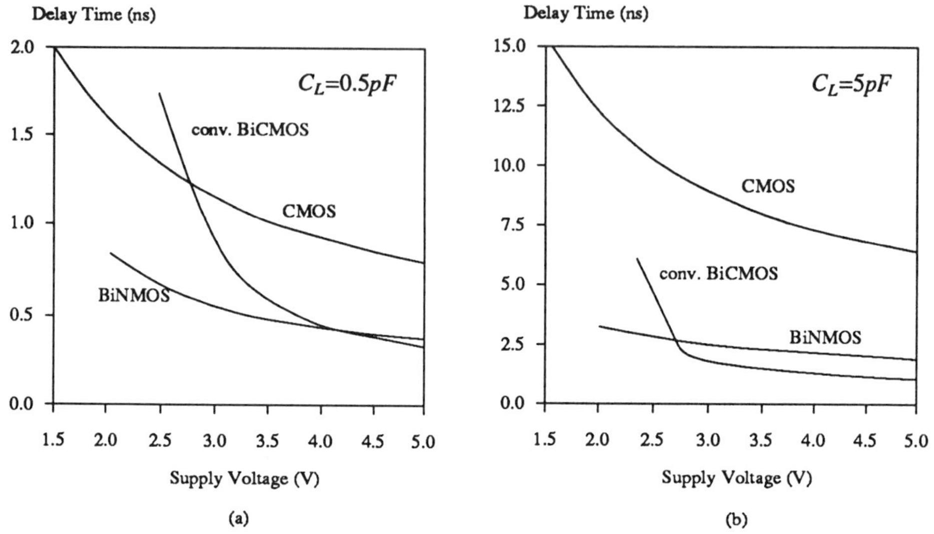

Figure 7.29 A comparison between the delay of the CMOS, conventional BiCMOS and BiNMOS inverters (all having the same input capacitance).

the output from $V_{H,RS}$ to V_{dd} is substantially increasing because the load is, in principle, charged through a resistance during this period. Under heavy loading and fast switching, the output node is prevented from swinging rail-to-rail. To reduce the time constant of the charging path when $V_o > V_{H,RS}$, the resistance (R) must be reduced. This, however, weakens the drive of the bipolar transistor (Q_1) and leads eventually to higher delays as shown in Fig. 7.31. However, as the supply voltage is reduced, the partial swing, that is exclusively charged or discharged through the resistors (R), increases. For example, at 2V supply voltage, about 50% of the swing is recovered by the resistors. It is therefore important to reduce the value of R. Fig. 7.31 shows the delay for R=3$k\Omega$ and R=50Ω. With large resistance, the R-type BiCMOS inverter has the same speed as conventional BiCMOS when the supply voltage is greater than 4V. Although the R-type BiCMOS operates at low supply voltages its delay is too large. By reducing the value of the bleeding resistors, the delay approaches that of the

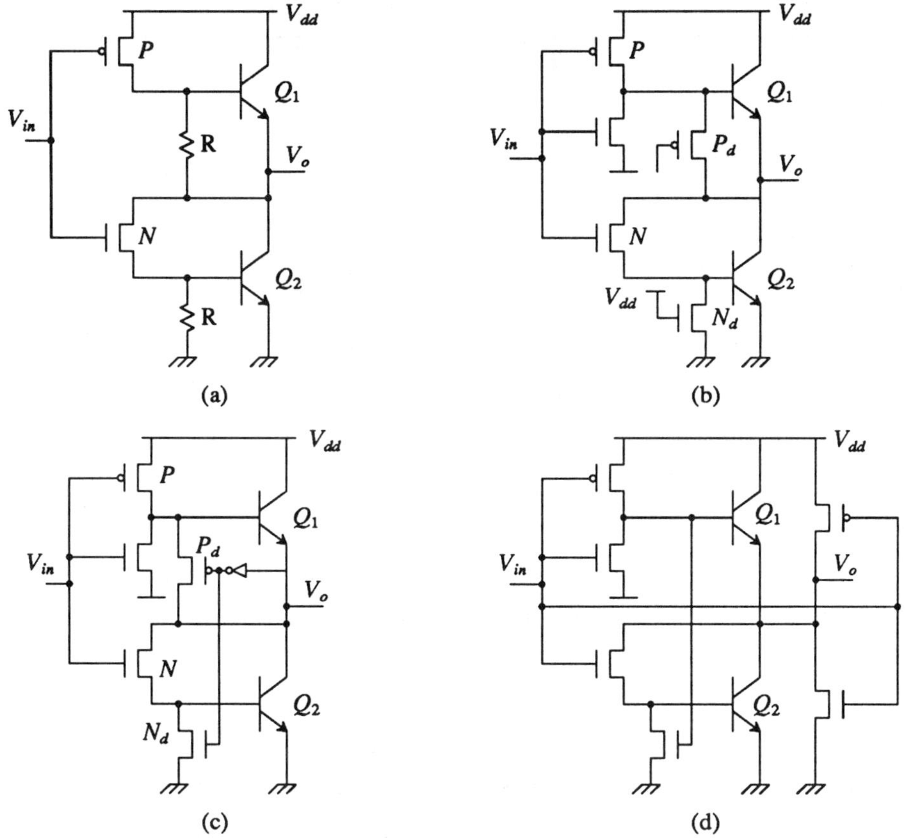

Figure 7.30 Full-swing BiCMOS Buffers

CMOS inverter as shown in Fig. 7.31. With 50Ω resistors, the bipolar transistors of the R-type BiCMOS are hardly active because of the poor base drive.

2- MOS-Type BiCMOS Circuit

The second circuit [Fig. 7.30(b)] is similar to that of Fig. 7.30(a) except that the bleeding resistors are replaced by MOS devices. P_d (N_d) is biased to operate in the linear region if the output voltage exceeds $V_{H,RS}$ (falls below $V_{L,RS}$). This

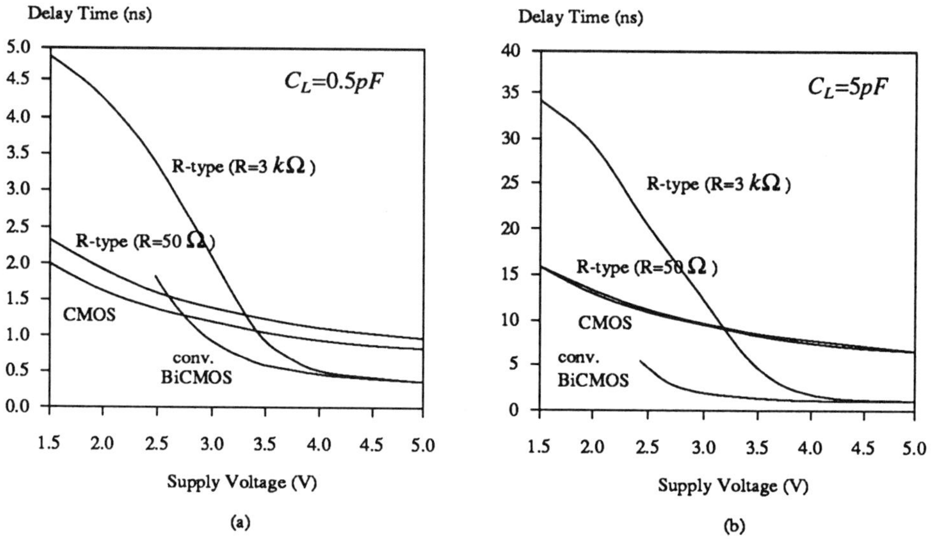

Figure 7.31 A comparison between the delay of the CMOS, conventional BiCMOS and R-type BiCMOS inverters (all having the same input capacitance).

circuit suffers from the same problems as the R-type. The size of the bleeding MOSFETs (P_d and N_d) cannot be increased to achieve full-swing under heavy loading because the base drive of Q_1 and Q_2 is jeopardized. Minimum sized P_d and N_d lead to high delays at low voltage as shown in Fig. 7.32. If they are sized up, the performance of the MOS-type BiCMOS approaches that of the CMOS. This corresponds to the case of reduced resistors of the R-type BiCMOS.

3- Feedback Type BiCMOS Circuit

The problems of the base drive and the voltage swing inherent in the R- and MOS-types can be decoupled by using a feedback CMOS inverter as shown in Fig. 7.30(c)[20]. During the pull-up transition, P_d is initially OFF and the current supplied by P is fully used to drive Q_1. P_d turns ON when V_o approaches $V_{H,RS}$. This allows the output node to rise to V_{dd}. It also serves to bleed the base charge of Q_1 when the input goes high. The operation in the pull-down phase

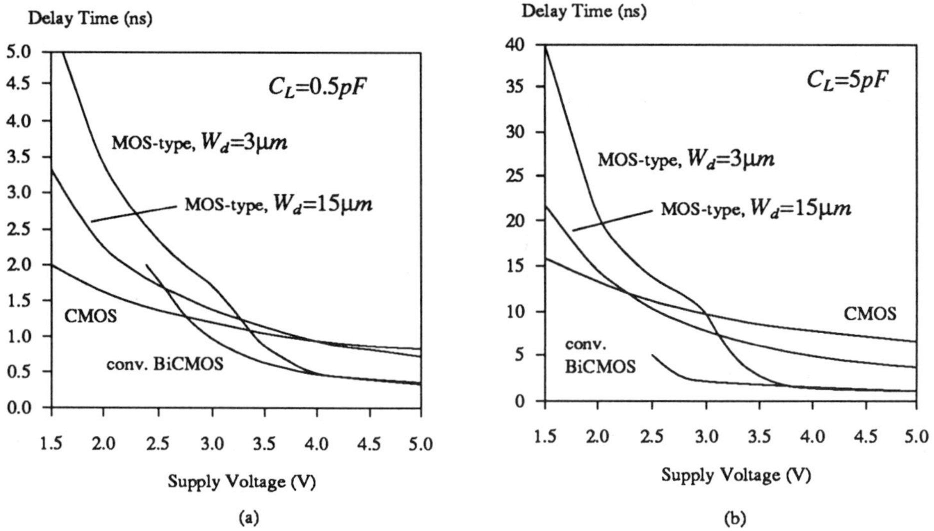

Figure 7.32 A comparison between the delay of the CMOS, conventional BiCMOS and MOS-type BiCMOS inverters (all having the same input capacitance).

can be similarly explained. Both transistors N_d and P_d can now be sized up to improve the charging or discharging times to the rail voltages without sacrificing the base drive of Q_1 and Q_2.

4- Collector-Emitter Shunt BiCMOS Circuit

The collectors and the emitters of Q_1 and Q_2 can be shunted by a PMOS transistor and an NMOS transistor respectively, whose gates are driven by the input signal as shown in Fig. 7.30(d). The shunting transistors keep sourcing (sinking) current to (from) the output load capacitance after Q_1 (Q_2) turns OFF. The output swing is pulled all the way up (down) to V_{dd} (ground).

If we compare the above full-swing circuits, we see that the R-type, MOS-type and collector-emitter shunt circuits are not useful for low voltage/high capacitance applications because they use passive elements (resistors or MOS in the linear region) to complement the swing. Although the feedback type is also using passive devices, the improved base drive allows it to operate at higher speed down to about 3V (compare Fig. 7.33 to Fig. 7.31 and Fig. 7.32). However, as the supply voltage falls below 3V, where the contribution of P_d and N_d as current source and current sink grows, the delay is substantially degraded.

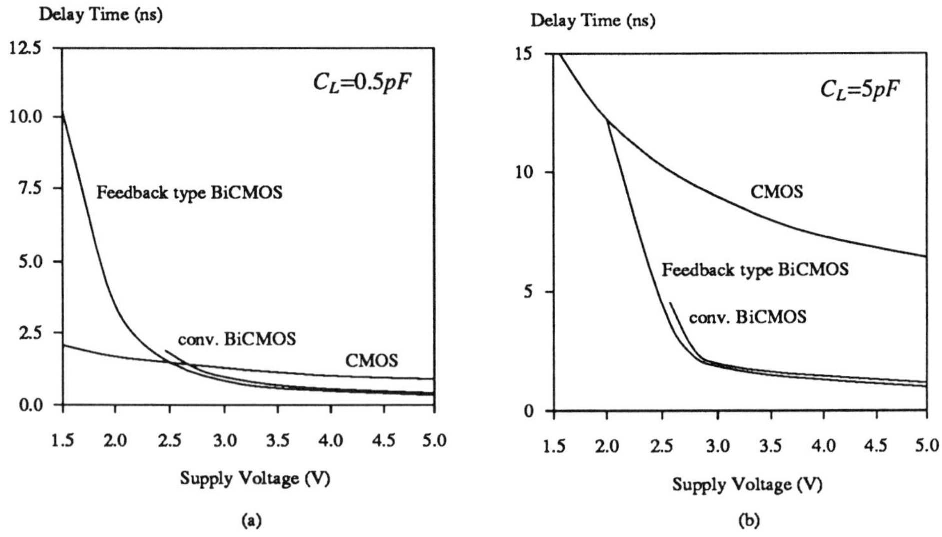

Figure 7.33 A comparison between the delay of the CMOS, conventional BiCMOS and feedback type BiCMOS inverters (all having the same input capacitance).

7.3.4. Full-Swing Common-Follower Complementary BiCMOS Circuits

Attempts to use complementary BiCMOS circuits have been reported in [2,21]. These circuits are based on the emitter-follower (EF) configuration shown in Fig. 7.34[22] .

The emitter-follower complementary BiCMOS inverter consists of NMOS/NPN pull-up and PMOS/PNP pull-down sections. The pull-up section is similar to that used in the conventional BiCMOS circuits [Fig. 7.10]. The swing of this circuit is similar to the swing of the conventional BiCMOS circuit. The static swing varies between V_{BEon} and $V_{dd}-V_{BEon}$. However, due to the bootstrapping effect explained in

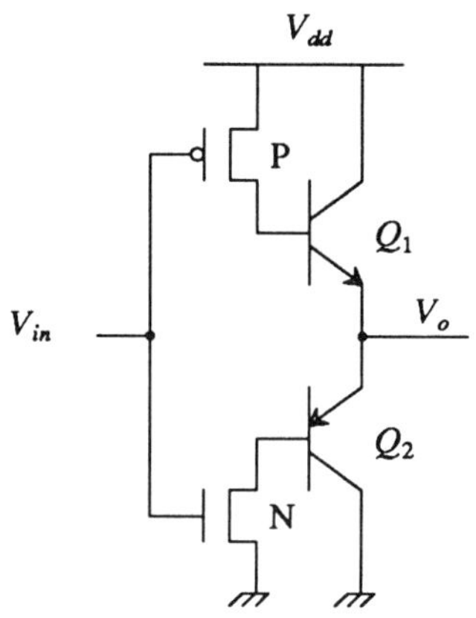

Figure 7.34 The circuit schematic of the emitter-follower BiCMOS inverter

Section 7.3, the actual swing might be larger than the static swing. The balanced transconductance of the PMOS/NPN and NMOS/PNP makes it easier to obtain symmetrical fall and rise times. The base drive of the pull-down transistor (Q_2) is greater than that of the conventional BiCMOS inverter because the gate-source voltage of transistor N is equal to the input voltage. While in the case of the conventional BiCMOS, it is equal to the input voltage less V_{BEon}. Note that the difference in the base drive of Q_2 becomes more significant as the supply voltage is scaled down, while the base-emitter built-in voltage is not scalable.

The emitter-follower complementary BiCMOS circuit shown in Fig. 7.34 can be modified to achieve full-swing operation by using base-emitter and/or collector-emitter shunting techniques[23]. Fig. 7.35(a) shows an emitter-follower complementary BiCMOS inverter where the transistors P_d and N_d have the function of bleeding the base charge of Q_1 and Q_2 respectively and complementing the output swing. Fig. 7.36 shows a comparison between the EF complementary BiCMOS (EF-CBiCMOS) [Fig. 7.35(a)], conventional BiCMOS and CMOS. The EF complementary BiCMOS offers better performance than the other two circuits. This is due mainly to the superior base drive. However, for low supply voltages, the delay increases significantly, and its performance is similar to the other full-swing

BiCMOS circuits, the R- and the MOS- types.

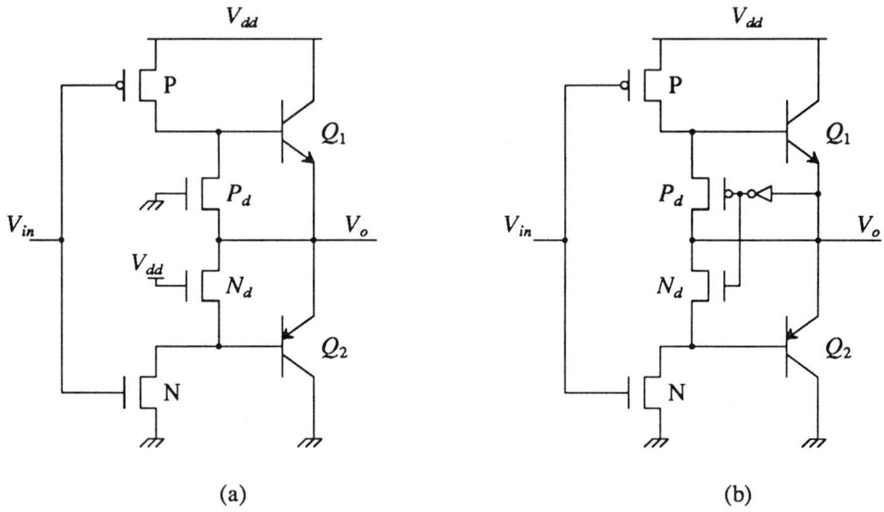

Figure 7.35 Schematic of full swing EF complementary BiCMOS inverters.

When the input of the circuit shown in Fig. 7.35(a) is LOW, transistor P turns ON. As the voltage of the base of Q_1 rises above the threshold voltage of P_d, it turns ON. Part of the drain current of P is diverted from the base of Q_1 to P_d, which weakens the drive of Q_1. The turning ON of P_d can be delayed by using a feedback inverter as shown in Fig. 7.35(b).

The full-swing circuits presented so far achieve the rail-to-rail swing by using resistors or MOSFETs that operate in the linear region. Because of the poor current drive of the passive elements used to complement the swing, these circuits are capable of achieving full-swing only in the case of small load and/or low operating frequencies.

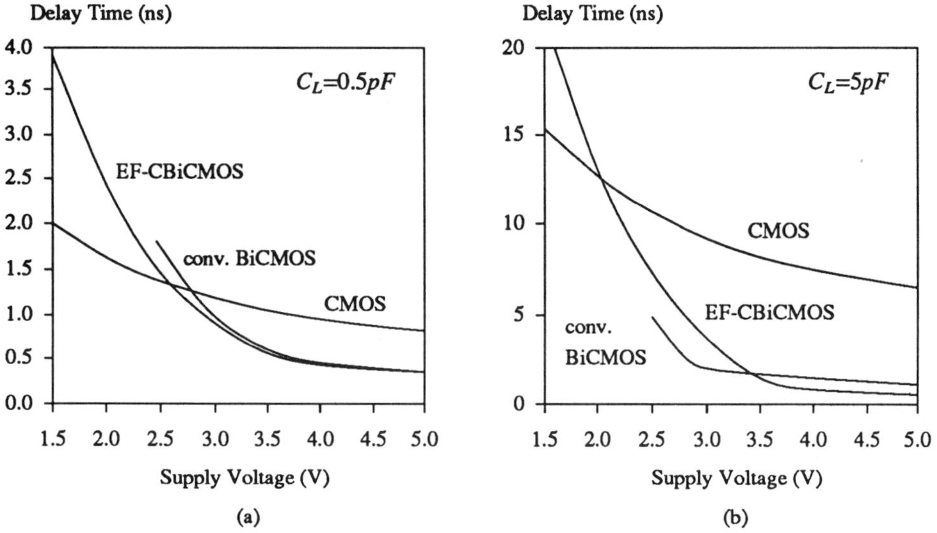

Figure 7.36 A comparison between the delay of the CMOS, conventional BiCMOS and EF-CBiCMOS inverters (all having the same input capacitance).

7.3.5. Full-Swing Common-Emitter Complementary BiCMOS Circuits

The basic common-emitter complementary BiCMOS (CE CBiCMOS) circuit is shown in Fig. 7.37. The pull-up section is made of the NMOSFET (*N*) and the PNP transistor (Q_1). The pull-down section consists of the PMOSFET (*P*) and the NPN transistor (Q_2). When the input rises to the HIGH level, *N* turns ON to sink current from the base of Q_1. When the base voltage of Q_1 falls to $V_{dd}-V_{BEon}$, Q_1 turns ON to source current to the output node. Q_1 eventually saturates and the output node is pulled-up to $V_{dd}-V_{CEsat}$. The pull-down process can be similarly explained. The main advantage of this circuit is that the output voltage of this circuit has near rail-to-rail swing (V_{CEsat} to $V_{dd}-V_{CEsat}$).

In this section, we present circuit techniques for increasing the voltage swing of BiCMOS buffers through active charging and discharging using complementary bipolar drivers. The concept of the operation of this type of circuits differs from the

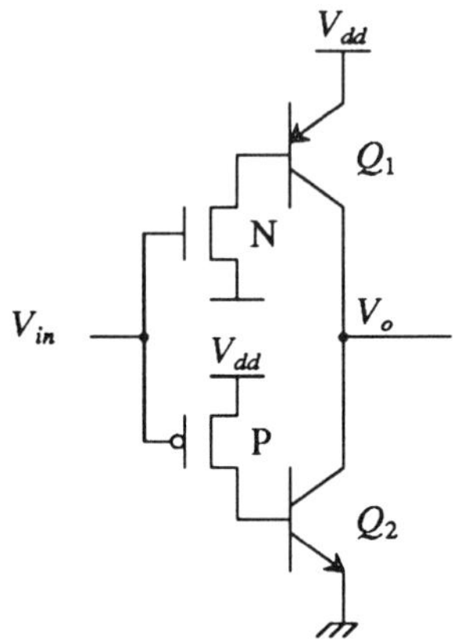

Figure 7.37 The schematic of the common-emitter complementary BiCMOS buffer.

above mentioned circuits. We therefore introduce the basic concept before discussing the circuit implementation.

Basic Concept

The common-emitter complementary buffer shown in Fig. 7.37 has two disadvantages; (i) the high static power dissipation due to the DC current flowing through the bases of either Q_1 or Q_2, and (ii) the excess delay due to the slow process of turning the saturated BJTs OFF. These problems can be solved by driving the bipolar transistors into saturation by using a base current pulse instead of the DC base current as in the circuit of Fig. 7.37. During the duration of the base current pulse, the bipolar drivers operate in the normal active region. At the end of the pulse, the base is isolated from the rest of the circuit to confine the excess minority charge in the base. This base charge will force the bipolar drivers to continue sourcing or sinking current to or from the output node until they saturate.

The advantage of this technique over the conventional common-emitter configuration [Fig. 7.37] is that there is no DC power consumption and the turn OFF time of the bipolar transistors is substantially reduced because most of the base charge would be recombined by the time the output voltage has reached the rail voltage.

To ensure full-swing operation, a minimum amount of the minority charge stored in the base must be guaranteed. To derive the condition that has to be satisfied to ensure the saturation of the bipolar drivers, we consider the pull-up section of the buffers shown in Fig. 7.38.

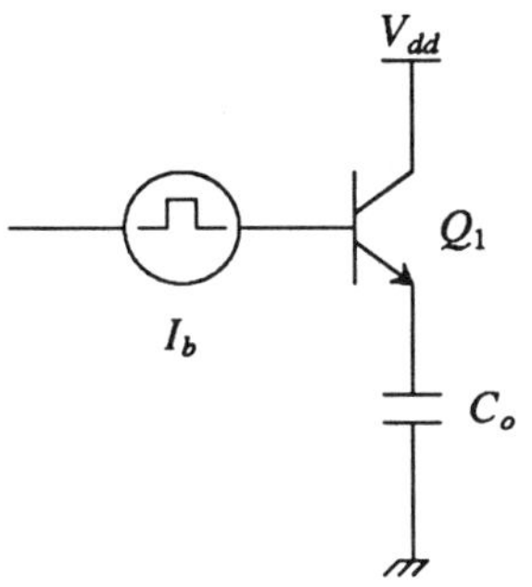

Figure 7.38 Pull-up section of a BiCMOS circuit driven by a base current pulse.

The minimum requirement to guarantee a full voltage swing is that the base charge q_b approaches zero only when transistor Q_1 enters the saturation region. Because q_b is an exponential function of time [Equation (7.60)], we derive a condition for the base charge to be, at least, equal to 10% of q_b^o when Q_l saturates. At the same instant $I_{C,Ql}$ has to be equal to 10% of $I_{C,Ql}^o$. By substituting for V_o in Equation (7.64) by $V_{dd}-V_{CEsat}-0.1I_{C,Ql}^oR_C$ and for $I_{C,Ql}$ in Equation (7.62) by $0.1I_{C,Ql}^o$ and eliminating the time from Equations (7.62) and (7.64) we get the minimum required value for $I_{C,Ql}^o$.

$$I_{C,Q1,min}^o = \frac{C_o \Delta V}{0.9\beta_P \tau_P + 0.1 R_C C_o} \qquad (7.65)$$

where τ_P and β_P are the forward transit time and current gain of the PNP transistor respectively, and $\Delta V = V_{dd} - V_{CEsat} - V_{H,RS}$. We refer to ΔV as the differential swing, which is generally defined as the change in the output voltage from the time the base current goes to zero to the time when the BJT saturates. A similar condition can be derived for the minimum required collector current level of Q_2. Equation (7.65) helps to explain the significance of the differential swing. By reducing the differential swing it is easier to force the output voltage to reach the rail voltages.

Circuit Implementations

In this section, we present three circuit implementations for the full-swing common-emitter CBiCMOS inverters. The circuits shown in Fig. 7.39 to 7.41 illustrate different ways of generating the base current pulse required to drive Q_1 and Q_2.

A. First Implementation:

The circuit shown in Fig. 7.39 consists of a multi-drain[24] CMOS input stage (P, N). The two discharging elements P_d and N_d are essential to discharge the bases of Q_1 and Q_2, respectively, to reduce their turn OFF time. Transistors N_1 and P_1 act as input/output controlled switches. The complementary pair Q_1 and Q_2 serve as output drivers. To avoid any DC level shifts due to leakage currents, the buffer has to be shunted by a minimum size CMOS inverter for which we use the second drain of the multi-drain input CMOS.

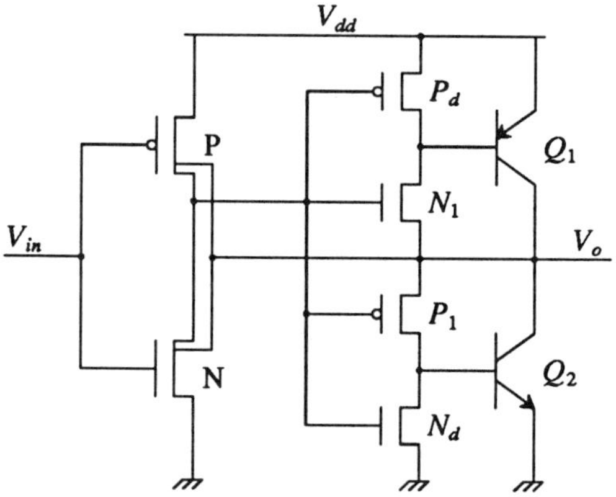

Figure 7.39 The circuit schematic of the MOS-type CE CBiCMOS inverter.

To understand the circuit operation, let us consider the circuit response to a falling transition at the input. Transistor P turns ON causing N_1 to turn ON. When the base of Q_l is pulled down to $V_{dd}-V_{BEon}$, Q_l turns ON. The output node is hence being pulled up. When the output voltage reaches $V_{dd}-V_{TN}$, transistor N_1 turns OFF provided that the threshold of N_1 is greater than or equal to V_{BEon}. The base of Q_l is thus isolated and the base charge, if large enough, forces Q_l into saturation. The pull-down operation is, in principle, similar to that of the pull-up.

It turns out that it is very difficult to design the circuit of Fig. 7.39 to swing rail-to-rail for the following reasons. The base drive is poor, because the gate-source and drain-source voltages of N_1 (P_1) decrease as the output rises (falls). Therefore, it is difficult to accumulate the base charge required for full-swing operation. Moreover, the differential swing (ΔV), which is equal to V_{TN} for the rising transition and V_{TP} for the falling transition, is relatively large.

B. Second Implementation:

To improve the drive of the output drivers, Q_l and Q_2, N_1 and P_1 can be replaced by a NPN (Q_{11}) and a PNP (Q_{22}) respectively as shown in Fig. 7.40[25]. The circuit is similar to the one shown in Fig. 7.39. In this case, Q_{11} and Q_{22} act as an input/output controlled switch. Because the bases of Q_{11} and Q_{22} should not be tied to the same node, they are connected to separate drains of the multi-drain P/N CMOS input stage. The circuit operation is also similar to that of the first implementation. If the input voltage is "0", the transistor P turns ON causing Q_{11} and in turn Q_1 to enter the active region. The output node is hence being pulled up. When the output voltage reaches $V_{dd}-V_{BEon}$, Q_{11} turns OFF. The base of Q_1 is hence isolated, preventing any leakage of its charge and thus allowing the output voltage to increase until Q_1 saturates.

The disadvantage of this circuit is that the differential swing (ΔV) is fixed and is equal to V_{BEon}. It is therefore difficult to satisfy the full-swing condition especially at very low power supply voltages (< 2V), unless the areas of Q_1 and Q_2 are substantially increased.

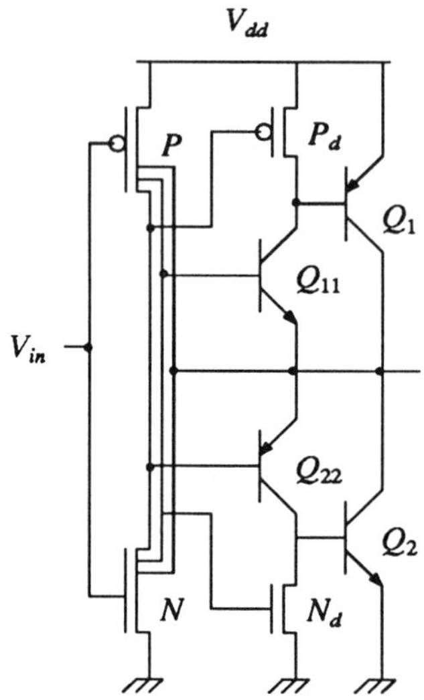

Figure 7.40 Schematic of the semi-full-swing CE CBiCMOS inverter.

C. Third Implementation:

Fig. 7.41 shows another implementation[25] of a complementary BiCMOS buffer. It consists of a multi-drain CMOS input stage. The first drain is used to compensate for any leakage of the charge stored in the load capacitance during steady state. The second drain drives the discharging transistors P_d and N_d. The inverter composed of P_1, P_i and N_1 controls the base current of Q_{11}. P_i is driven by the input, while P_1 and N_1 are driven by the output signal. The current amplifier, Q_{11}, serves as the base driver for Q_1. The pull down section is the complement of the pull up section.

The operating principle of this circuit is similar to the one of Fig. 7.40. When the output is LOW and the input falls from HIGH to LOW, the switches P_i and P_1 are closed to turn Q_{11} and Q_1 ON. Consequently, the output node is being pulled up. As the output voltage reaches the threshold voltage of the inverter P_1, P_i and N_1, the voltage of the base of Q_{11} falls below V_{BEon} causing Q_{11} to turn OFF. The time period required to turn Q_{11} OFF can be tailored by sizing the transistors P_1, N_1 and

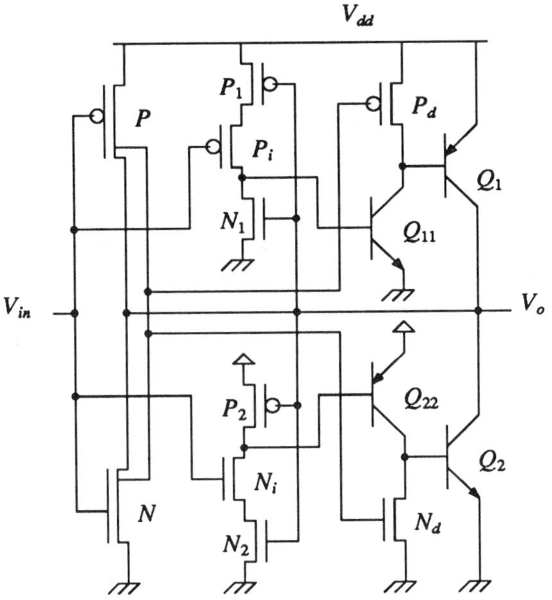

Figure 7.41 Schematic of the full-swing CE CBiCMOS inverter.

P_i. As soon as Q_{11} cuts OFF, the base minority charge is trapped in the base of Q_1. Hence, Q_1 stays ON supplying the output load with the charging current until Q_1 saturates. In this implementation, the conduction periods of Q_{11} (Q_{22}) can be extended to allow the output voltage to get as close as possible to $V_{dd}-V_{CEsat}$ (V_{CEsat}) before the base current of Q_1 (Q_2) falls to zero. As a result, the differential swing (ΔV) can be reduced. Hence, the collector currents required to achieve full-swing are lower than those required in the case of the circuit of Fig. 7.40, as explained in Section 7.4.3.1. In fact, simulations have shown that the circuit shown in Fig. 7.41 gives rail-to-rail voltage swing while the output voltage of the circuit shown in Fig. 7.40 is reduced by about 0.2V at 5V supply. We will, therefore, refer to the second implementation [Fig. 7.40] as the semi-full-swing common-emitter complementary BiCMOS (SFS-CE-CBiCMOS) and to the third implementation [Fig. 7.41] as the full-swing common-emitter complementary BiCMOS (FS-CE-CBiCMOS).

Circuit Simulations

The speed performance of the CE CBiCMOS circuits was evaluated using circuit simulation (HSPICE). The delay was extracted from the simulation of 5-stage ring oscillators for two loads 0.5pF and 5pF over a wide range of supply voltages from 1.5V to 5V. The delay curves in Fig. 7.42 indicate that the SFS-CE-CBiCMOS buffer is almost insensitive to supply voltage scaling down to 1.5V. This can be explained by the severe swing losses at such low voltages. The FS-CE-CBiCMOS driver also demonstrates low sensitivity to supply voltage as depicted in Fig. 7.43. It is also interesting to note that both circuits offer low delay-to-load sensitivity. The FS-CE-CBiCMOS circuit, though having the disadvantage of large area, has two advantages over the semi-full-swing type. It offers a rail-to-rail swing, which allows it to operate at low supply voltages (1.5V). Its performance can be further improved as the cutoff frequencies of the bipolar transistors are increased [see Fig. 7.43]. The voltage swing of the SFS-CE-CBiCMOS buffer is sensitive to the forward transit time. It is less probable that the full-swing can be achieved if τ_P or in general τ_{fo} is decreased as indicated by Equation (7.65). The swing of the FS-CE-CBiCMOS buffer is almost independent on the transit time, because ΔV is small.

7.3.6. CMOS versus BiCMOS under Supply Voltage Scaling

In comparing the performance of the various circuits; CMOS, conventional BiCMOS, BiNMOS, EF-CBiCMOS and CE-CBiCMOS, we need to establish a methodology by which these circuits can be relatively evaluated. The performance of logic gates is typically measured by their delay, area and power dissipation. For the circuits of interest the power dissipation is purely dynamic. If we exclude the power dissipation due to short circuit current, which can be substantially reduced by good design, the dynamic power dissipation of all circuits is quite similar. We therefore limit the comparison on the basis of delay and area. However, if the delay is compared while the areas are kept constant, the input capacitances may not be equal. This means that the delay of the previous stage is unequal. On the other hand, if the delay is compared for constant input capacitance, the areas will be unequal. Ideally, one should compare the performance for the same area and the same input capacitance. Therefore, in comparing a CMOS buffer with a BiCMOS buffer, multi-

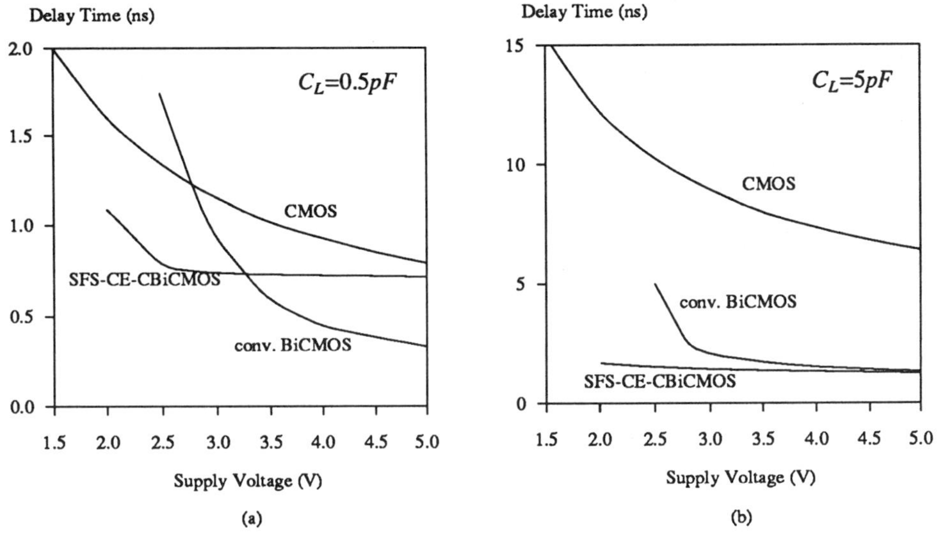

Figure 7.42 A comparison between the delay of the CMOS, conventional BiCMOS and SFS-CE-CBiCMOS inverters (all having the same input capacitance).

stage CMOS buffer should be considered[26].

Another criterion for comparison would be the speed alone. In such a case, the minima of the delay of all circuits is compared regardless of the area, while both the input and output capacitances are kept constant. This comparison is more meaningful for the selection of driver circuits. In such an application, speed has higher priority over area.

The following comparisons are based on delay times extracted from 5-stage ring oscillators simulated using HSPICE. Each of the circuits was first optimized for certain input capacitance and loading conditions. Some circuits have been excluded from the comparison. For example, the conventional, R- and MOS- type BiCMOS buffers are outperformed by other circuits over the whole range of supply voltage and load capacitance.

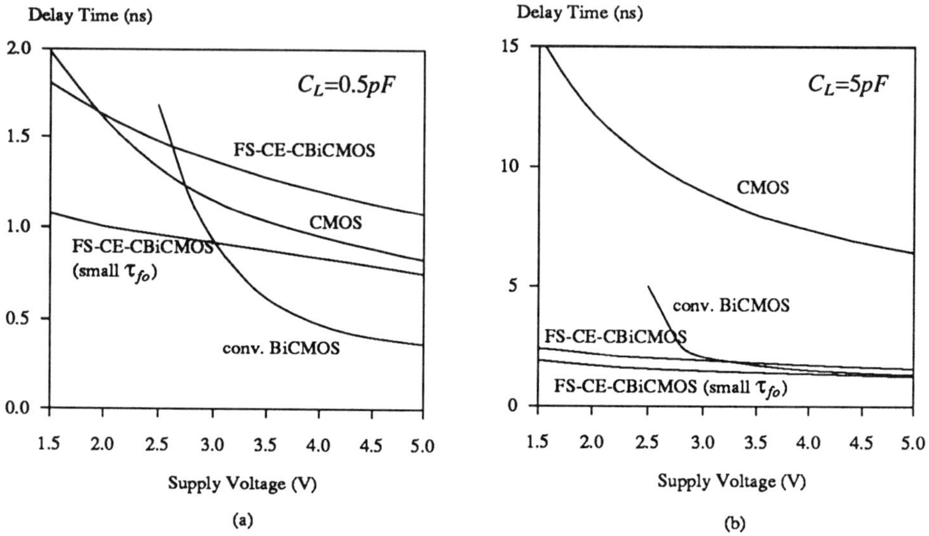

Figure 7.43 A comparison between the delay of the CMOS, conventional BiCMOS and FS-CE-CBiCMOS inverters (all having the same input capacitance).

In the case of small load capacitances (e.g. 0.5pf), we compared three candidates, the BiNMOS, the feedback type and the SFS-CE-CBiCMOS inverters. Fig. 7.44(a) shows the delay of these circuits and how they compare to a single stage CMOS driver. The feedback type is an efficient driver for high supply voltage (> 3V). The BiNMOS seems to be the best candidate in the low voltage range. As the loading increases [see Fig. 7.44(b)], the EF-CBiCMOS outperforms the feedback type in the high voltage domain. Below 3V supply voltage, only the 3-stage CMOS buffer and the FS-CE-CBiCMOS are in competition. The FS-CE-CBiCMOS buffer is superior to the CMOS and has almost double the speed compared to CMOS for high performance PNP and NPN bipolar transistors (small τ_{fo}).

In conclusion, we can say that the application (i.e. the loading specification and the supply voltage etc.), plus the quality of the devices will determine which type of circuit to be used. However, we have observed that for high supply voltages, simple circuits such as the BiNMOS or the feedback type are the most appropriate circuits.

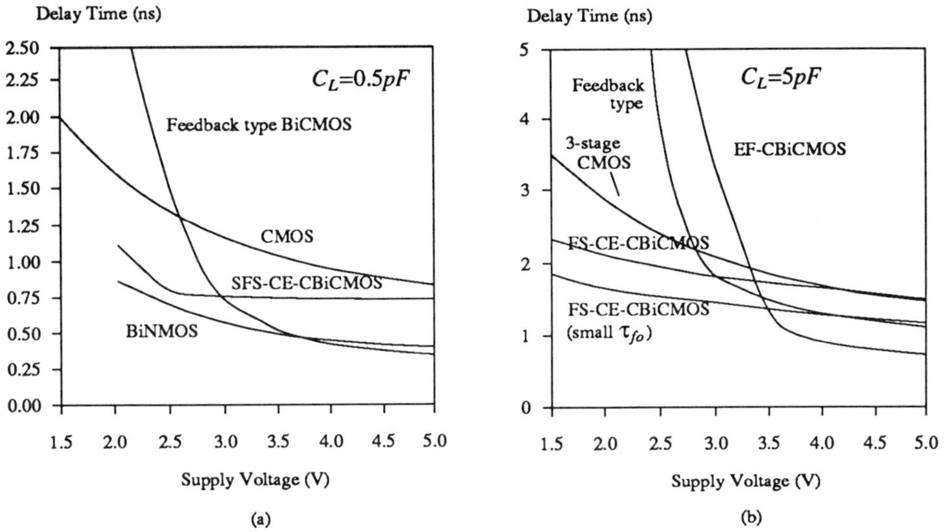

Figure 7.44 A comparison between the delay of the different candidates for small and large loading.

In the case of low supply voltage, the full-swing BiCMOS circuits using passive charging elements fail and more complex circuitry, such as the FS-CE-CBiCMOS, has to be used especially when heavy loads are to be driven.

7.4. Current Mode BiCMOS Circuits

So far we have considered BiCMOS circuits which are used for driving applications. In this section, we look at a different application, in which we exploit the availability of the CMOS and bipolar devices to integrate different logic families with different voltage swings, e.g. CMOS, ECL and BiCMOS in a single system. Combining high- and low- voltage-swing circuits allows for better power and speed mangement. Fig. 7.45 shows how a mixed system, which combines different logic families, can fill the performance gap that exists between the low-speed/low-power CMOS and high-speed/high-power ECL systems. Such a mixed system can only be realized in a BiCMOS technology.

It is expected that in future BiCMOS VLSI systems, the dense logic will be predominantly CMOS. BiCMOS buffers are to be employed for driving the heavily loaded nodes or off chip load. The ECL or ECL-like (low-voltage-swing logic) circuits can be used to realize a small portion of the logic that would be a bottle neck if implemented in CMOS. For example the delay time of the critical paths can be shortened by replacing a CMOS block by an equivalent ECL or ECL-like block. Provided that the extra power dissipation is tolerable, the overall speed of the system can be enhanced.

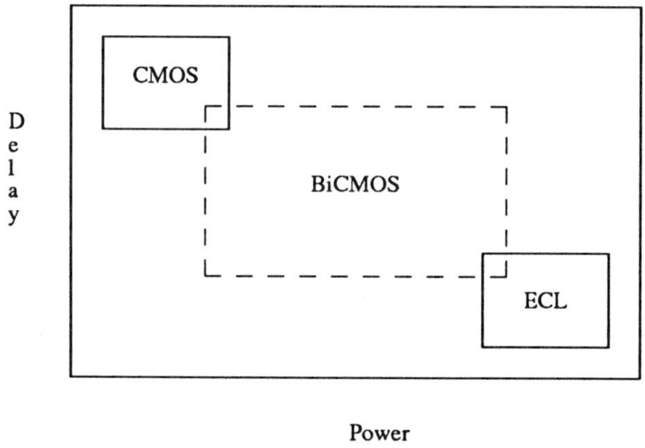

Figure 7.45 BiCMOS can be used to fill the gap between CMOS and ECL.

In the following sections, we introduce variations of the MOS/bipolar merged CML circuits.

7.4.1. The Basic MOS/Bipolar Merged CMS Cell

The first attempt of a MOS/bipolar merged current switch was reported by Heimsch et al[27]. Their proposed switch is shown in Fig. 7.46.

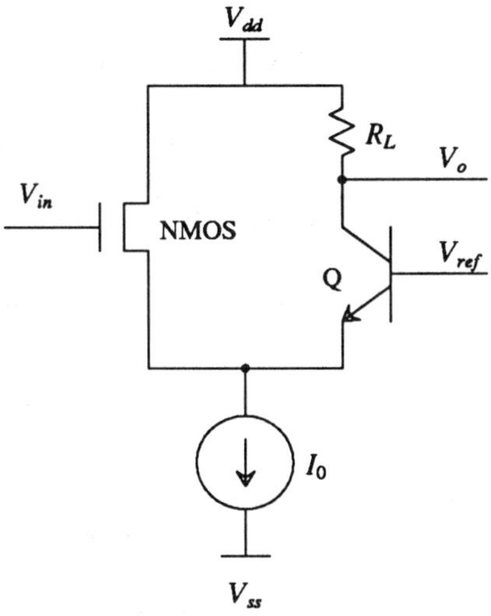

Figure 7.46 The merged MOS/bipolar current mode switch.

This CMS is basically a differential pair with a MOS transistor on one side of the differential pair and a bipolar transistor on the other side. The swing of the input signal can be as large as the CMOS swing.

The NMOS device in the circuit shown in Fig. 7.46 can be replaced by an PMOS transistor to perform an inverting function.

(i) DC Characteristics of the MOS/Bipolar CMS

As the input voltage increases above $V_{ref}-V_{BEon}+V_{TN}$ the NMOS transistor turns ON and its current increases, causing the collector current of the bipolar transistor to fall gradually, which in turn forces the output voltage to rise. The final value of the output voltage depends on the percentage of I_0 steered to the NMOS side. If the current is fully switched to the NMOS branch, the output voltage is pulled up to V_{dd}. To achieve symmetrical noise margins, the reference voltage has to be around $(V_{dd}+V_{ss})/2$ provided that $V_{TN} \approx V_{BEon}$. In general, the output voltage of the merged CML gate is given by :

$$V_o = V_{dd} - (I_0 - I_n)R_L \tag{7.66}$$

where the current through the NMOS (I_n) is given by,

$$I_n = K_N C_{ox} v_{sat} W_N (V_{in} - V_{ref} + V_{BEon} - V_{TN}) \tag{7.67}$$

Hence the noise margin, when the output is HIGH, is given by :

$$NM_H = V_{dd} - \frac{I_0}{K_N C_{ox} v_{sat} W_N} - V_{ref} - V_{TN} + V_{BEon} \tag{7.68}$$

Assuming that $V_{ref}=(V_{dd}+V_{ss})/2$ and $V_{TN}=V_{BEon}$, the noise margins reduce to :

$$NM_H = \frac{V_{dd} - V_{ss}}{2} - \frac{I_0}{K_N C_{ox} v_{sat} W_N} \tag{7.69.a}$$

$$NM_L = \frac{V_{dd} - V_{ss}}{2} \tag{7.69.b}$$

From Equation (7.66) and (7.67), one can prove that the gain dV_o/dV_{in} can be increased by increasing W_N. This means that NM_H is stretched and becomes comparable to NM_L [see Equation (7.69)]. Equation (7.69.a) also shows that W_N must be increased if I_0 is increased to keep NM_H comparable to NM_L.

(ii) Transient Behaviour of the MOS/Bipolar CMS

In this section, we discuss the transient behaviour of this circuit qualitatively. The input voltage has to rise from its LOW level to $V_{ref}-V_{BEon}+V_{TN}$ to turn the NMOS ON. As the input voltage increases, the emitter voltage of the bipolar transistor rises causing the base-emitter voltage to be reduced. Consequently, the collector current drops quickly and the output rises abruptly to V_{dd}. The capacitive coupling between the gate of the NMOS transistor and the emitter node enhances the turn OFF of the bipolar transistor especially if the input voltage swing is large. As the input signal is falling the drain current decreases gradually until the NMOS cuts off. Ideally, the current I_0 is steered to the bipolar side in a short time as in the bipolar CML circuits. However, because the voltage swing of the emitter may be large compared to the case of bipolar CML, the parasitical capacitance at the emitter node may be discharged by I_0 causing the switching of the current to the bipolar branch to be slow. As a result, the fall of the output voltage is delayed. Therefore, the capacitance of the current source has to minimized. This can be achieved by using a MOS current source instead of a bipolar source as shown in Fig. 7.47.

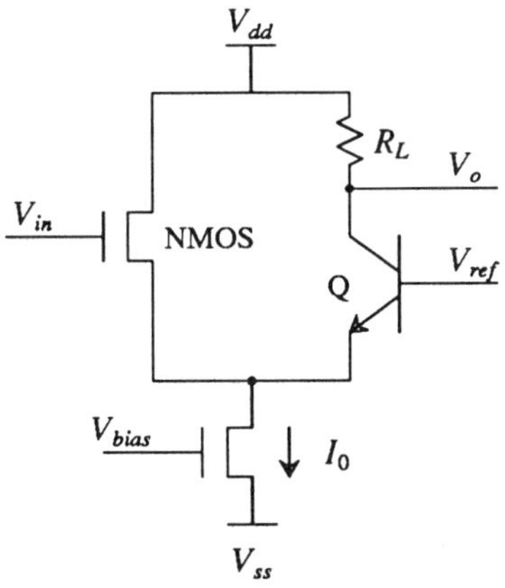

Figure 7.47 MOS/bipolar CMS with an NMOS current source.

7.4.2. MOS/Bipolar Merged CML Gates

Fig. 7.48 shows an OR gate implementation. If one or both of the NMOS devices turn ON, the collector current drops to zero and the output voltage is hence pulled up to V_{dd}. Each of the NMOS devices have to be sized such that their saturation current is equal to I_0. When both NMOS transistors are on, they share the current I_0. On the other hand if both NMOSFETs are OFF, when A and B are LOW, the output falls to $V_{dd}-I_0R_L$. An AND gate is shown in Fig. 7.49(a). The ANDing operation is performed by two series connected transistors. Only if A and B are HIGH, the current I_0 is steered away from the bipolar branch, and the output voltage rises to V_{dd}. Any other combination of A and B results in a low level at the output.

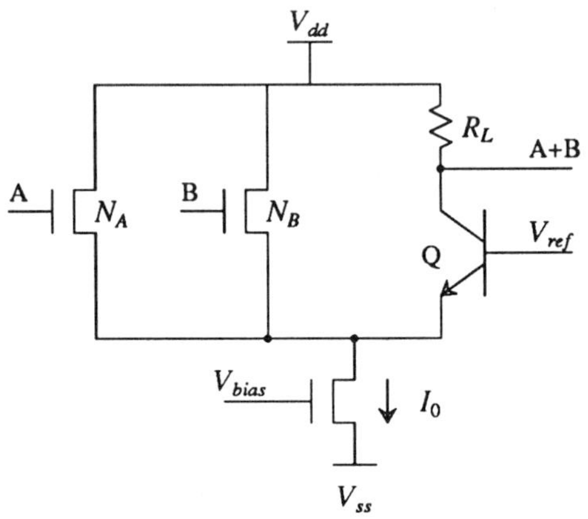

Figure 7.48 MOS/bipolar current mode OR gate.

When A='0' and the B is in a transition from '0' to '1', a glitch can be observed at the output. This is due to the capacitive coupling between the input of N_B and the emitter, which causes the bipolar transistor to turn off temporarily during the transition of B. In addition, when N_B turns ON while N_A is OFF, the parasitic capacitance at the intermediate node between N_A and N_B is discharged through the current source causing a transient decrease in the collector current and hence an enhancement of the induced glitch at the output. One way of solving this problem is to bypass the output current using a small NMOS transistor connected between the output node and V_{ss} [Fig. 7.49(b)]. So if N_A is OFF, the output node is held at its LOW level through N_A'.

In general, the ANDing operation can be realized by stacking NMOS transistors in one branch, while ORing translates into parallel NMOS branches. Any complex function that can be described by sum of products can be easily implemented by parallel branches of series connected NMOS transistors [Fig. 7.50]. To reduce the width and height of a glitch at the output node, the intermediate nodes between the stacked NMOS transistors must be discharged. This can be done by using small size

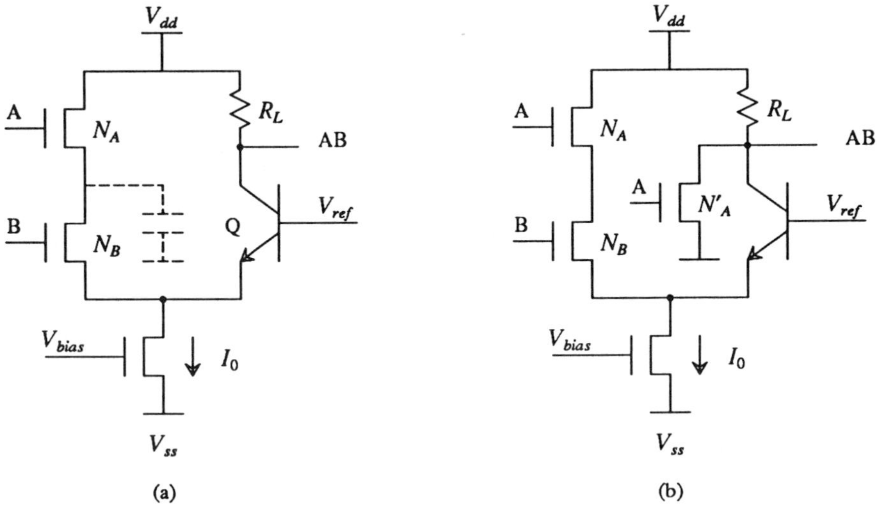

Figure 7.49 MOS/bipolar current mode AND gate.

NMOS switches as shown in Fig. 7.50. If for example, $X1$ is LOW, the first branch is OFF and the intermediate node between N_1 and N_2 can be discharged through N_1' which is driven by the complement of $X1$. The voltage of the source of N_1' must be equal to V_x, otherwise if N_2, N_3 and N_1' are ON, these three transistors act as a current source parallel to I_0. This may cause faulty logic. Therefore, the source of N_1' is connected to V_x' which has the same potential as V_x. Although these discharging elements improve the speed, they cause higher power dissipation, need more area and more complex wiring.

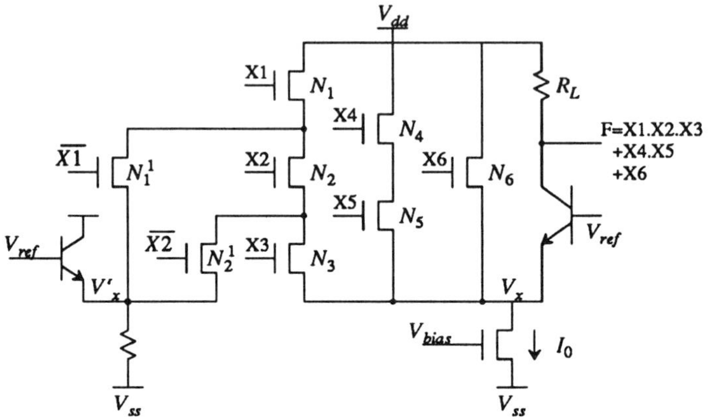

Figure 7.50 Complex merged MOS/bipolar current mode gate.

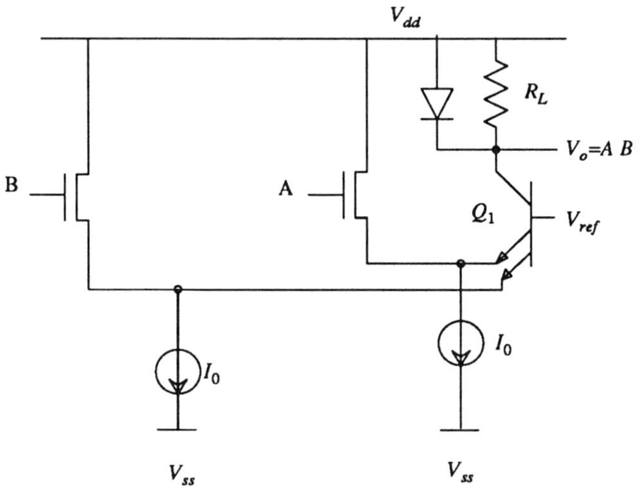

Figure 7.51 Multi-emitter MOS/bipolar merged CML AND gate.

7.4.3. Multi-Emitter MOS/Bipolar Merged CML Gates

The multi-emitter technique allows for implementing a wide variety of logic functions without the need for series gating [28]. It is therefore attractive for low voltage applications. Fig. 7.51 shows a multi-emitter MOS/bipolar merged CML AND gate. Its operation can be described as follows :

(i) If both input signals (A and B) are HIGH, the input MOS transistors are turned ON causing the CML currents to be steered away from the output branch. The output voltage is hence equal to V_{dd}.

(ii) If one of the inputs is HIGH and the other is LOW, a diode drop is developed across R_L (provided that $I_0R_L > V_{BEon}$, where V_{BEon} is the diode drop). The output level is thus $V_{dd}-V_{BEon}$, which corresponds to a logic LOW.

(iii) If both inputs are LOW, then both NMOS transistors are OFF. The output voltage is clamped at $V_{dd}-V_{BEon}$.

If the input NMOS transistors are replaced by PMOS transistors the function becomes $\bar{A}.\bar{B}$. One of the advantages of the multi-emitter technique is that it can be used for ANDing CMOS and ECL signals as shown in Fig. 7.52.

The consumed power of the multi-emitter MOS/bipolar merged CML gate can be reduced by replacing the current source of the NMOS current switch [Fig. 7.51] by a PMOS transistor as shown in Fig. 7.53. The average power of the low power multi-emitter CML is 25% less than that of the conventional multi-emitter CML of Fig. 7.51. Therefore, for the same power dissipation the low power multi-emitter CML is faster than the conventional multi-emitter CML[28].

The multi-emitter MOS/bipolar merged CML is a flexible structure that allows for implementing a variety of logic functions without increasing the circuit complexity or area. For example, interchanging the PMOS and NMOS transistors in the circuit of Fig. 7.53, the function becomes $\bar{A}.B$. Fig. 7.54 illustrates the simplicity of implementing an XOR gate using the multi-emitter approach.

It is of interest to compare the multi-emitter technique to series gating. MOS/bipolar merged series gating has been used to implement AND operations as shown in Fig. 7.55[27]. The disadvantage of series gating, in comparison to multi-emitter

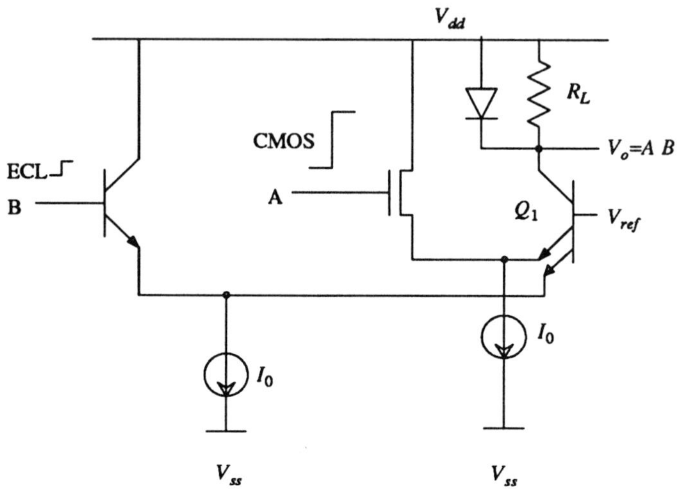

Figure 7.52 Multi-emitter MOS/bipolar merged CML AND gate with mixed input signals.

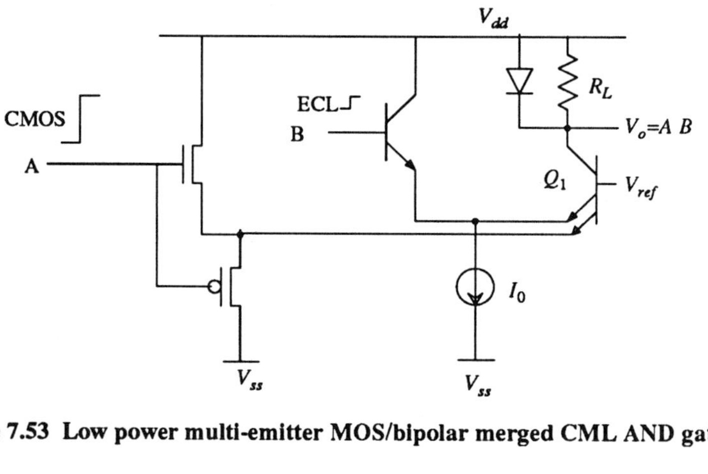

Figure 7.53 Low power multi-emitter MOS/bipolar merged CML AND gate.

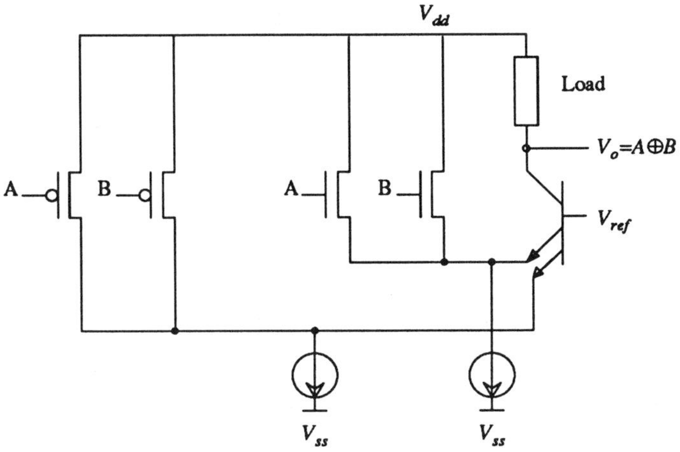

Figure 7.54 Multi-emitter MOS/bipolar merged CML XOR gate.

approach, is that the power supply scaling is limited by the number of inputs. In addition, the parasitical capacitances (collector-substrate and collector-base) of Q_1 and Q_2 in the series gate [Fig. 7.55] have a significant contribution to the total output capacitance. In the multi-emitter circuits, the parasitics of only one transistor add to the external loading. Therefore, the multi-emitter circuits are faster than series gates in the cases where the external loading is small.

7.5. Chapter Summary

In this chapter, we have demonstrated the advantage of using bipolar over MOS for driving large capacitive loads. We have shown the historical evolution of the BiCMOS gate. DC and transient analyses were derived. From the transient analysis we have derived a delay model for the conventional BiCMOS inverter. This model was used to demonstrate the impact of the device parameters on the circuit's delay. In this Chapter, we have discussed the origin of the voltage swing loss of the conventional BiCMOS circuit and its effect on the circuit performance. A variety of alternative circuit techniques for low voltage systems have been presented and compared to the conventional BiCMOS and CMOS circuits. It is shown that the

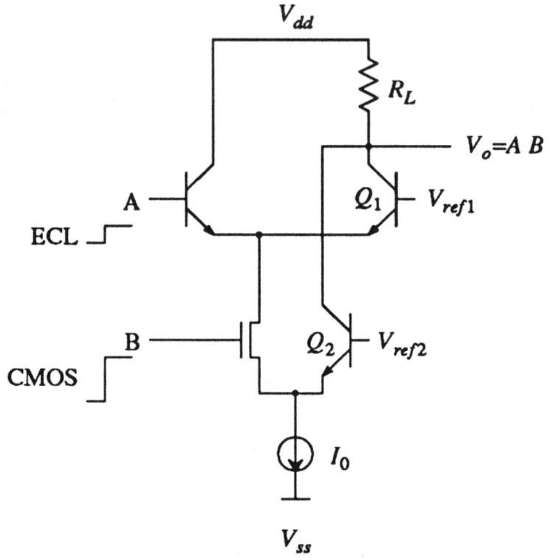

Figure 7.55 MOS/bipolar merged CML AND gate.

choice of the BiCMOS gate or driver must be based on the supply voltage and the loading conditions. In Section 7.4, we introduced different techniques for merging MOS and bipolar transistors in current mode configurations. They have the capability to process different swings without level conversion.

References

1. R.M. Warner, Jr and R.D. Schrimpf, "BJT-MOSFET Transconductance Comparisons," *IEEE Trans. on Electron Devices*, pp. 1061-1065 (May 1987).

2. W.R. Burger, C. Lage, B. Landau, M. DeLong, and J. Small, "An Advanced 0.8 Micron Complementary BiCMOS Technology for Ultra-High Speed Circuit Performance," *Bipolar Circuits and Technology Meeting Tech. Dig.*, pp. 78-81 (1990).

3. H.C. Lin, J.C. Ho, R.R. Iyer, and K. Kwong, "CMOS-Bipolar Transistor Structure," *IEEE Trans. on Electron Devices*, **Vol. ED-6, No. 11** pp. 945-951 (November 1969).

4. Y.Nishio, et al., "0.45ns 7K Hi-BiCMOS Gate Array with Configurable 3-Port 4.6K SRAM," *IEEE Custom Integrated Circuits Conference Tech. Dig.*, pp. 203-204 (1987).

5. T. Ikeda, A. Watanabe, Y. Nishio, I. Masuda, N. Tamba, M. Okada, and K. Ogiue, "High-Speed BiCMOS Technology with a Buried Twin Well Structure," *IEEE Trans. on Electron Devices*, **Vol. ED-34, No. 6** pp. 1304-1309 (June 1987).

6. H. De Los Santos and B. Hoefflinger, "Optimization and Scaling of CMOS-Bipolar Drivers for VLSI Interconnects," *IEEE Trans. on Electron Devices*, **ED-33, No. 11** pp. 1722-1730 (Nov. 1986).

7. N. Weste and K. Eshraghian, *Principles of CMOS VLSI Design*, Addison-Wesley, Reading, MA (1985).

8. E.W. Greenich and K.L. McLaughlin, "Analysis and Characterization of BiCMOS for High Speed Digital Logic," *IEEE Journal of Solid-State Circuits*, **SC-23,No 2** pp. 566-572 (April 1988).

9. G.P. Rosseel and R.W. Dutton, "Infleunce of Device Parameters on the Switching Speed of BiCMOS Buffers," *IEEE Journal of Solid-State Circuits*, **24, No 1** pp. 90-99 (Feb. 1989).

10. P.L. Heedley and R.C. Jaeger, "An Analytical Model for BiCMOS Logic Transient Response Allowing Parameter Variations," *IEEE Custom Integrated Circuits Conference Tech. Dig.*, (1989).

11. P. Raje, K. Cham and K. Saraswat, "BiCMOS Gate Performance Optimization using Unified Delay Model," *Symposium on VLSI Technology Tech. Dig.*, pp. 91-92 (1990).

12. C.H. Diaz, S-M. Kang and Y. Leblebici, "An Accurate Analytical Delay Model for BiCMOS Driver Circuits," *IEEE Trans. on Computer-Aided Design*, pp. 577-588 (1991).

13. M. Fujishima, K. Asada and T. Sugano, "Appraisal of BiCMOS from Circuit Voltage and Delay Time," *Symposium on VLSI Circuits Technology Tech. Dig.*, pp. 91-92 (1990).

14. S.H.K. Embabi, A. Bellaouar and M. I. Elmasry, "Analysis and Optimization of BiCMOS Digital Circuit Structures," *IEEE Journal of Solid-State Circuits*, pp. 676-679 (April 1991).

15. J. Warnock et al., "A 27 GHz 20 ps PNP Technology," *International Electron Devices Meeting Tech. Dig.*, (1989).

16. D. de Lang, E. Bladt, A. v.d. Goor and W. Josquin, "Integration of Vertical PNP Transistors in a Double-Polysilicon BI-CMOS Process," *Bipolar Circuit and Technology Meeting Tech. Dig.*, (1989).

17. C. Yamaguchi, Y. Kobayashi and T. Sakai, "A 7GHz PNP Transistor for Complementary Bipolar LSI," *Symposium on VLSI Technology Tech. Dig.*, pp. 39-40 (1987).

18. P-F Lu and D.D. Tang, "Simulations of Collector resistance of PNP Transistors for Complementary Bipolar Technology," *Solid State Electronics*, **32 no.8** pp. 675-678 (1989).

19. A.R. Alvarez, *BiCMOS Technology and Applications*, Kluwer Academic Pub. (1989).

20. Y. Nishio, et al., "A BiCMOS Logic Gate with Positive Feedback," *International Solid-State Circuits Conference Tech. Dig.*, pp. 116-117 (Feb. 1989).

21. H.J. Shin, "Full-Swing Logic Circuits in a Complementary BiCMOS Technology," *Symposium on VLSI Circuits Tech. Dig.*, pp. 89-90 (1990).

22. N. Rovedo et al., "Process Design for Mergerd Complementary BiCMOS," *International Electron Devices Meeting Tech. Dig.*, pp. 485-488 (1990).

23. H.J. Shin, "Performance Comparison of Driver Configurations and Full-Swing Techniques for BiCMOS Logic Circuits," *IEEE Journal of Solid-State Circuit*, **Vol. 25, No.3** pp. 863-865 (June 1990).

24. M.I. Elmasry, "Multidrain NMOS for VLSI Logic Design," *IEEE Journal of Solid-State Circuits*, pp. 779-781 (April, 1982).

25. S.H.K. Embabi, A. Bellaouar, M.I. Elmasry, and R.A. Hadaway, "New Full-Voltage-Swing BiCMOS Buffers," *IEEE Journal of Solid-State Circuits*, **Vol. SC-26** pp. 150-153 (Feb. 1991).

26. P. Raje, K. Saraswat and K. Cham, "A New BiCMOS/CMOS Gate Comparison/Design Methodology and Supply Voltage Scaling Model," *International Electron Devices Meeting Tech. Dig.*, pp. 433-436 (Dec. 1989).

27. W. Heimsch, et al., "Merged CMOS/Bipolar Current Switch Logic (MCSL)," *IEEE Journal of Solid-State Circuits*, **Vol. SC-24** pp. 1307-1311 (Oct. 1989).

28. M. Elrabaa and M.I. Elmasry, "Multi-Emitter BiCMOS CML Circuits," *IEEE Journal of Solid-State Circuits*, pp. 454-458 (March 1992).

CHAPTER 8

BiCMOS DIGITAL CIRCUIT APPLICATIONS

In this chapter we study the application of BiCMOS digital circuits in the implementation of building blocks such as adders, ALU's, memories, PLA's and standard cells. The objective is to identify some of the applications where the BiCMOS can be useful. We will consider different design styles for BiCMOS circuits.

8.1. Adders

An adder is an important component of a digital VLSI system. Many types of adders have been implemented in CMOS technology. The application of BiCMOS technology to adders has resulted in high performance circuits. In this section, we study different adder implementations for CMOS, bipolar and BiCMOS technologies.

8.1.1. Combinational Full-Adder

The Boolean functions of a full-adder are:

$$S_{out} = A \oplus B \oplus C_{in} \tag{8.1}$$

$$C_{out} = A.B + C_{in}(A+B) \tag{8.2}$$

A and B are the inputs, C_{in} is the carry input, S_{out} is the sum output, and C_{out} is the carry output. The truth table of an adder is shown in Table 8.1.

The CMOS implementation of a one-bit full-adder is shown in Fig. 8.1[1]. It requires 28 devices and has two gate delays. In this circuit the transistors controlled by the carry signal C_{in} should be placed close to the output. This will offset the body effect

Table 8.1 Adder truth table

A	B	C_{in}	S_{out}	C_{out}
0	0	0	0	0
0	1	0	1	0
1	0	0	1	0
1	1	0	0	1
0	0	1	1	0
0	1	1	0	1
1	0	1	0	1
1	1	1	1	1

problem, since the carry is the latest arriving signal.

The delay of an n-bit ripple adder is determined by the carry path which represents this critical path. Because the carry "ripples" through the n-stages as shown in Fig. 8.2, the sum of the nth bit cannot be performed until the carry C_{n-1} is evaluated. The delay of n-bit addition is given by :

$$t_a = (n-1)t_c + t_s \tag{8.3}$$

where t_c is the carry delay and t_s is the sum delay. The delay of an adder can be reduced by using low voltage swing circuits (e.g. ECL or ECL-like) for its critical path. Heimish et al[2] proposed a full adder [Fig. 8.3] which is based on the merged CMOS/bipolar current switch logic (MCSL). The inputs A and B have CMOS swings and the carry input C_{in} has an ECL-like swing. Three-level series gating is employed. The NMOS transistors are used in the first level and the PMOS transistors are used in the second level to minimize the circuit area. In principle, it is possible to use either NMOS or PMOS transistors in both levels, but with the smaller gate-source voltage and body effect, the area of the MOS transistors would be too large. In the third level, only bipolar transistors with small input swings are used to allow for fast carry propagation. The low swing of the sum outputs can be converted

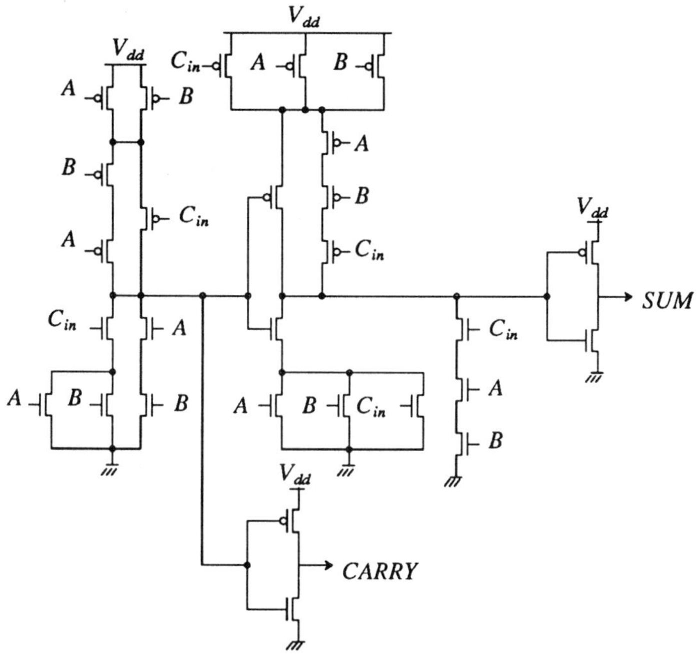

Figure 8.1 CMOS full-adder.

to CMOS swing using ECL/CMOS level converters. Although the MCSL operates at speeds comparable to those of ECL, it consumes less area and power than ECL. In comparison to CMOS, the MCSL circuits are faster but more power and area consuming. The pure ECL implementation with emitter followers is shown in Fig. 8.4. This configuration uses only two-level series gating.

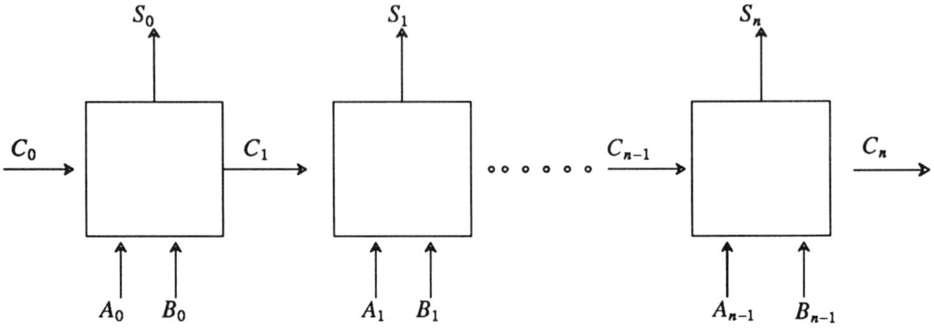

Figure 8.2 n-bit ripple adder.

8.1.2. Carry Lookahead Adder

To avoid the linear growth of the carry delay, we use the carry lookahead adders (CLA) in which the carrys can be generated in parallel. The carry of each bit is generated from the propagate and the generate signals (P_i,G_i) as well as the input carry (C_0) [1]. The propagate and the generate signals (P_i,G_i) are derived from the operands A_i and B_i.

$$G_i = A_i B_i \tag{8.4}$$

$$P_i = A_i + B_i \tag{8.5}$$

The carries of the four stages are given by :

$$C_1 = G_1 + P_1 C_0 \tag{8.6.a}$$

$$C_2 = G_2 + P_2 G_1 + P_2 P_1 C_0 \tag{8.6.b}$$

$$C_3 = G_3 + P_3 G_2 + P_3 P_2 G_1 + P_3 P_2 P_1 C_0 \tag{8.6.c}$$

$$C_4 = G_4 + P_4 G_3 + P_4 P_3 G_2 + P_4 P_3 P_2 G_1 + P_4 P_3 P_2 P_1 C_0 \tag{8.6.d}$$

Fig. 8.5 shows the block diagram of the adder. The carry generator blocks (*CLG*1 to *CLG*4) generate the carries C_1 to C_4. The sum is generated by :

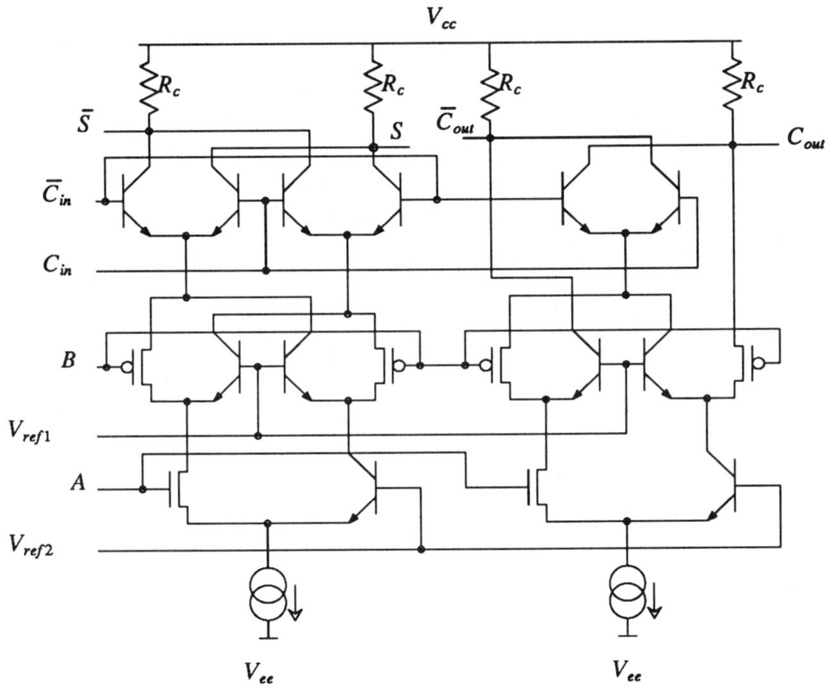

Figure 8.3 MCSL full-adder cell.

$$S_i = C_{i-1} \oplus A_i \oplus B_i \tag{8.7}$$

This function is realized by the sum generators ($SG1$ to $SG4$).

The concept of carry lookahead adders can be implemented in CMOS, ECL or BiCMOS technologies. An example of a CMOS carry lookahead adder is considered first. Fig. 8.6(a) and 8.6(b) show the first and the fourth CMOS carry lookahead circuits respectively[1]. The generate and propagate signals are generated in parallel and are fed to all carry generators with the input carry signal C_0. The carry signals are generated simultaneously. However, because the number of stacked MOS transistors increase, the delay of the fourth carry is greater than that of the first and limits the adder speed. The sum generator of the CMOS adder is shown in Fig. 8.7.

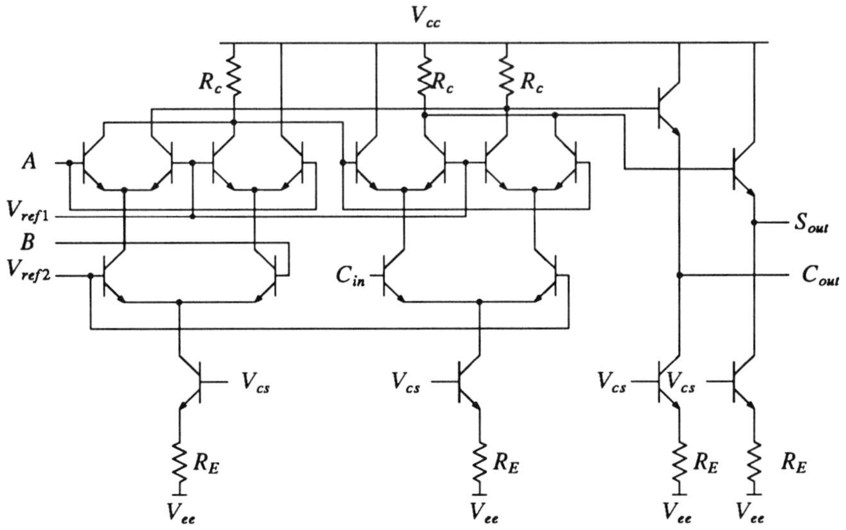

Figure 8.4 ECL full-adder.

The same circuit is used for all four bits. The CMOS implementation is slow because of the large numbers of stacked MOS transistors which represent a high resistance pull-up (pull-down) path.

The speed of the carry and sum generators can be enhanced by reducing the voltage swing of the carry signals. Merging MOS with bipolar in a current mode configuration can be used to implement a mixed-voltage-swing BiCMOS carry lookahead adder. The swing of the operands and the input carry are CMOS like, while that of the propagate, generate, carry (C_1–C_4) and sum (S_1–S_4) signals are reduced. Such an adder can be used in a predominant CMOS system.

The first carry, given by Equation (8.6.a), can be realized by using two parallel branches. One driven by G_l and the other is made of two series NMOS transistors which are driven by P_l and C_0 as shown by Fig. 8.8. If G_l or P_l and C_0 or all of them are HIGH, then the current through R_L goes to zero and the output voltage goes HIGH. It is recommended to keep the swing of the input signals as small as possible to avoid large fluctuations of V_x due to capacitive coupling. If we assume -5V power

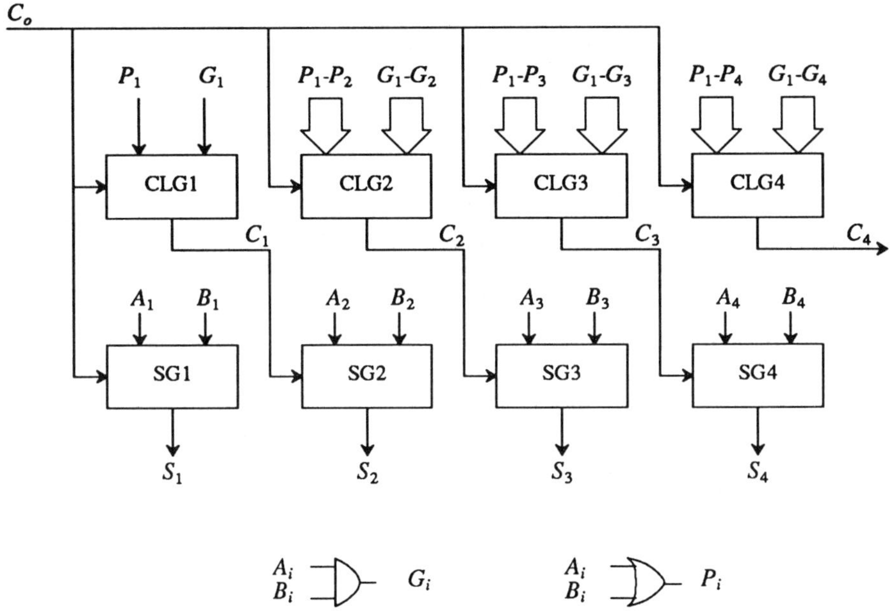

Figure 8.5 Block diagram of the carry lookahead adder.

supply and -3V reference voltage, the voltage of the emitter node (V_x) would be -3.8V. Assuming that the threshold voltage of the NMOSFET is 0.8V, the input signals can be switched between -3V and 0V. If, for example, the voltage of G_I is -3V, N_{G1} turns OFF. When G_I goes to zero, the gate to source voltage is around 3.8V. N_{G1} is hence saturated and the current I_{ss} is steered to N_{G1}. By limiting the swing of the input signals, the time required to turn the NMOS devices ON is reduced. The intermediate node between N_{P_I} and N_{C0} is discharged through the N_1 - N_2 branch, which is only conducting when N_{G1} is OFF.

The same technique used for the CLG1 can be extended to implement CLG2 to CLG4 as shown in Fig. 8.9.

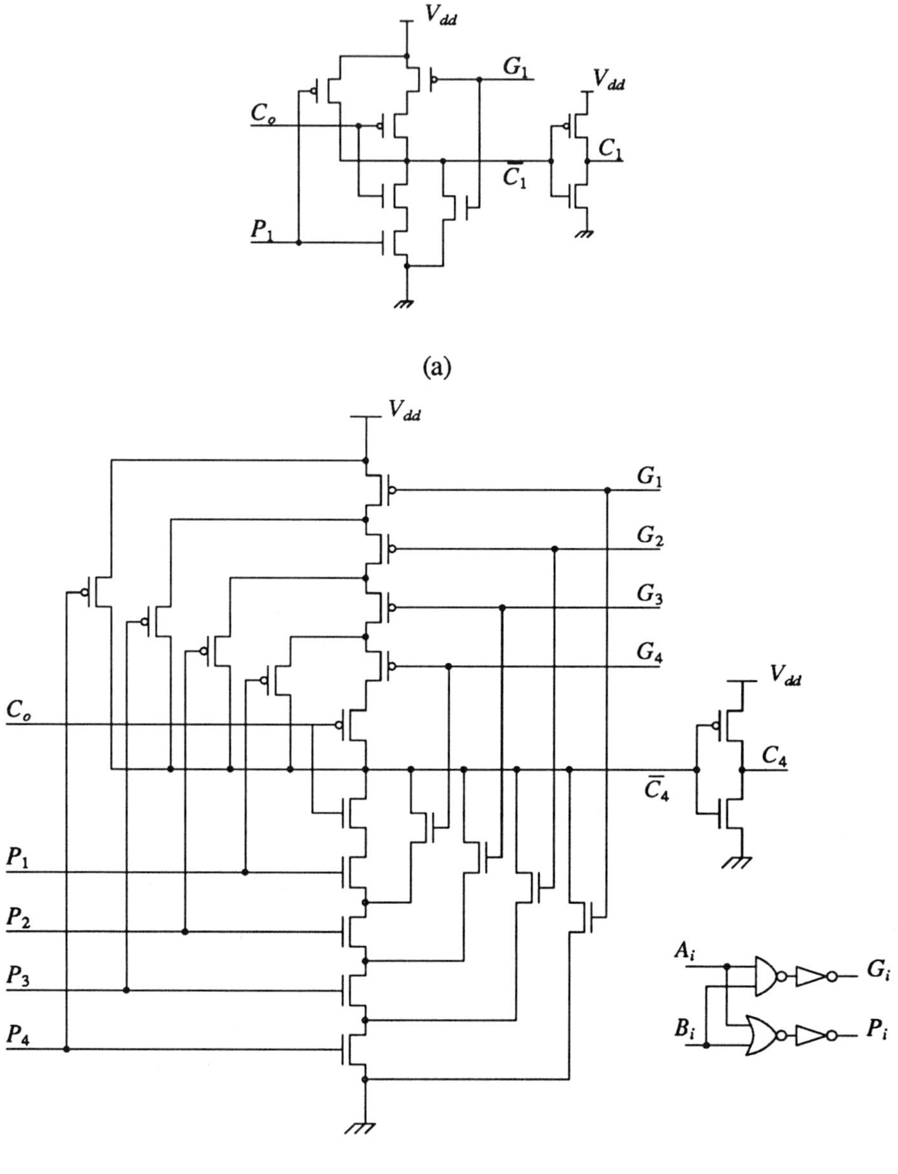

Figure 8.6 CMOS carry lookahead generator: (a) First bit; (b) Fourth bit.

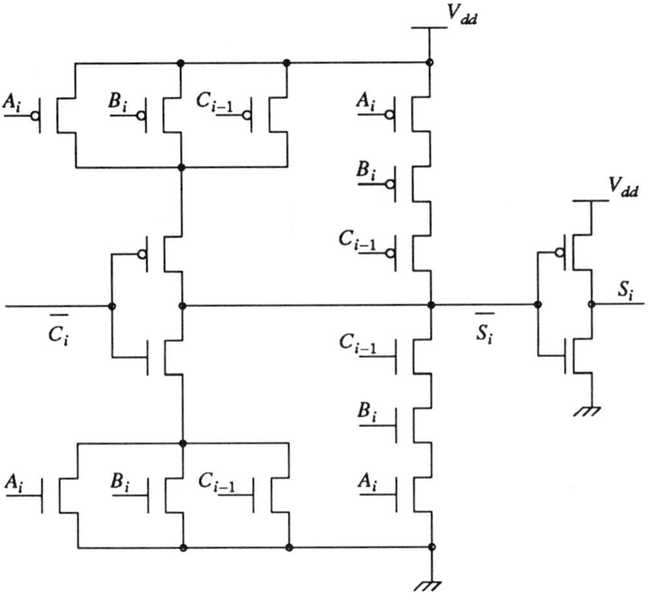

Figure 8.7 CMOS sum generator.

As the number of series connected transistors grows, the width of the stacked devices has to be increased which leads to large parasitic capacitances at the intermediate nodes. To achieve maximum speed, these capacitances must be discharged using additional circuits as shown in Fig. 8.8 and 8.9(a).

The speed of the CLG4 is limited because it has five stacked NMOS devices. The fourth carry generator may be replaced by a CLG2 type where P_1, P_2, G_1 and G_2 are replaced by P_3, P_4, G_3 and G_4 respectively as shown in Fig. 8.10. The input carry, C_0, is replaced by C_2 after amplifying its swing (-3V to 0V).

The sum is equal to the XOR of A_i and B_i when the input carry is LOW and is equal to the XNOR of A_i and B_i when the carry is HIGH. Fig. 8.11(a) shows the schematic of the sum generator. The circuit consists of a XNOR gate which drives two current switches. The first, which consists of Q_2 and Q_3, is active when the carry input signal $C_{i-1,s}$ is HIGH and acts as a noninverting buffer. Therefore, when $C_{i-1,s}$ goes

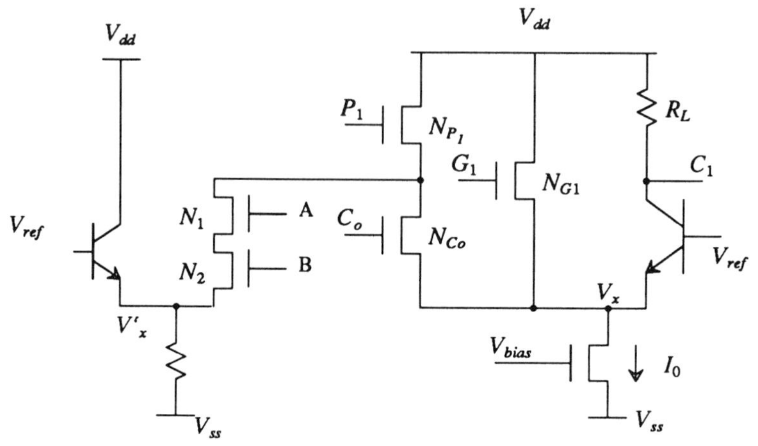

Figure 8.8 Circuit schematic of the first carry generator.

HIGH, the sum (S_i) follows the output of the XNOR gate. When it is LOW, the second current switch (Q_4 and Q_5) becomes active, and the sum generated is the complement of the output of the XNOR gate. The carry signal which is generated by the CLGs has to be shifted down by three diode voltage drops, using the emitter follower shown in Fig. 8.11(b), before being fed to the sum generators. The levels of C_{i-1} are shifted from -1V and 0V to about -3.5 and -2.5 respectively. Hence the reference voltage, at the base of Q_6 [Fig. 8.11(a)], should be -3V which is the same voltage used for the CLGs, AND, OR and XNOR gates. The reference voltage, V_{refs}, is determined by the voltage swing of the output of the XNOR gate, which is typically around 500mV. Thus, V_{refs} has to be set to -250mV.

The first sum generator, which receives an input carry with CMOS levels, uses the same circuit shown in Fig. 8.11, except that the transistor Q_{C_i} is replaced by a NMOSFET.

Comparing the MOS/bipolar (BiCMOS) adder (4-bit) with the CMOS and the ECL carry lookahead adders in terms of speed, power and area, shows that the BiCMOS adder is three times faster than its CMOS counterpart and offers half the speed of

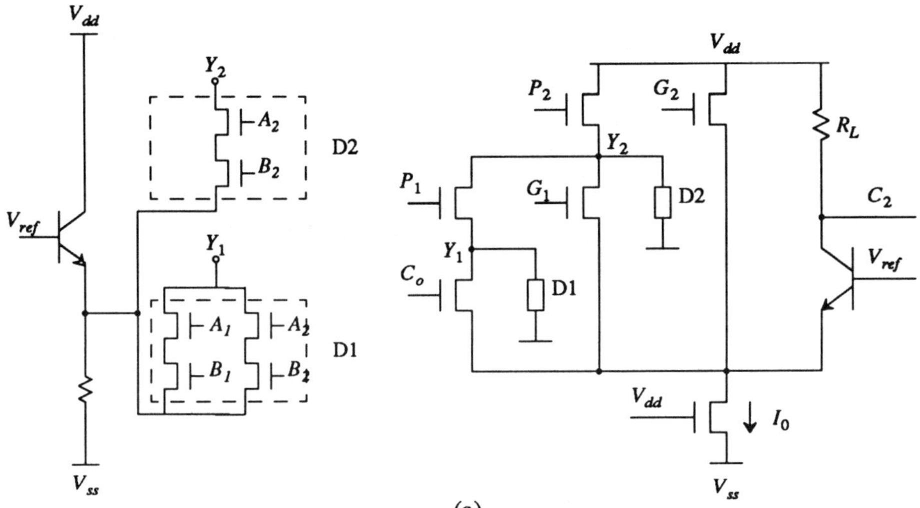

(a)

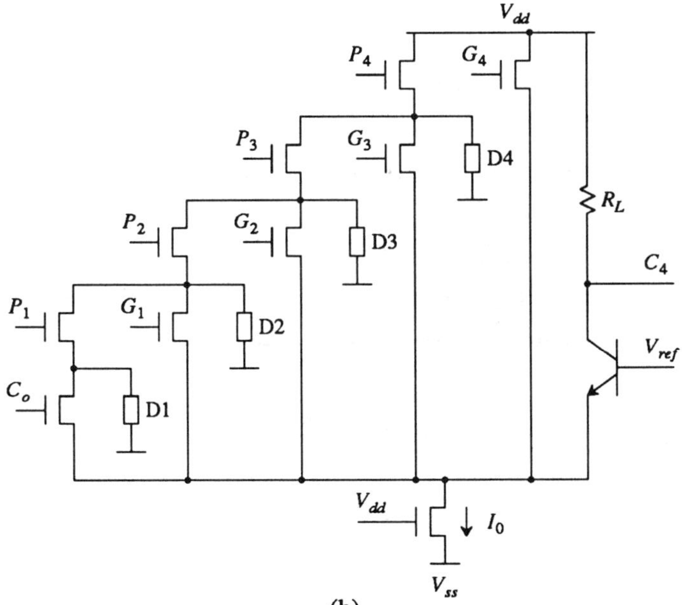

(b)

Figure 8.9 The second and fourth carry generators in (a) and (b) respectively.

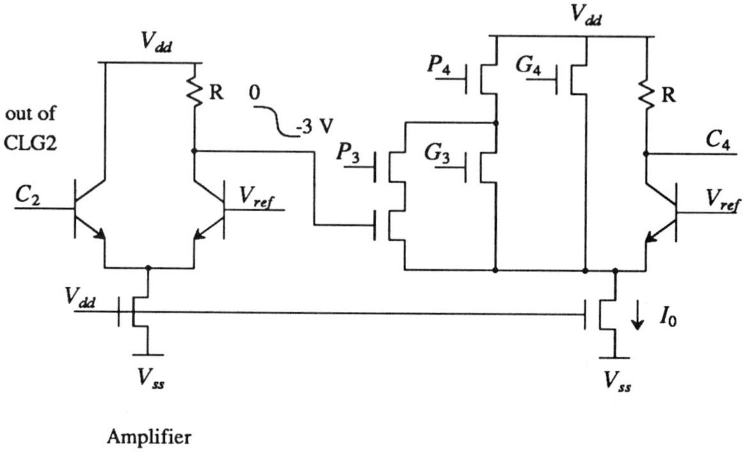

Figure 8.10 The second stage of the fourth ripple carry generator.

ECL. The BiCMOS adder consumes more power than the CMOS adder. The area of the BiCMOS adder is only 20% larger than that of the CMOS adder and half of that of the ECL adder. Fig. 8.12 summarizes the results obtained from the simulation of the CMOS, ECL and BiCMOS implementations of a 4-bit adder. These results support the principle of using BiCMOS to fill the gap between CMOS and ECL. The arrows in the Fig. 8.12 indicate that the performance of the BiCMOS adder can shift up or down towards CMOS or ECL, respectively, by trading off speed for power or vice versa.

It is interesting to compare the speed of the same three adder implementations as a function of the number of bits. In order to do that let us consider 8-bit adders, which are typically made up of two 4-bit carry lookahead adders as shown in Fig. 8.13. The simulations indicate that the delays are approximately proportional to the number of bits. Hence, the ratios of the delay times of the three implementations remain almost constant. This means that as the number of bits increases the difference between the delay of the CMOS and BiCMOS is increasing as demonstrated by Fig. 8.14. In other words, the speed of the BiCMOS adder is less sensitive to the number of bits than the CMOS adder.

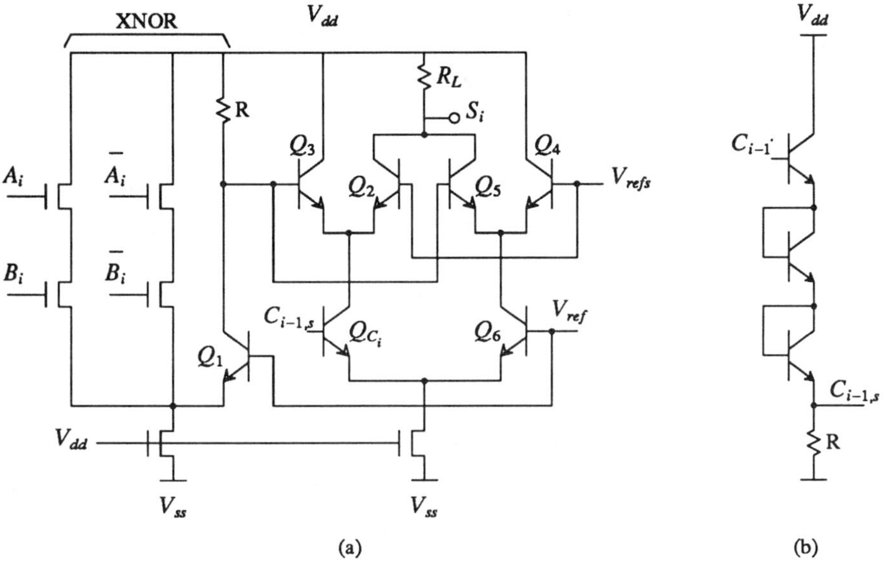

Figure 8.11 The circuit schematic of the sum generator.

8.1.3. Carry Propagation Adder

Fig. 8.15(a) shows a fast dynamic 4-bit carry propagation adder. The circuit operates as follows. When ϕ is LOW, the nodes $\overline{C_i}$ (i=0,..3) are precharged and the output C_4 is precharged LOW. When ϕ goes HIGH, the precharge transistors turn OFF and the pull-down NMOS transistors are ON to allow for the C_4 to be evaluated. The critical path in this circuit is defined by six series NMOS transistors with the input signals ϕ, C_0, P_1, P_2, P_3, and P_4 as shown in Fig. 8.15(a). One way to speed up the delay of this path is to use a small voltage swing with a sense circuit. Fig. 8.15(b) shows the 4-bit BiCMOS carry propagation circuit, which uses a common emitter sense circuit [3,4]. When ϕ_1 is LOW and ϕ_2 is HIGH (precharge cycle), the MOS transistor M_1 is conducting and supplies the base current required to turn Q_1 ON. As a result C_4 is pulled down to the LOW level. The transistor M_3 acts as clamp to prevent Q_1 from saturation. When ϕ_1 is HIGH and ϕ_2 is LOW (evaluation cycle), the transistor M_1 is OFF. The MOS carry lookahead circuit determines the

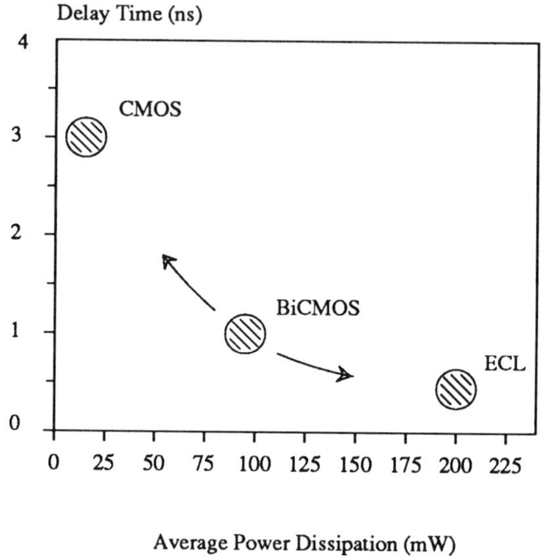

Figure 8.12 A comparison between CMOS, ECL and BiCMOS adders. The areas of the CMOS, BiCMOS and ECL implementations are 44, 51 and 97 mil^2.

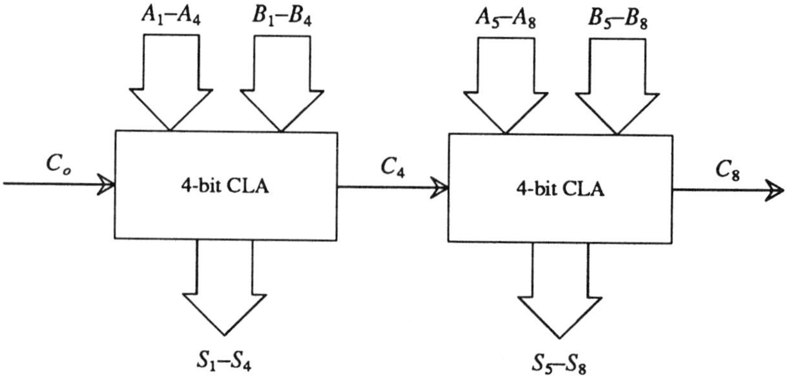

Figure 8.13 The organization of an 8-bit adder of 2 4-bit carry lookahead adders.

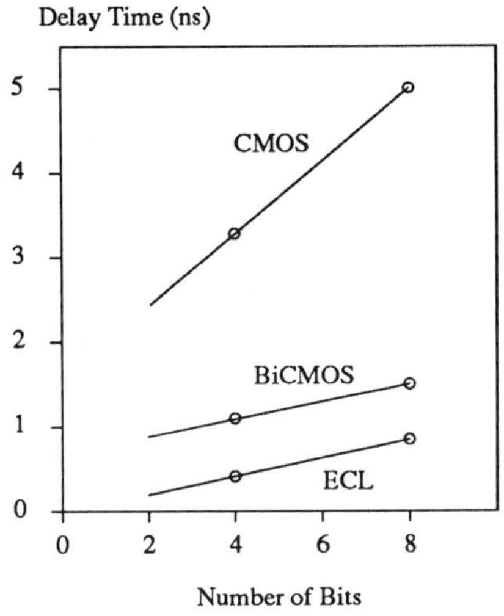

Figure 8.14 A comparison between the simulated delay of the CMOS, ECL and BiCMOS adders as function of the number of bits.

voltage of node A. If A is LOW, the output C_4 is pulled up to the HIGH level. If A is not LOW, the transistor Q_1 will be conducting such that the ouput node C_4 stays at its LOW level. The voltage swing of node A is within $0.8V$. This low voltage swing results in high speed operation of the carry propagation circuit. The signal levels of C_4 are CMOS compatible and hence C_4 can be connected directly to C_0 of the next stage of the carry propagation circuit.

The common emitter sense circuit, shown in Fig. 8.15(b), has low noise immunity. A small change in the base voltage swing causes large variations in the collector current and the output voltage level. Another problem with this circuit is the possibility that the bipolar transistor could saturate. The BiCMOS sense circuit shown in Fig. 8.16 was developed to overcome these two problems by using a current mirror[5].

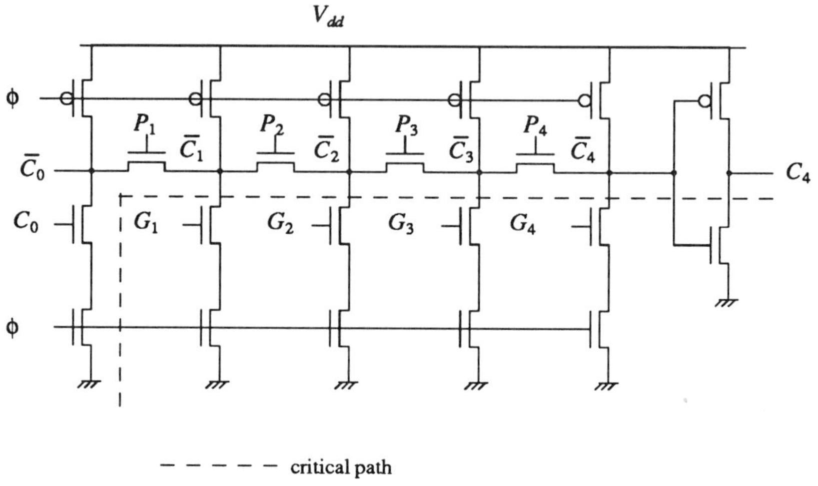

(a)

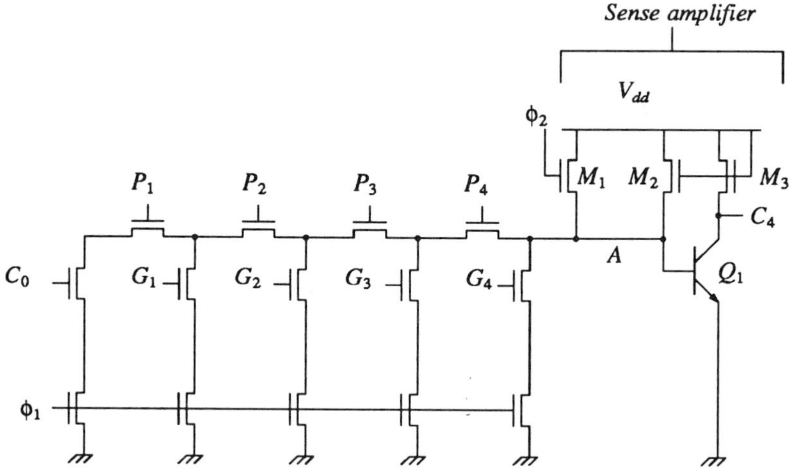

(b)

Figure 8.15 4-bit carry propagation: (a) CMOS; (b) BiCMOS.

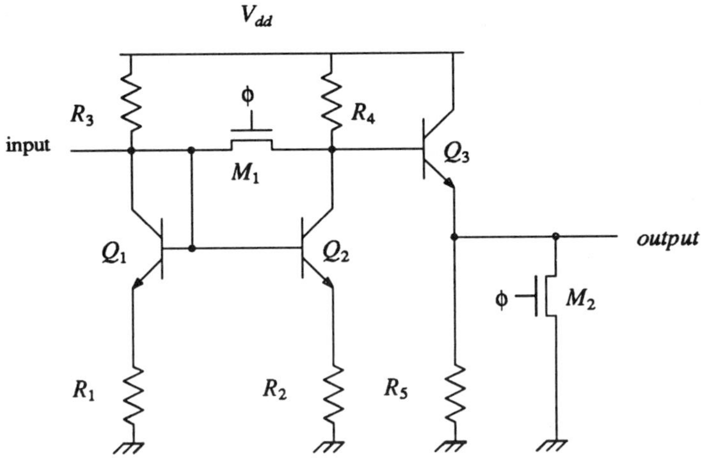

Figure 8.16 BiCMOS sensing circuit.

During the precharging phase (ϕ=1), M_1 and M_2 are turned ON. The output voltage is pulled down by M_2. At the same time, M_1 pulls the bases of Q_1 and Q_2 through R_4, which is typically smaller than R_3 to reduce the precharging time. During the evaluation phase (ϕ=0), M_1 and M_2 are OFF. When the input is HIGH, the collector currents of Q_1 and Q_2 increase causing the drop across R_4 to increase which, in turn, causes Q_3 to turn OFF, and the output voltage remains at its LOW level (as in the precharge phase). If, however, the input is pulled down, the collector currents of Q_1 and Q_2 decrease. Hence, the collector of Q_2 is pulled up to turn Q_3 ON and charge the output node to its HIGH level.

The emitter follower (Q_3 and R_5) is used so that the HIGH and LOW levels of the output are CMOS compatible and to isolate the current mirror from the load. To reduce the delay time of this sense circuit the RC constant at the collector of Q_2 must be minimized. Therefore, Q_2 must be of minimum size. To achieve the current amplification the ratio $R_1/R_2 > 1$. Another advantage of R_1 and R_2 is the reduction of the sensitivity of the collector currents of Q_1 and Q_2 to input voltage fluctuations.

This can be explained by the negative feedback effect due to R_1 and R_2. An incremental increase in the input voltage is followed by an increase in the collector currents and the voltage drops across R_1 and R_2. This results in a reduction in the base-emitter voltage of Q_1 and Q_2, which causes the collector currents to decrease and eventually remain stable. Hence, the circuit of Figure 8.16 is more immune to input noise than the common emitter sense circuit.

8.2. Multiplier

Multipliers are used in data processors as well as in digital signal processors to perform binary multiplication. They can be classified into two categories. The first one includes serial multipliers and the second includes parallel multipliers. Many algorithms have been developed for parallel multiplication. In this section we discuss a simple parallel array multiplier for unsigned binary numbers and the application of BiCMOS to speed-up the multiplication time. The reader can consult the references[6, 7] for more details.

As an example for parallel array multiplication, we choose the 4-by-4 unsigned array multiplier. Consider two unsigned numbers $X=x_{n-1}\cdots x_1x_0$ and $Y=y_{n-1}\cdots y_1y_0$

$$X = \sum_{i=0}^{i=n-1} x_i 2^i$$

$$Y = \sum_{i=0}^{i=n-1} y_i 2^i$$

The product $P=P_{2n-1}\cdots P_1P_0$, which results from multiplying the multiplicand X by the multiplier Y, can be written in the following form:

$$P = \sum_{i=0}^{i=n-1}\sum_{j=0}^{j=n-1} (x_i b_j) 2^{i+j}$$

Each of the partial product terms $P_k=x_iy_j$ is called summand. The $n \times n$ summands are generated in parallel by n^2 AND gates. Fig. 8.17 shows the implementation of this parallel array. It needs $n(n-1)$ full adders, n^2 AND gates and n half adders.

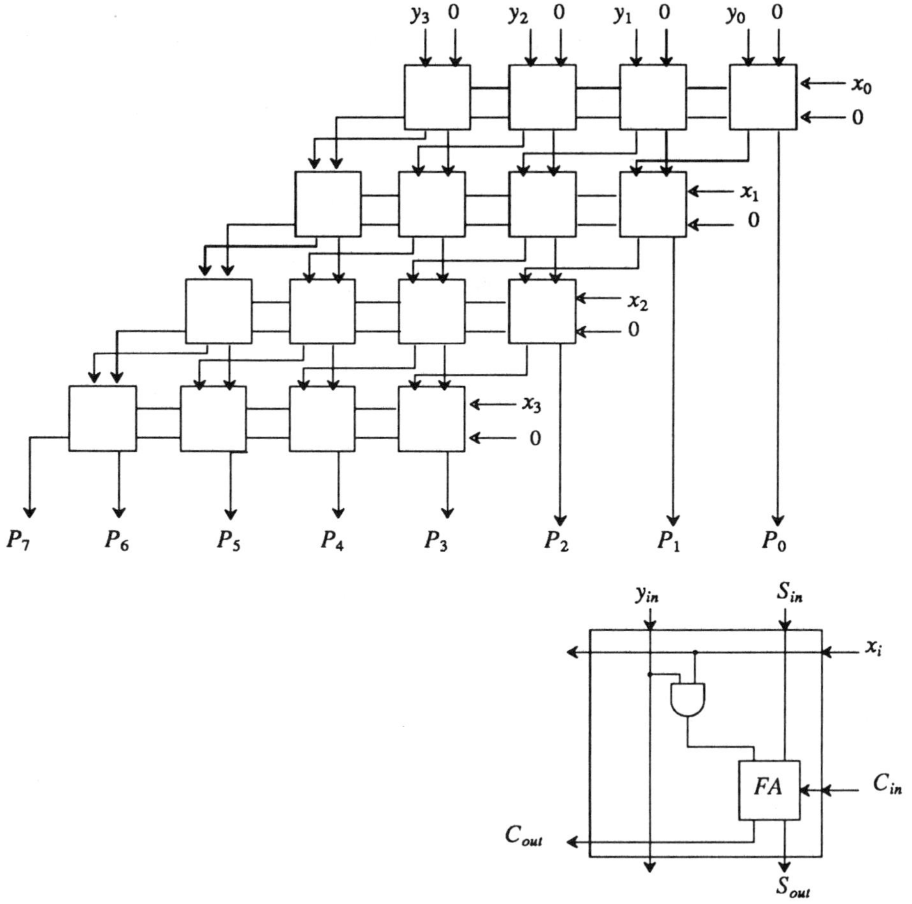

Figure 8.17 4 × 4 bit unsigned array multiplier.

In general the multiplicand and the multiplier are stored in the registers. Because each digit x_i or y_j is connected to a great number of gates, BiCMOS buffers for the X and Y can be used to drive the cell arrays as illustrated in Fig. 8.18.

In practice, modified Booth algorithm is used to halve the number of rows at the expense of additional hardware.

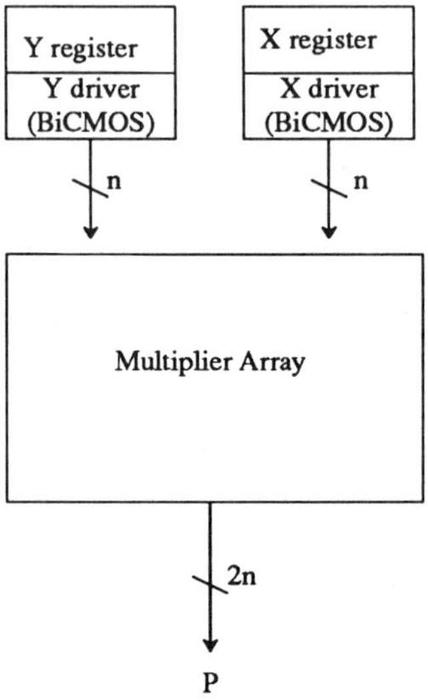

Figure 8.18 Block diagram of a n × n array multiplier using BiCMOS drivers.

8.3. Random Access Memory

In this section, we investigate the use of BiCMOS for memory applications such as read only memories, static and dynamic memories.

8.3.1. Read Only Memory

The read only memory (ROM) circuit is often used in VLSI processors as microcode controller. In this case, the ROM contains the microprogram instructions. The read-out cycle of the ROM limits the speed of the processor. The time required for the read-out cycle can be reduced by using BiCMOS circuits[8]. A typical micro-ROM size is 2k words of 64 bits. Fig. 8.19 shows a simple ROM circuit configuration. BiCMOS circuits are used in the ROM for the decoder driver, the precharge circuit and the sense circuit, to improve the speed of operation.

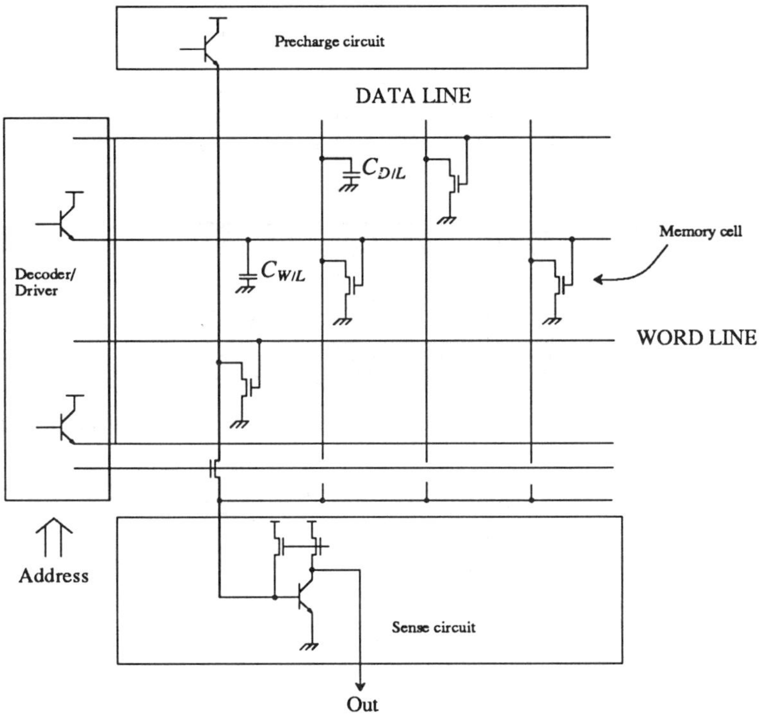

Figure 8.19 BiCMOS ROM memory.

(a) ROM Memory Cell

The memory cell is the same as the one used in CMOS technology. It is implemented with only one transistor which stores the state indefinitely. The W/L of the memory cell NMOS transistor is small to obtain dense layout. Fig. 8.20 illustrates the layout of the memory cell. Metal1 and metal2 are used to reduce the resistance of lines, hence the delay caused by this latter is reduced.

(b) Address Decoder

The decoder is necessary to select one of 2^n words by decoding the n-bit address. An example of a 2-bit row decoder truth table is shown in Table 8.2. N-bit address can be decoded using a set of CMOS NOR or NAND gates in both static and dynamic

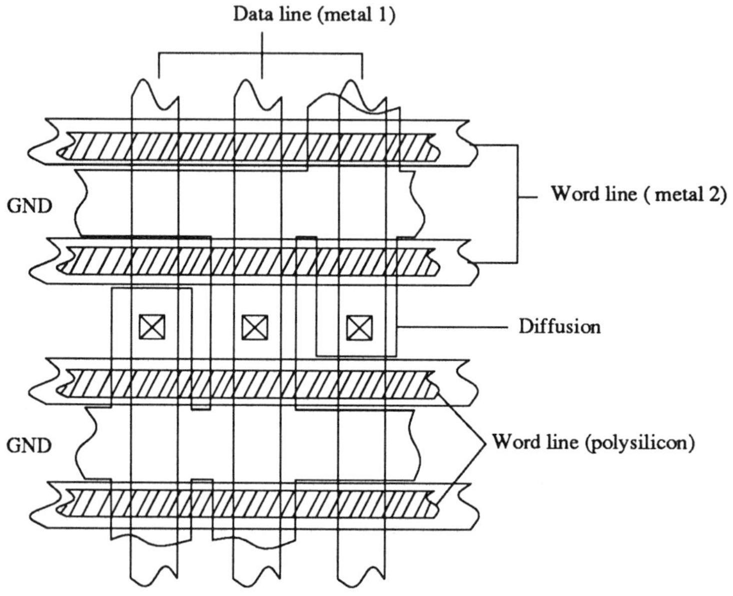

Figure 8.20 Layout of a ROM memory cell.

styles. Decoders can be single-level circuits, but if the bit number of the address word is HIGH, then multi-level circuits are used. A CMOS implementation of a single-level NOR-based decoder is shown in Fig. 8.21. It is pseudo NMOS-like style which is usually preferred, because the NMOS devices are in parallel. Compared to complementary CMOS, the input capacitance is cut in half. The address for the different NOR gates is provided with both signals and their complements. This pseudo NMOS-like implementation has the disadvantage of consuming DC power. To optimize the area and speed of decoders, multi-level circuits are more often used. This implementation simplifies the pitch matching with the ROM cell. A two-level decoder using the combination of CMOS NAND and NOR gates is shown in Fig. 8.22.

Table 8.2 Decoder truth table

A_0	A_1	O_0	O_1	O_2	O_3
0	0	1	0	0	0
0	1	0	1	0	0
1	0	0	0	1	0
1	1	0	0	0	1

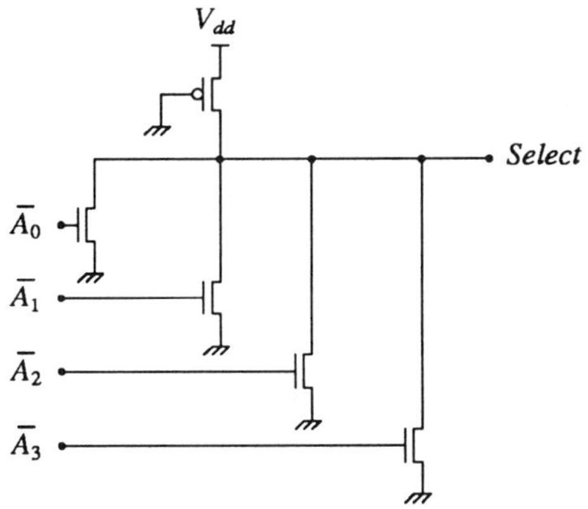

Figure 8.21 Pseudo NMOS-like CMOS row decoder with NOR gate (single level).

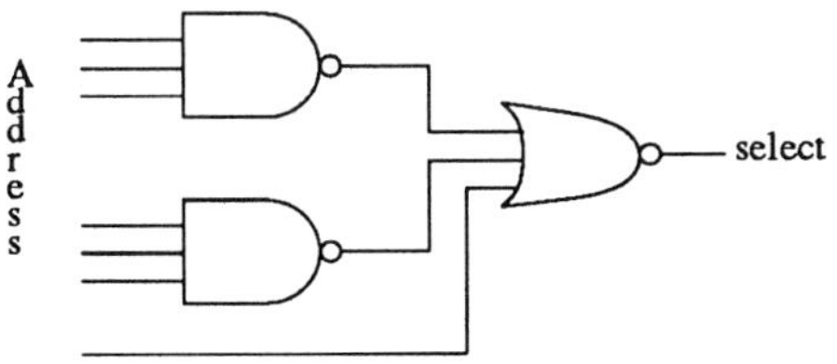

Figure 8.22 Two level decoder.

(c) Word-Line Driver

The parasitic capacitances of the word-line of a memory array is large (few pF). The delay of the output of the CMOS decoder can be reduced by using conventional BiCMOS drivers. However, since the LOW level of the output of the driver is V_{BEon}, the MOS transistor of the memory cell may not turn OFF causing high power dissipation and possible malfunction. One solution to this problem is to use a BiNMOS driver as shown in Fig. 8.23.

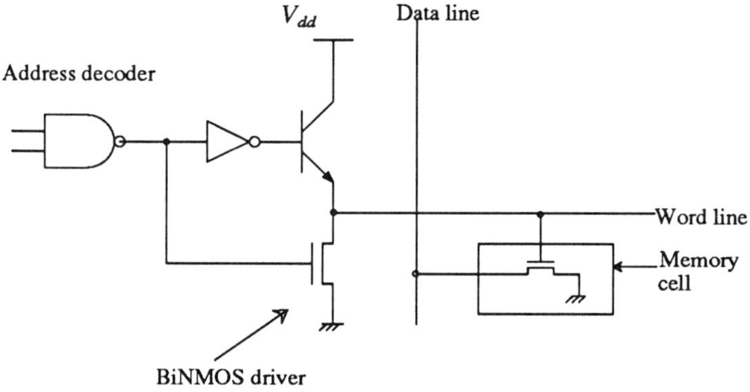

Figure 8.23 Static BiCMOS word line driver.

(d) Data-line Sense Amplifier

The MOS transistor of the memory cell is slow in pulling the heavily loaded data-line to ground. Hence, a sense amplifier, which detects small variations on the data-line, is needed. Fig. 8.24 shows the sensing circuit of the ROM. This circuit is similar to that used for the carry propagation sense amplifier. During the read operation the data-line is selected. The MOS transistor of the memory cell shunts the base current of the bipolar transistor so that the output goes HIGH. The voltage swing in the data-line is within the base-emitter voltage of the bipolar transistor, hence the circuit is fast.

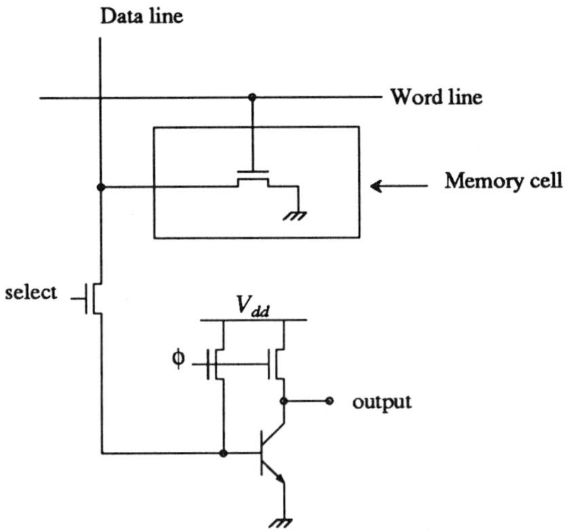

Figure 8.24 Data-line sense circuit.

(e) Data-line Precharge Circuit

A bipolar transistor is used to precharge the data-line at a voltage equal to V_{BEon} as mentioned before. Fig. 8.25 shows the precharge circuit with a Darlington connection. The base of Q_2 is clamped by three diodes to assure a V_{BEon} voltage on the data-line. If ϕ_1 is HIGH, the data-line goes HIGH only if selected, otherwise the data-line stays LOW.

8.3.2. Static RAM

The ECL static RAMs (SRAMs) developed with bipolar technology have serious problems in power dissipation and production yield when their bit-capacity is increased. On the other hand, they have low access time compared to CMOS SRAMs. But CMOS allows high density memory. In recent years BiCMOS SRAMs composed of ECL interface bipolar circuits and high-density MOS memory cell arrays, have resulted in high-speed high-density memory[9-14]. A typical SRAM architecture is shown in Fig. 8.26. The memory array contains the memory

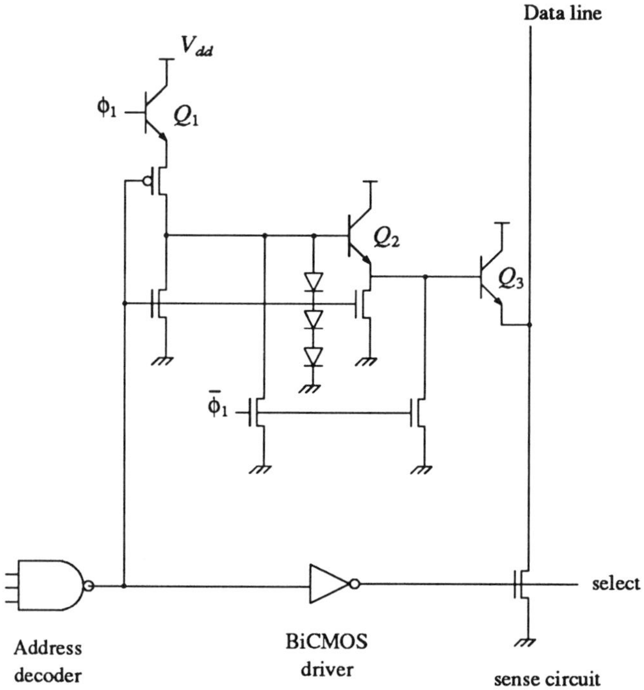

Figure 8.25 BiCMOS precharge circuit.

cells which are readable and writable. In general, the memory array is implemented in CMOS. The address is available in ECL levels. The row decoder (X-decoder) selects 1 out of 2^n rows, while the column decoder (Y-decoder) selects 2^l out of 2^m columns. Sense amplifiers detect small voltage variations on the memory bit-line which reduces the reading time. An example of SRAM organization is illustrated in Fig. 8.27. It has 64-kbit, organized in 16-k word-by-4-bit. The 16-k words are in the form of 512 rows-by-32 columns. Hence, 14-bit address word is needed to decode 512 rows (9-bit) and 32 columns (5-bit). This configuration reduces the length of bit and control lines. Moreover, many row and column decoders can be used. Fig. 8.28 shows the critical path for read access in conventional SRAMs.

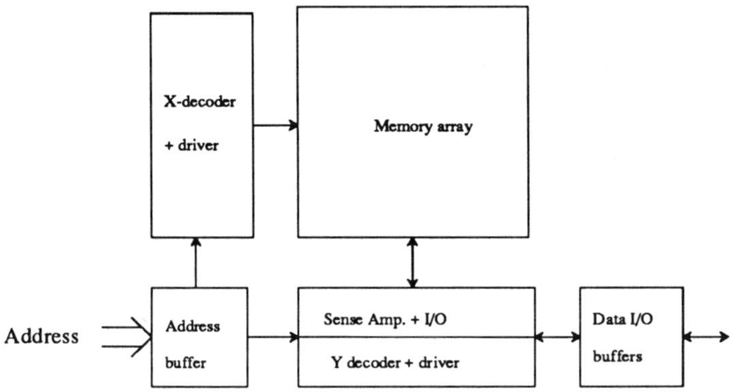

Figure 8.26 A typical SRAM architecture.

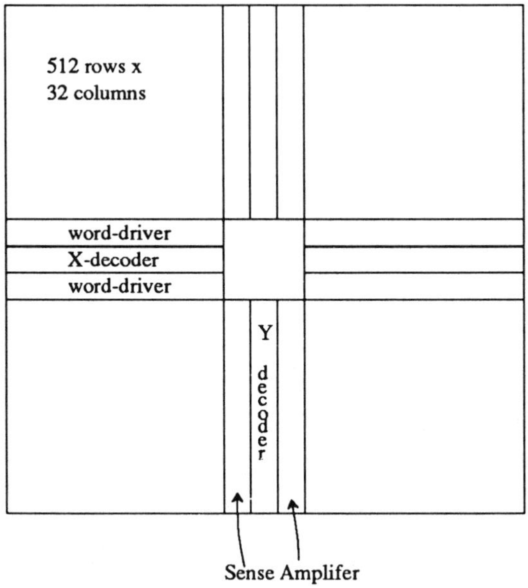

Figure 8.27 Floor plan of a 64 kbit SRAM.

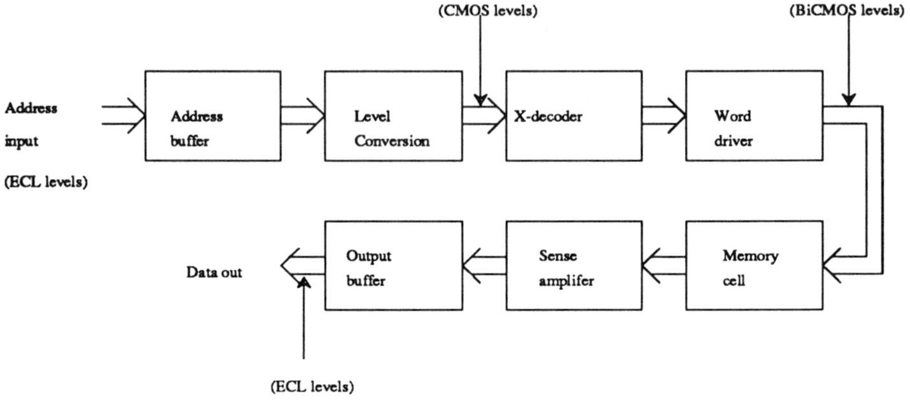

Figure 8.28 The critical path for read access in a conventional BiCMOS SRAM with ECL I/O.

In the following subsections, we present the different circuits and cells of a BiCMOS SRAM with ECL I/O.

(a) Static RAM Cells

The BiCMOS SRAMs use CMOS for cell memory array, as it permits high capacity and low power dissipation. There are two types of CMOS memory cells (MC). Fig. 8.29 shows the CMOS MC with PMOS transistor as a load. The cell has six transistors in the form of two inverters crosscoupled with two pass-transistors connected to two complementary bit-lines BL and $\overline{BL}$. The pass-transistors are controlled by the signal WL (word-line).

During the read cycle, the bit-lines are held HIGH (precharged). Assume that a "0" is stored at node A and a "1" is stored at node B. When the cell is selected; i.e. WL set to "1", $\overline{BL}$ is discharged through $MN1$ and $MN3$.

To write in the cell, one of the bit-lines is pulled LOW and the other HIGH and then the cell is selected by WL. Assume that $\overline{BL}$ is set to "0" while initially a "1" is stored at node A ("0" at B). $MN1$ and $MP1$ should be sized such that node A is pulled down enough to turn $MP2$ ON. This in turn causes node B to be pulled up. The

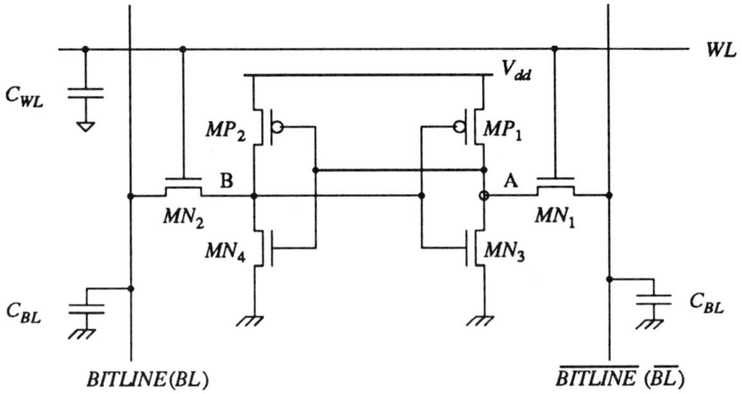

Figure 8.29 PMOS load CMOS memory cell.

crosscoupled inverter pair has a high gain to cause the nodes *A* and *B* to switch to opposite voltages.

Another memory cell configuration is shown in Fig. 8.30. This cell is similar to the one presented before, except that the PMOS pull-up devices are replaced by high-resistance polysilicon loads. Typical values of resistance are in the $G\Omega$ range. This value is necessary to allow low standby current ($<0.1nA$). The memory cell area can be up to 50% smaller than the CMOS six-transistor MC, because the two polysilicon resistances can be formed on top of the two NMOS transistors. Moreover, it was shown that a polysilicon load cell has an excellent soft-error immunity and low-voltage operation capability.

(b) ECL Input Buffers

The availability of bipolar transistors in BiCMOS allows for using ECL circuits for high-speed operation in conjunction with high-density MOS memory cell arrays. Fig. 8.31 shows a typical input buffer, which receives ECL levels (i.e. H=-0.9V, L=-1.7V, for V_{cc}=0, and V_{ee}=−5.2V). These levels are shifted by $2V_{BEon}$ drops. The swing is then amplified using a differential pair. The resulting levels ($V_{cc}-V_{BEon}$ to $V_{cc}-V_{BEon}-R_cI$) are then converted to CMOS using a level shifter such as that shown

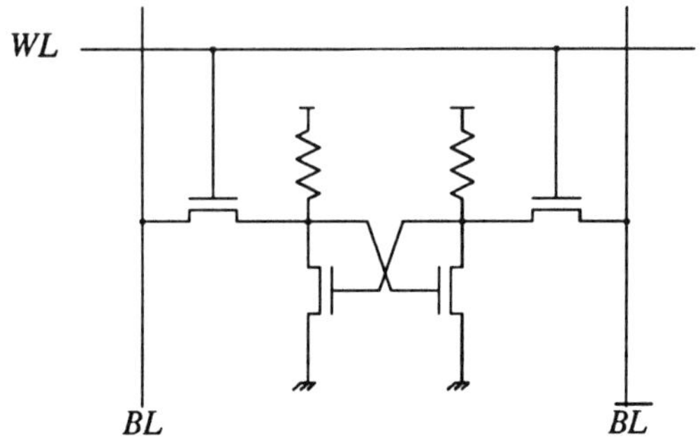

Figure 8.30 High-resistance polysilicon load memory cell.

in Fig. 8.32. The diode D_2 [Fig. 8.31] shifts the input levels of the translator down by V_{BEon} to increase the transconductance of the PMOS devices of the CMOS shifter. To avoid the saturation of Q_1 and Q_2 the ECL levels are shifted by $2V_{BEon}$ using the emitter follower with D_1. The input stage has a diode D_1. The current sources of the input buffer are composed of a bipolar transistor and series-resistance.

(c) Decoders and Word Drivers

In conventional SRAM the decoding is performed with CMOS logic as shown in Fig. 8.33. At the output of the CMOS decoder circuit, a BiCMOS buffer is used to drive the word-line. The decoding circuit can be implemented in pseudo-NMOS.

Another alternative, an ECL decoder or a predecoder can be placed between the ECL input address buffer and the level translator to speed up the critical path of read cycle. But this type of decoder is limited by the permissible power dissipation. Fig. 8.34 shows an ECL OR/NOR decoding circuit [15]. The swing at the ouput of the emitter follower is designed to be high for the CMOS level translator. Other predecoder configurations are proposed in [13, 14].

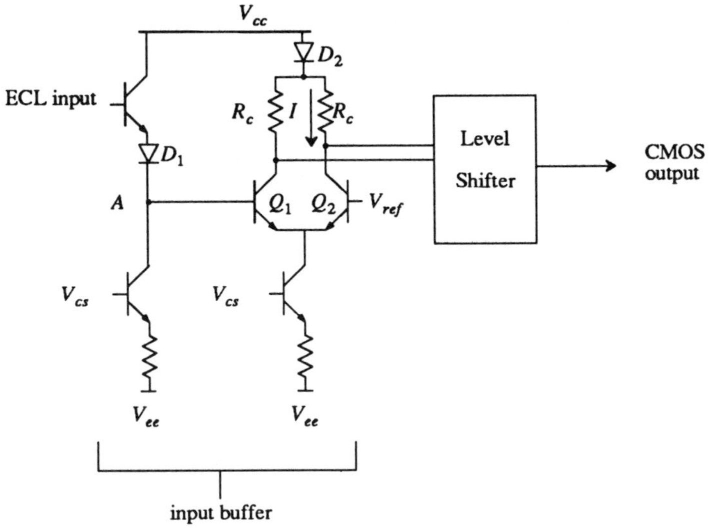

Figure 8.31 ECL input buffer circuit with level translator.

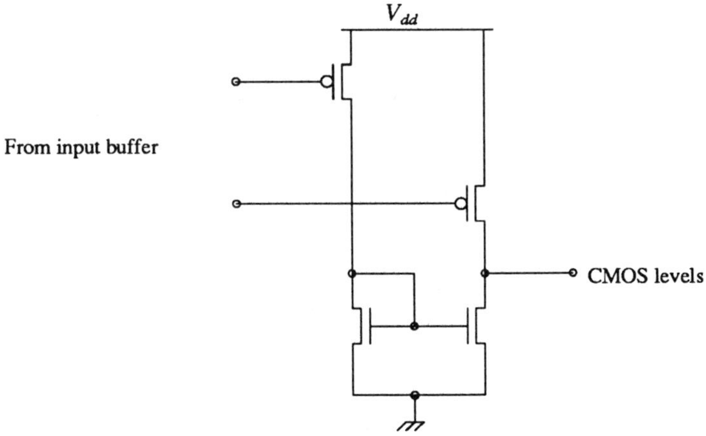

Figure 8.32 An example of a CMOS level translator.

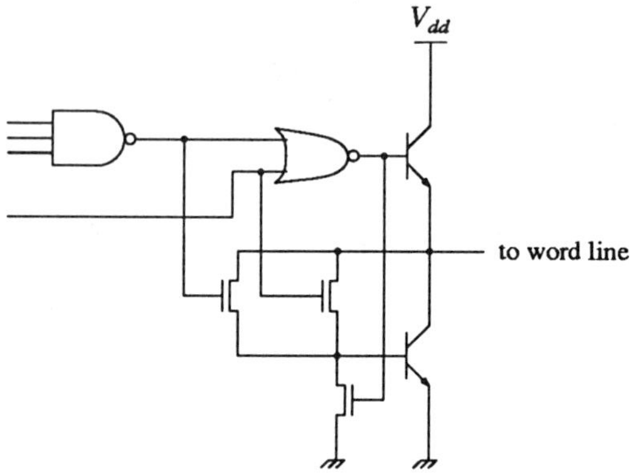

Figure 8.33 CMOS X-Decoder circuit with word line driver.

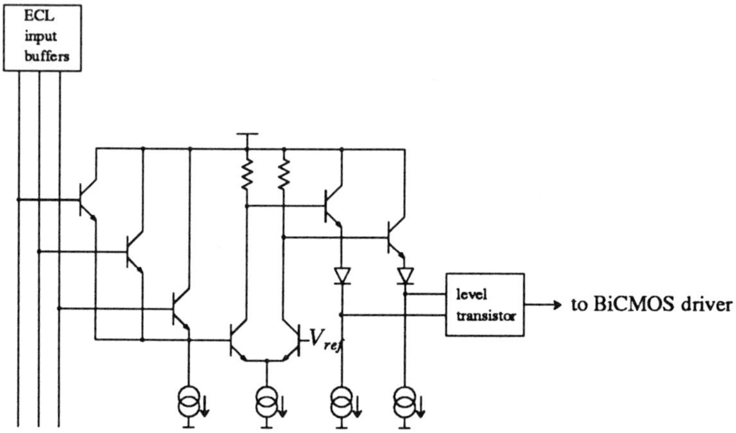

Figure 8.34 ECL wordline decoder.

(d) Read/Write Circuitry

Fig. 8.35 illustrates the peripheral circuit and the sense amplifier (SA) of bit-lines for common data lines (CDL) for read/write operation. A variable impedance bit-line load is employed in this circuit. It realizes fast sensing in the read cycle and a short write pulse width in the write cycle[16]. For fast sensing, the voltage swing of the bit-line should be small. To achieve this, the load impedance should be low. On the other hand, to obtain a short write pulse width, the load impedance of the bit-lines should be high. During the read cycle all four PMOS transistors $MP1$, $MP2$, $MP3$, and $MP4$ are turned ON. The bit-lines are switched into a low-impedance state so that the voltage swing of the bit-lines is limited to a small value (e.g., $50mV$). The signal WL is selected, and the small-signal from the memory cell is transfered to a local sense amplifier through the transfer gates $TG1$ and $TG2$. This small swing is sensed by a bipolar differential amplifier. The sense amplifier is composed of two stages. The first one is a simple differential pair formed by the bipolar transistors Q_3 and Q_4. The emitter followers formed by the transistors Q_1 and Q_2 and the diodes D_1 and D_2 shift the bit-lines voltage by $2V_{BEon}$ to prevent the sense amplifier bipolar transistors to enter saturation. The first stage is selected by the signal SAS. The collector lines drive the second stage of the sense amplifier. This latter is a dual cascode sense amplifier which uses a small voltage swing at the capacitively loaded collector-lines. Hence the delay of sensing is small. During the write cycle, the two PMOS $MP1$ and $MP2$ are turned OFF, and the bit-lines loads go to a high-impedance state. Therefore, the bit-line can be quickly discharged to V_{ee} by a write control circuit. In high speed SRAMs, write recovery time is an important component of the write cycle time. It is defined as the time necessary to recover from the write state to the read state after the $\overline{WE}$ signal rises. To obtain fast recovery time after the write cycle, a signal ϕ_{EQ} is sent to the common data-lines equalization circuit when the rising edge of $\overline{WE}$ is detected. The common data-lines are equalized to ground level.

Another bit-line technique[11] using $V_{BL}=V_{cc}-V_{BEon}$ is shown in Fig. 8.36. During the read operation, the bit-lines are set to $V_{cc}-V_{BEon}$ level by the bipolar transistors Q_1 and Q_2, which are controlled by the read/write control circuit. Two additional

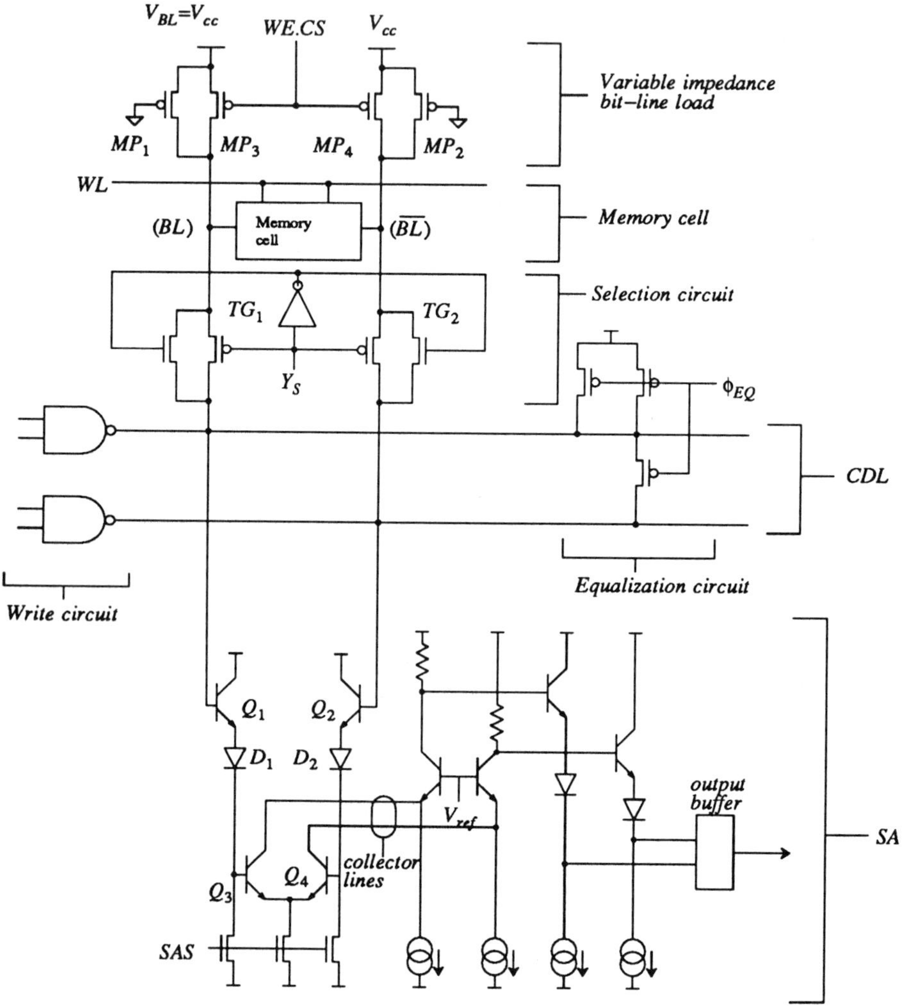

Figure 8.35 Read/Write circuitry in an SRAM.

diffusion resistances (R) are added in the bit-lines to control the bit-line's voltage swing. The bit-line voltages are sensed by a differential-pair called individual column amplification sensing (ICAS) sense amplifier. The collector of the bipolar transistors in the ICAS sense amplifier are connected to the local data-line pairs. Each local data-line pair is connected to l ICAS sense amplifiers (i.e., l=32), to minimize the capacitive load of the data-line. During the read operation, the input data D_{in} and $\overline{D_{in}}$ are LOW and the nodes A and B are set to V_{cc}. The ICAS sense amplifier is selected by the column select signal ($COLSEL$). The memory cell is selected and the bit-line voltage swing is detected by the sense amplifier. The unselected bit-lines have the signal $COLSEL$ LOW. Hence they are equalized by the PMOS transistor MP. During the write cycle, data input D_{in} and $\overline{D_{in}}$ are set to their values. If $\overline{D_{in}}$ is HIGH, $COLSEL$ is selected causing $MN1$ and $MN3$ to turn ON. The input data, through the read/write control circuit, turn the bipolar transistor Q_2 OFF while the bit-line is pulled down through $MN1$ and $MN3$. The data is stored in the memory cell when the WL signal is selected.

(e) Output Buffer

The small swing signal generated by the sense amplifiers can be fed directly to the standard ECL output buffer shown in Fig. 8.37.

Fig. 8.38 demonstrates the improvement in the SRAM access time when BiCMOS is used[17]. The delay of each stage of the BiCMOS SRAM is compared with its CMOS counterpart. Using BiCMOS reduces the delay of each stage except for the output buffer, however, the data sensing has benefited the most. The saving in the decoding delay time increases with the memory size. By using ECL predecoding schemes the decoding time can be cut down by a factor of two[18].

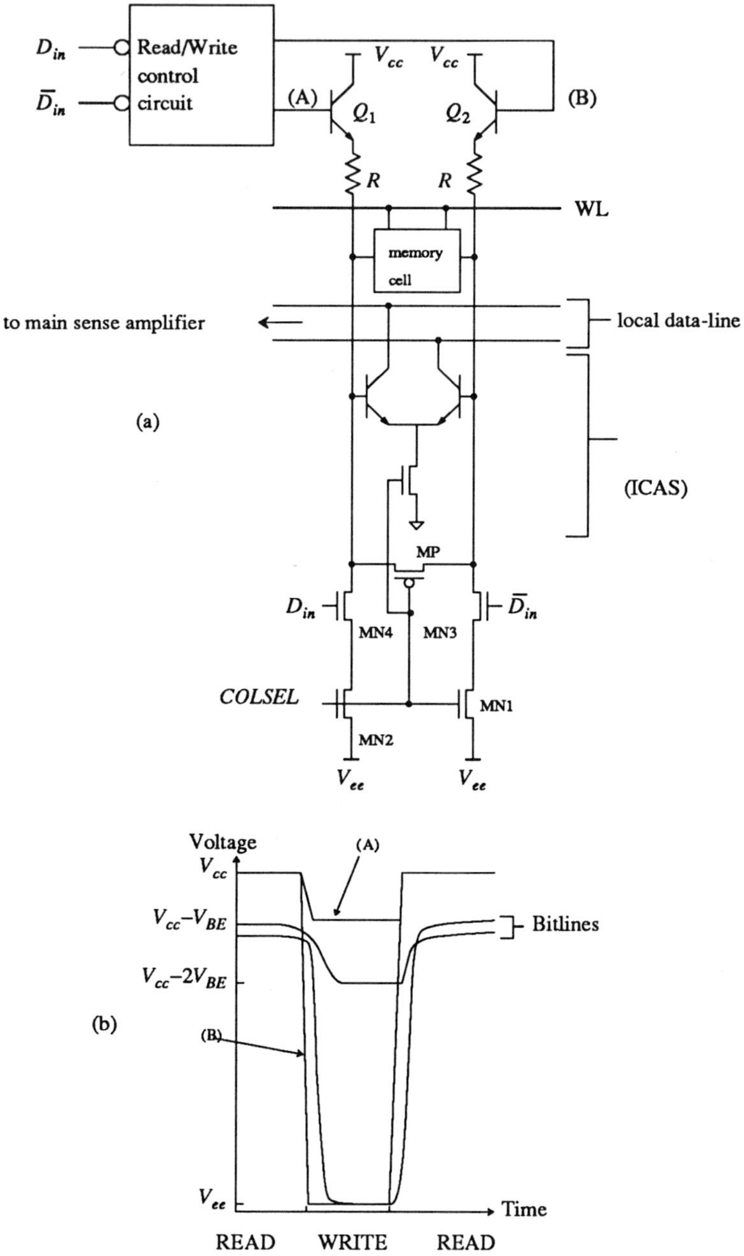

Figure 8.36 (a) Memory peripherals and sense amplifier; (b) Bit-line operation.

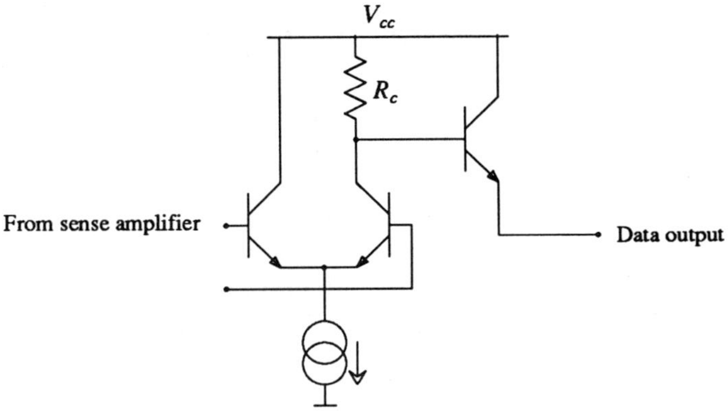

Figure 8.37 ECL output buffer.

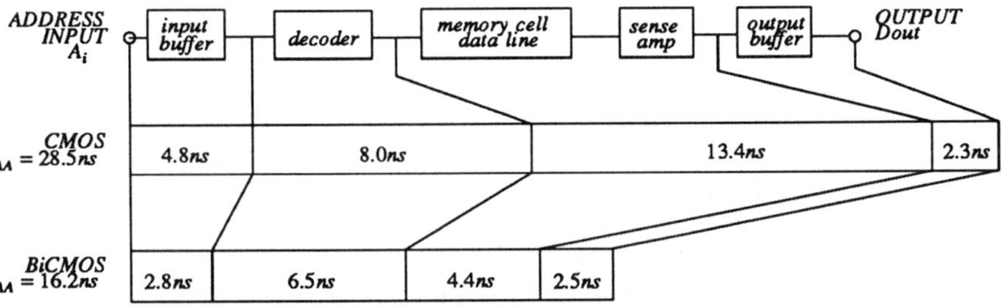

Figure 8.38 Comparison between the access times of CMOS and BiCMOS SRAMs.

8.3.3. Dynamic RAM

CMOS Dynamic RAMs (DRAMs) have reached high bit-density up to 16 Mb. But still the access time has to be improved to meet the requirements of high speed microprocessors systems and computers. The application of BiCMOS technology to DRAMs has allowed higher speed of operation[19-24].

(a) DRAM Memory Cell

CMOS DRAMs with three-transistor and four-transistor cells were used in 1- and 4-kb generations. One-transistor (1T) cell offers smaller chip size and low cost.

A schematic of a 1T DRAM cell is illustrated in Fig. 8.39. The charge is stored in capacitor C. To prevent loss of the stored information, the capacitor must be refreshed within a specific time with special circuitry. The bit line has a capacity C_{BL}, which can be reduced by dividing each bit line into k sections. Typical values for the storage and the bit-line capacitors are 40 fF and 350 fF respectively. The ratio C/C_{BL} is very important for the sensing operation. During the read operation the bit line voltage changes by

$$\Delta V_{BL} = (V_{MC}-V_{BL})\frac{C}{C_{BL}+C}$$

where $(V_{MC}-V_{BL})$ is the difference between the memory cell voltage and the bit line voltage before the selection of the cell. A typical value of the difference is 2.5 V. Hence, ΔV_{BL} is in the order of $100mV$. This small voltage change of the bit-line requires powerful sensing circuits.

(b) Read/Write Circuitry

Fig. 8.40 illustrates the different circuits for read, write and precharge functions. The read operation is performed as follows. The signal WL is selected by the X decoder. The HIGH level of the word-line voltage has to be greater than 5V to increase the stored charge in the memory cell. Both bit lines (BL and $\overline{BL}$) are precharged to V_p which is equal to $V_{dd}/2$. The selected memory cell is connected to one bit line. The bit line differential voltage signal is transferred to the differential read circuitry through two clocked CMOS inverters. Each inverter isolates the bit line from the base of the bipolar. Otherwise, the read information on the floating bit-line is

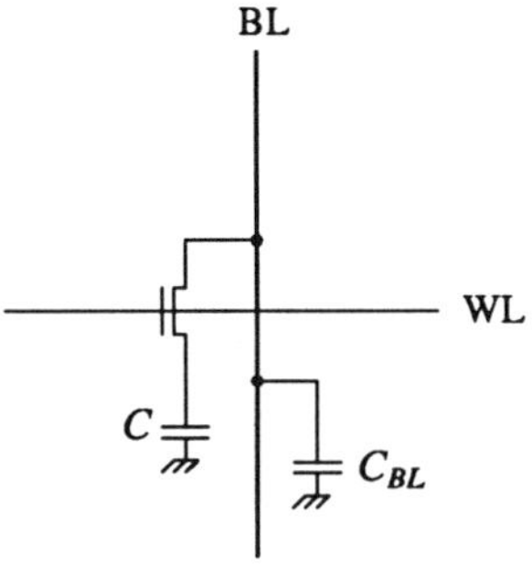

Figure 8.39 One-transistor memory cell.

quickly destroyed due to the base current. The read circuit has a bipolar differential amplifier, which has a high drivability and sensitivity for small signals compared to a CMOS amplifier. One disadvantage of the bipolar sensing circuit is the extra area compared to CMOS. Once a read circuit is selected by *YR*, a differential current appears on the read output lines ($RO, \overline{RO}$). An output amplifier is used for data output. Other improved BiCMOS sensing circuits exist in [23].

In the write mode, the *YW* signal is selected as shown in Fig. 8.40. The selected bit-lines are connected to a pair of write-lines W and $\overline{W}$ and the data are transferred to the memory cell when *WL* goes HIGH. Note that BiCMOS circuits are used as drivers for X and Y decoders and word-line drivers.

8.3.4. Content Addressable Memory

A content-addressable memory (CAM) is an important macrocell of a translation-lookaside buffer (TLB)[25] and cache memory [26] circuits of computer systems. The TLB permits the translation of the virtual address of a CPU to the physical address, and the cache memory from the physical address to the memory data.

A CAM stores tags which can be compared against an input address word as shown in Fig. 8.41(a). HIT detection signal is sent by the CAM if the values stored in the CAM array match with the input address word.

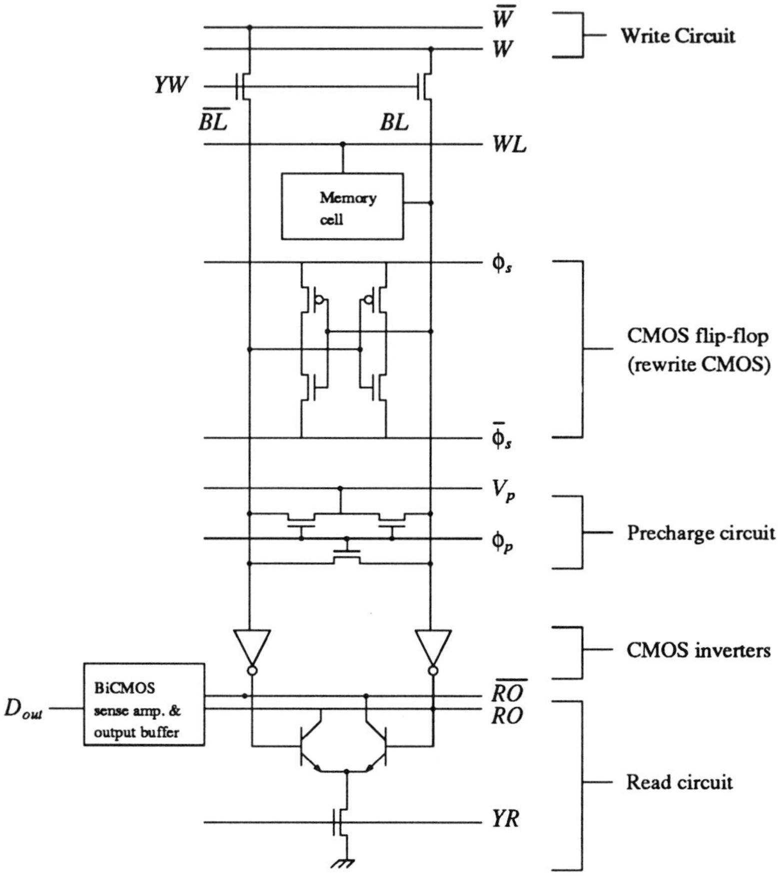

Figure 8.40 BiCMOS DRAM.

A CMOS implementation of the CAM cell is illustrated in Fig. 8.41(b). It can be readable and writable just as an ordinary memory cell. A tag word is formed by identical cells which are repeated in a horizontal array. The match line (*ML*) is precharged HIGH. For a comparison operation, suppose that a "1" is stored at node *A*. Assume that *CBL* line is held HIGH and $\overline{CBL}$ line is held LOW. In this case, the transistors *MN*3 and *MN*1 are OFF, hence *ML* line remains HIGH, indicating a match at this bit location. Assume now that *CBL* is driven LOW and $\overline{CBL}$ HIGH.

The transistor $MN4$ is OFF, but $MN1$ and $MN2$ are ON. Then ML line is discharged, indicating a mismatch at this bit location.

To improve the speed of the CAM many circuits techniques using BiCMOS can be implemented. One of these techniques has been successfully applied to implement a BiCMOS CAM array for TLB. Fig. 8.42 shows a BiCMOS CAM cell with the match line sense amplifier and the HIT detection circuit[27]. Prior to the comparison, the CBL, $\overline{CBL}$ and ML lines are set LOW. Assume that CBL is pulled up, and a "1" is stored at the node A. The transistor $MP1$ is OFF. Since $MP2$ is ON, then the voltage at the base of the bipolar transistor Q_1 remains LOW. In this case, the ML line remains LOW. This indicates that a match exists in this bit location. When $\overline{CBL}$ is driven HIGH (and CBL remains LOW), then the base of Q_1 is HIGH, and the ML line is pulled HIGH. In this case, a mismatch exists in this bit location.

This BiCMOS CAM cell uses a small swing (V_{BEon}) on the CAM bit-lines CBL, $\overline{CBL}$ and the match-lines during the comparison cycle. The CAM cell is written by holding $\overline{CWL}$ LOW and setting CBL and $\overline{CBL}$ differential lines (small differential voltage). The transistor $MN1$ is OFF and $MP3$ and $MP4$ are ON. The latch formed by the crosscoupled inverters is disabled. Nodes A and B of the cell are connected to the bit-lines. When $\overline{CWL}$ goes HIGH (CMOS level), the regeneration in the latch drives nodes A and B to full CMOS levels.

The CAM array is organized so that the row cells have a common match-line which forms a wired-OR of the tag. If there is a mismatch in any bit of the tag word, the match line is pulled HIGH. If all the m bits match, the common match-line remains LOW. The match-line is connected to a sense amplifier which detects the small swing on the match-line. Since initially all the match-lines are LOW (match) then the sense amplifier should be disabled by the signals $ENABLE$ and $\overline{ENABLE}$ to inhibit their outputs. The vertical organization of the CAM array has n match-lines connected to emitter followers. These emitter followers form a wired-OR signal as shown in Fig. 8.42. When at least one of the match-lines is raised, the output of the match-line high detection circuit is HIGH. A special circuit sets the $ENABLE$ and $\overline{ENABLE}$ signal to "1" and "0" respectively. These signals enable the sense amplifier, then the output is evaluated. A HIT detection circuit can easily be implemented.

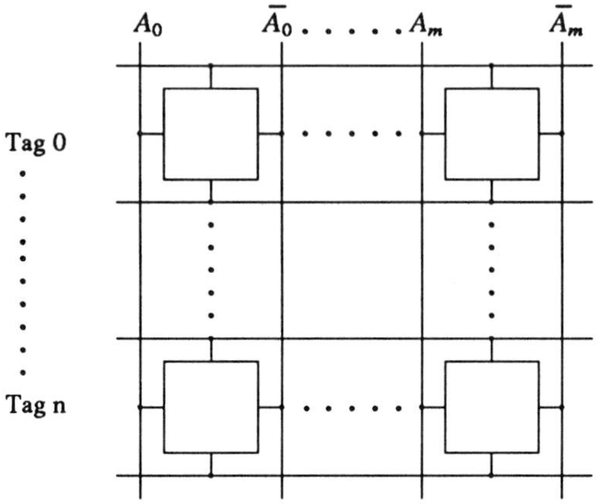

(a)

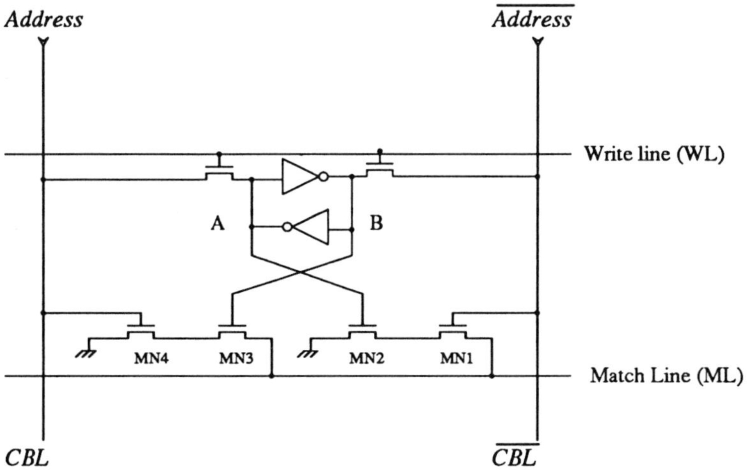

(b)

Figure 8.41 (a) CAM array; (b) CMOS CAM cell.

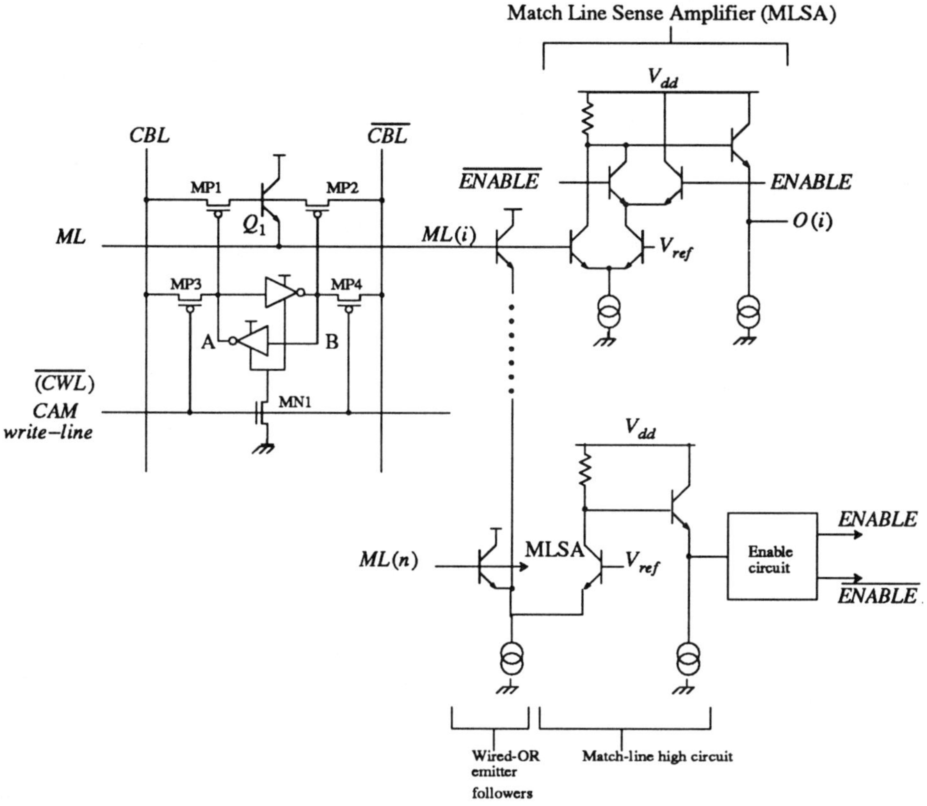

Figure 8.42 BiCMOS CAM cell with sense amplifier and match-line detection circuit.

8.4. Programmable Logic Arrays

Logic functions such as those used in the control units of VLSI processors or in finite-state machines are hard to implement in random logic. One way of implementing these functions in a regular structure is the use of Programmable Logic Array (PLA).

PLAs have a regular architecture divided mainly in two planes as shown in Fig. 8.43. These planes perform a specific function such as OR and AND. CMOS PLAs can be implemented in both static and dynamic styles. The style is chosen depending on the timing strategy in the chip. Other factors such as the availability of clock, speed,

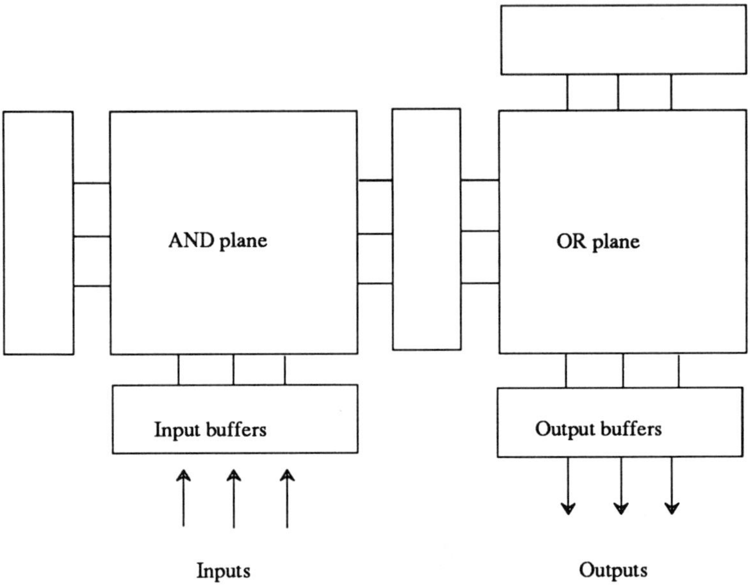

Figure 8.43 AND-OR PLA architecture.

power dissipation, and the allowed area play an important role in the PLA design style. A CMOS PLA example using pseudo-NMOS like style is shown in Fig. 8.44. The output OR functions are realized with NOR gates. From Fig. 8.44(a), we have

$$P_1 = \overline{\bar{A}+B+\bar{C}} = A.\bar{B}.C$$

$$P_2 = \overline{A+C} = \bar{A}.\bar{C}$$

$$P_3 = \overline{\bar{B}+C} = B.\bar{C}$$

$$P_4 = \overline{A+\bar{C}} = \bar{A}.C$$

The OR plane is in principle similar to the AND plane [Fig. 8.44(b)]. From Fig. 8.44(b), we have

(a)

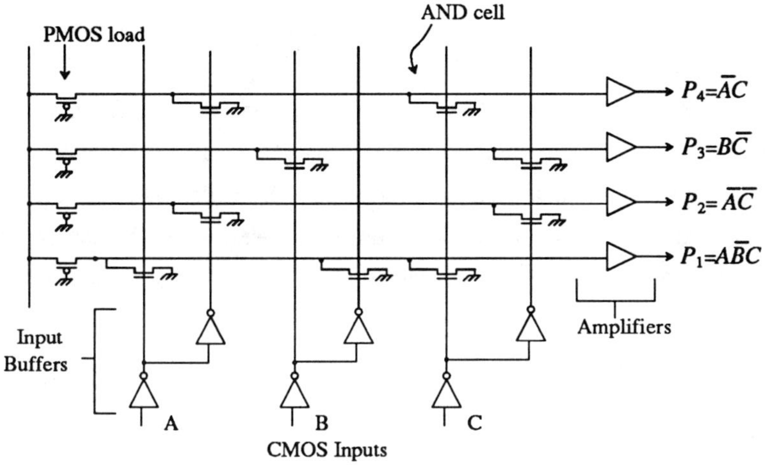

(b)

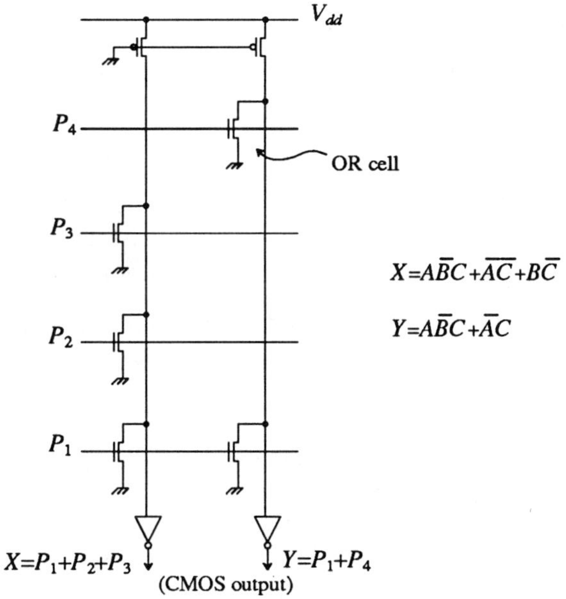

Figure 8.44 Pseudo-NMOS like CMOS PLA: (a) AND plane; (b) OR plane.

$$X = P_1 + P_2 + P_3$$

$$Y = P_1 + P_4$$

This example shows that the PLA organization is useful for implementing sum of products (SOP) functions. Hence any SOP functions can be realized by programming the array with the AND and OR cells. This design style of PLAs is preferred when the area is an important constraint, while the number of inputs, speed and power dissipation are not critical constraints. If the number of inputs is HIGH, this style is not preferred because of the increase in power dissipation. Usually very large PLAs are implemented in dynamic style[1].

The critical path of a PLA is composed of an input buffer, AND array, AND amplifier, OR array, and output amplifier. In CMOS technology the most critical sections are the AND and OR arrays. In order to achieve high speed operation, the delay of these arrays should be minimized. This can be achieved by using bipolar transistors. Fig. 8.45 shows a BiCMOS PLA using ECL levels in the AND and OR arrays. If the inputs and outputs are CMOS, then CMOS/ECL and ECL/CMOS converters/buffers are needed for level translation. The example of logic functions implemented in CMOS is used in this BiCMOS diagram. Fig. 8.45(a) illustrates the BiCMOS AND plane. The inputs are CMOS and they are translated to ECL level and then buffered. These inputs then drive bipolar wired-OR and inverted, to obtain the products P_i. The OR functions are realized by bipolar wired-OR with ECL level. ECL/CMOS converters/buffers are needed at the ouput to translate the signals to CMOS level. This type of BiCMOS PLA offers higher speed operation compared to CMOS PLA, but at the cost of more power dissipation and area.

Different types of BiCMOS PLAs can be designed, depending on the type of inputs available and outputs needed and the area reserved for the PLA. If the inputs and outputs are ECL the extra delay and area of CMOS/ECL and ECL/CMOS converters are saved. Another type of PLA can be designed by implementing the AND plane in CMOS (with CMOS levels) and the OR plane in ECL.

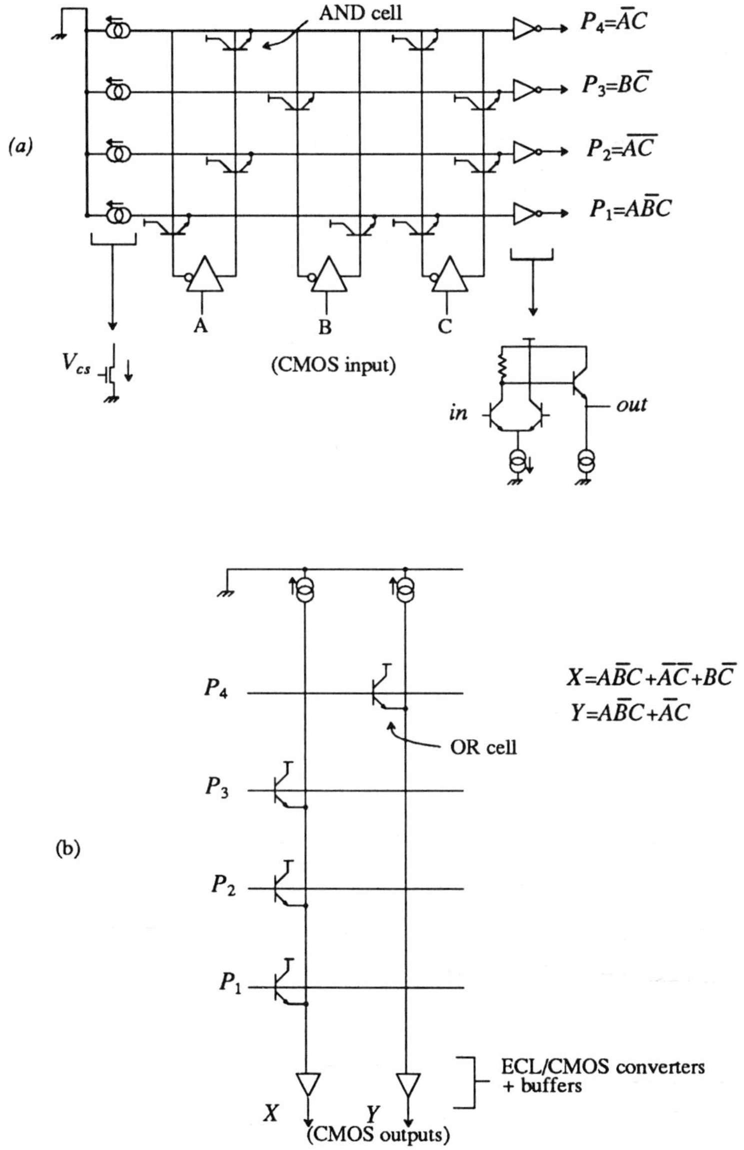

Figure 8.45 BiCMOS PLA with mixed CMOS/ECL levels: (a) AND plane; (b) OR plane.

8.5. Data Path

A VLSI chip can be partitioned in two parts; the data path (or execution unit) and the control unit. The data path consists of an arithmetic logic unit (ALU), a shifter, file registers, I/O ports, multipliers, data busses, etc. It performs many operations on the data in the register file, to which the results are sent back. The data busses permit the communication between the different units of the data path. The control unit delivers the instructions to the data path. These instructions determine the operations that the data path has to perform. The control unit can be implemented using random logic, micro-ROM, PLAs or a combination of these three implementations. Other macrocells such as TLB, cache memory, etc. can be added to the data path and the control unit.

The use of BiCMOS technology allows the data path to operate with clock frequency over 100 MHz. In this section, we will focus on the use of BiCMOS for data path implementation.

8.5.1. Arithmetic Logic Unit

ALU is an important part of a data path. It is a macrocell which executes arithmetic operations such as multiplication, addition, subtraction, negation, and logic operations such as AND, OR, XOR, comparison, etc. It performs the operation on two operands stored in latch A and latch B and puts the result in latch C as shown in Fig. 8.46. The operation code (op code) selects the operation of the ALU to be executed. The flags indicate the status of the ALU such as overflow, zero-result, carry generation, etc. The input latches A and B are in general connected to two parallel data busses. The result latch is connected to one of the busses or to a third one.

The maximum clock frequency of a VLSI circuit may be limited by the ALU operations; especially the arithmetic ones. The critical delay of an arithmetic operation is due mainly to the carry propagation along the width of the carry. In CMOS ALU, a 4-bit carry propagation circuit is used in the carry path. BiCMOS is employed in the ALU to speed-up the carry path by using the techniques discussed in Section 8.1.3. The rest of the ALU is implemented in CMOS.

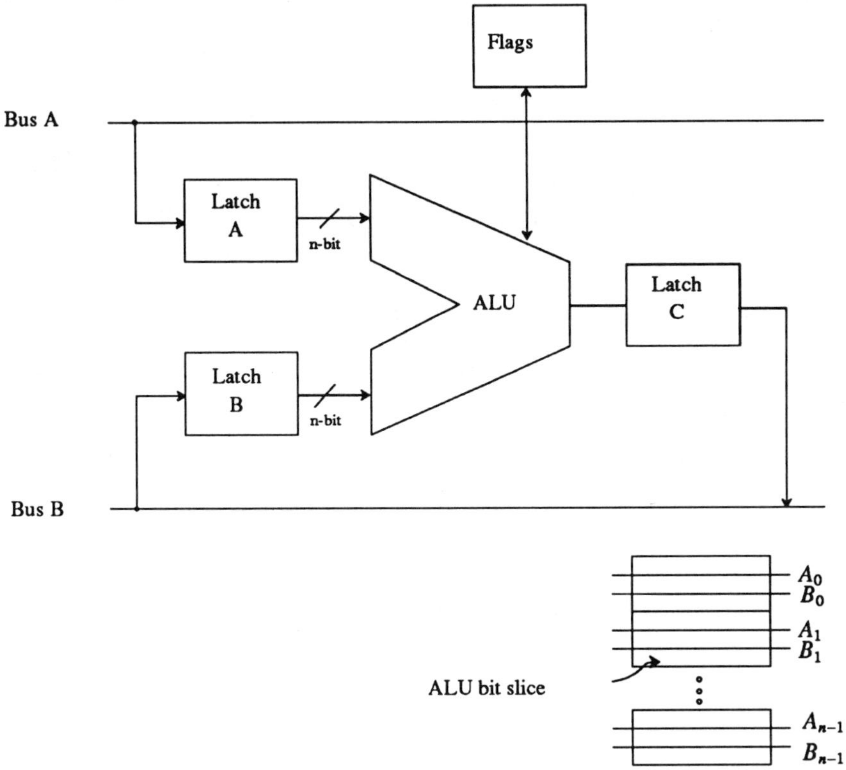

Figure 8.46 Arithmetic logic unit.

8.5.2. Shifter

Another macrocell of the data path is the shifter. It performs shift or rotate operations on the data. If the number of bits to be shifted is arbitrary, then a barrel shifter is used [28,29]. Fig. 8.47 shows the CMOS implementation of a 4-bit barrel shifter. NMOS transistors are used as switches in the array. The input busses (D_o–D_3) can be connected to the output busses (L_o–L_6) via the pass transistors. The control signal S_0-S_3 selects the pass transistors to be switched. These signals are generated by a 2-bit decoder.

Since the barrel shifter is not a critical unit for the delay, no bipolar transistors are required to speed up its operation. The CMOS implementation can be used in BiCMOS design.

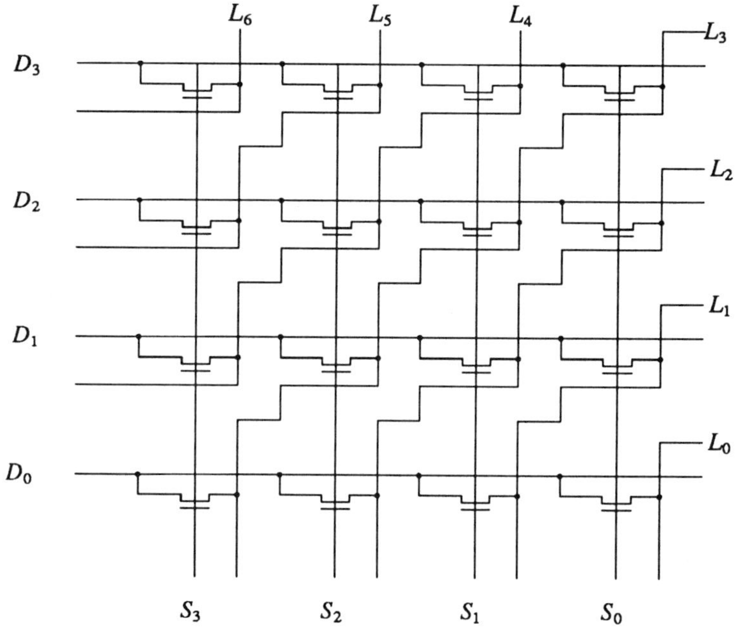

Figure 8.47 Barrel shifter circuit.

8.5.3. Register File

A register file is a set of registers which store data. It consists of a small array of static memory cells. A typical array is 32 registers of 32-bit. Since the ALU needs two pieces of data from the register file, the array has dual-read single-write architecture. The memory cell schematic with dual port [30] is shown in Fig. 8.48. During the write operation the data and it's complement value are held on the bus A and bus B respectively. The signals $WL1$ and $WL2$ are selected to store the data in the flipflop. In the read operation, the data is put in the data bus A, while the complement data from another register (in the same row) is put on the bus B.

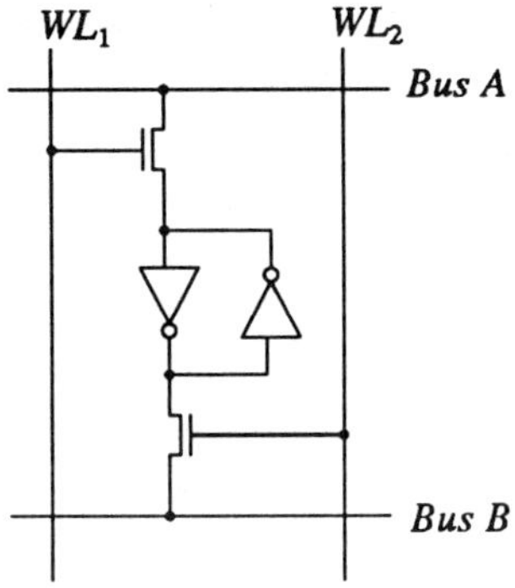

Figure 8.48 CMOS dual-port register cell.

The data busses are the communication means for the data transfer between the ALU, shifter, file register, etc. These busses have a heavy load (few *pF*). In CMOS design, dynamic techniques are used to allow fast operation. The bus shown in Fig. 8.49 is precharged by the clock ϕ_1. During the evaluation (phase ϕ_2), if the data to be sent to the bus is "1" then the bus is discharged, otherwise it is left HIGH (negative logic on busses). The dynamic techniques permit the use of small area destination and source circuits connected to the busses, because the operation is based only on discharging the busses.

In CMOS the busses are precharged to V_{dd}. Using BiCMOS sensing circuits allows for low precharging voltage levels. This leads to faster operation speed.

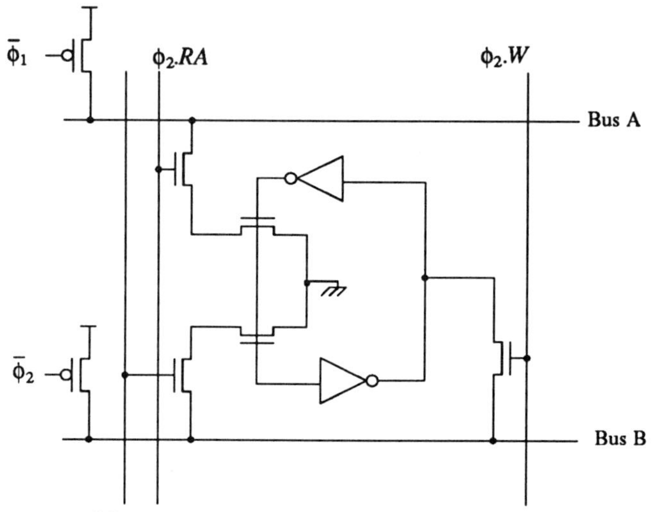

Figure 8.49 A register file memory cell with bus precharge circuit.

8.6. BiCMOS Logic Cells

The design cycle of custom VLSI Application Specific ICs (ASIC's) can be shortened by using macrocells. A set of standard cells behind the macrocells also provide a fast means for semi-custom design. The standard cells library contains inverters, buffers, n-input NOR gates (n=2..4), n-input NAND gates, n-input OR gates, multiplexers, XOR gate, flipflops, etc. The cells must have the same height so they can be butted together.

In BiCMOS technology, the cells should be designed in CMOS and BiCMOS for the same logic functions. The BiCMOS gates are faster than CMOS gates when the load capacitance is greater than C_x as shown in Fig. 7.24. A typical value of C_x is $0.1pF$. Hence, in standard cells approach, the BiCMOS gates are used when the load is high. Fig. 8.50 illustrates an example of how to combine BiCMOS and CMOS cells in a logic block. The load of a cell is greatly affected by the interconnection. When BiCMOS gate is used it consumes more area than an equivalent CMOS gate. But since in standard cells approach the interconnections consume a fair amount of area, the extra area required for BiCMOS gates is relatively small.

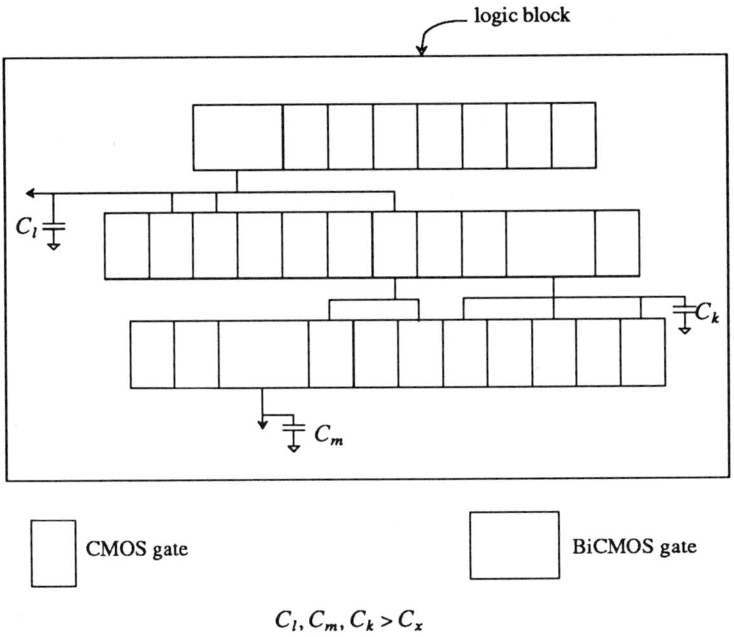

Figure 8.50 BiCMOS standard cell block.

8.7. BiCMOS Gate Arrays

Gate arrays became very popular for a wide spectrum of applications because of their low cost and short turn-around time. Gate array chips consist of a large number of identical *sites* or *basic cells* which are usually placed in rows. The rows are separated by routing channels. The core of rows and channels is surrounded by I/O cells at the chip periphery as illustrated in Fig. 8.51[1].

Each of the basic cells is typically made up of a number of transistors which can be connected to form a two input NAND or NOR gate or a simple latch. The only processing step that can be customized is the metallization. The user of a gate array can implement his or her system by specifying the required connections between the devices in each cell and then the connection between the various cells. This is done

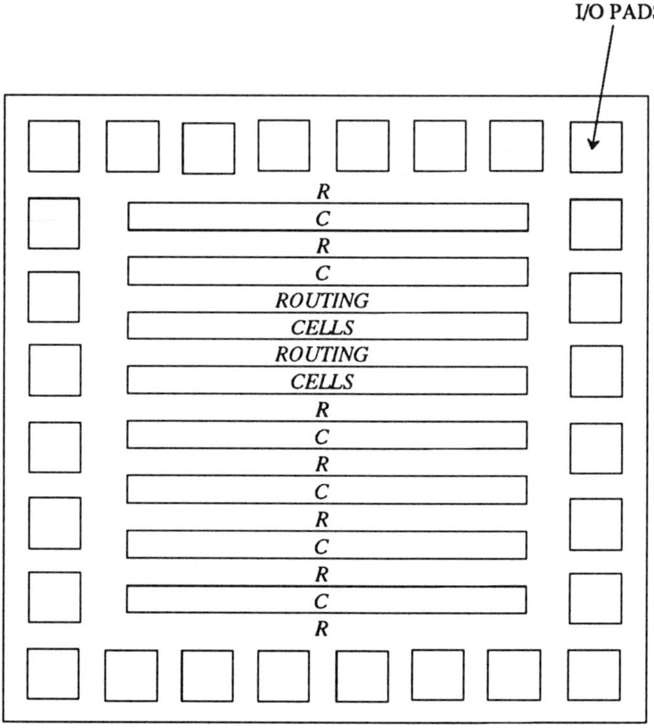

Figure 8.51 Gate Array chip floor plan.

automatically using CAD tools. The number of metal levels used for wiring varies from 2 to 4. The first one or two levels are used for internal wiring of the cell and the upper levels (e.g. third and fourth) for wiring between the cells in the horizontal and vertical directions[31].

BiCMOS technology has been used extensively for building gate array and channelless gate arrays (sea-of-gates) [31-35]. Having NMOS, PMOS and bipolar transistors in the basic cell allows for mixed CMOS, BiCMOS and ECL designs, and offers enhanced current drive capability. Measured data show that for a fan-out of 10, the delay of a 2-input BiCMOS NAND gate is half that of a 2-input CMOS NAND gate[32]. In addition, the shallow slope of the BiCMOS delay curve (17 ps/fan-out vs 60 ps/fan-out for CMOS [32]) makes it less sensitive to load variations (fan-out or wire length) and hence allows for greater tolerance on timing-driven route tools[32].

There are many problems associated with the design of BiCMOS gate arrays. For example, the basic cell structure which includes issues such as the ratio of MOS to bipolar devices, the sizing of both types, the floor plan of the cell, merging PMOS with bipolar, etc. All these issues are strongly related to the applications and the nature of the systems to be implemented. In this section, we will address some of these topics briefly. Our treatment is supposed to shed some light on the important considerations. Let us first look at the typical cell floor planning. Various basic cell styles have been proposed. The conventional basic cell consists of a number of NMOS and PMOS transistors and two bipolar devices. An example of such a cell is shown in Fig. 8.52[32].

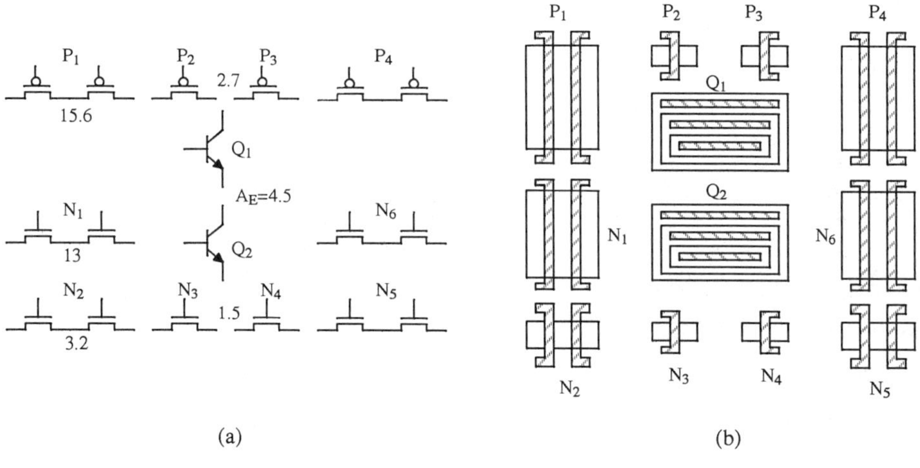

Figure 8.52 Schematic and layout of a BiCMOS gate array cell with two center bipolar transistors.

There are four large PMOS transistors (W/L=15.6/0.8) and four large NMOS transistors (W/L=13/0.8). These transistors can be used as the base drivers of the BiCMOS gates (e.g. transistors N and P in the circuit of Fig. 7.10). They can also be used for CMOS gates. The cell includes additional small size NMOS and PMOS transistors which can be employed as bleeding transistors (e.g. N_{d1} and N_{d2} in the circuit of Fig. 7.10). They can also be used to build the feedback inverters required for the feedback type BiCMOS circuits [Fig. 7.30(c)]. The centered bipolar transistor pair are obviously used for the bipolar output drivers of a BiCMOS gate. The bipolar transistors could also be placed on the top and the bottom of the NMOS and PMOS transistors[31]. The cell area can be reduced by using only one bipolar transistor per cell as shown in Fig. 8.53(a). In this case, a single cell can realize CMOS or BiNMOS gates only. For building a BiCMOS gate two cells have to used, a full cell and the bipolar transistor of an adjacent cell[35]. The concept of sharing bipolar transistor between adjacent cells limits the flexibility of macrocell placement. Another way of saving area is to use bipolar-PMOS merged transistors [Fig. 8.53(b)][34]. The merged bipolar-PMOS transistor can be used to realize the pull-up section of the BiCMOS gate.

The sizing of the bipolar and MOS devices is important because it will eventually determine the total cell area. However, the optimum emitter area is a function of the load capacitance and/or fan-out. There is an obvious trade-off between the delay and the area. Increasing the emitter size improves the delay in the case of heavy loading but increases the cell area. It is also not clear how many of the cells will be required to drive heavy loads and where they will be placed. The common approach that has been adopted so far is to assume that the number and the location of the BiCMOS drivers is unknown, hence all cells have to be identical. Therefore, the selection of the emitter area must be based on statistical information on the distribution functions of the load and fan-out. The sizing of the MOSFETs is more complicated because it should account for effect of the input capacitance on the delay of the previous stages. This makes the problem more complicated than that of sizing the bipolars. However, the final decision will be based on the statistical data and the simulations which show the best compromise for the expected range of loads and fan-outs[32].

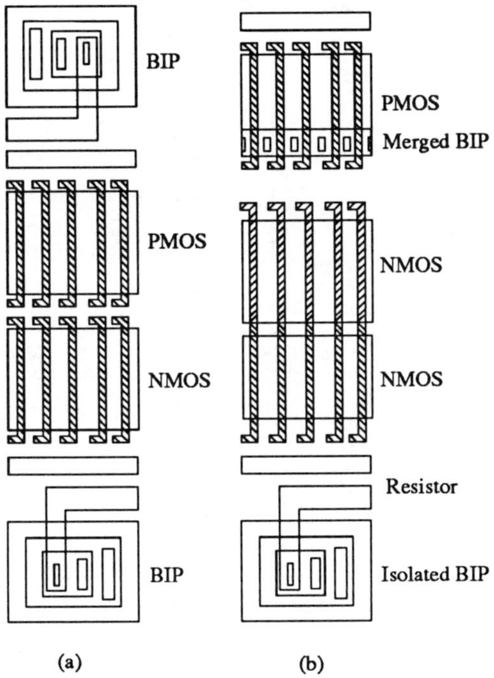

Figure 8.53 (a) Conventional BiCMOS basic cell. (b) Bipolar-PMOS merged basic cell.

In most of the reports on BiCMOS gate arrays it is not clear whether any methodology has been used to determine the structure of the basic cell (e.g. the number of bipolar transistors per cell). There is little information in the literature about techniques to determine the ratio of BiCMOS to CMOS in gate arrays. However, Duchene and Declercq[36] have presented an analysis that can be used to estimate the optimal bipolar-to-MOS count ratio of BiCMOS sea-of-gates. Their analysis might not be general enough to apply for all applications as is, however, the concepts can be extended to other areas with some modifications.

We will briefly describe the analysis which might serve as a basis for a design methodology of BiCMOS gate arrays and sea-of-gates. The analysis is based on the assumption that the simple gates can be realized in several ways such as single-stage

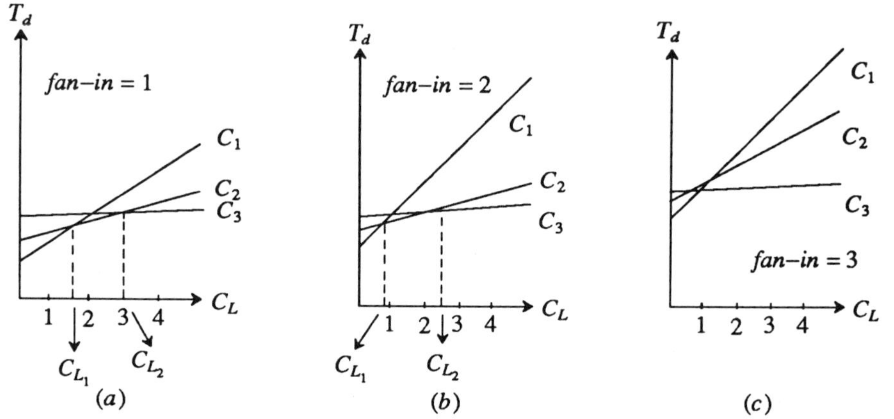

Figure 8.54 The delay of the three configurations vs C_L for FI=1,2 and 3.

CMOS, single-stage BiCMOS (conventional BiCMOS, feedback type BiCMOS, BiNMOS, etc.), single-stage CMOS with CMOS buffer or single-stage CMOS with BiCMOS buffer. These basic configurations will be represented by the symbol C_i, where the subscript i indicates the configuration type. A strategy is defined as a set which includes either a single configuration or a combination of these configurations. Now the question is, how can the designer determine which strategy to choose and what will be ratios of the configurations to each other ? To answer this question we must first define the criterion by which we compare the different strategies. In this analysis, the criterion was the average delay of a given strategy. The exact definition of this average delay will be derived. To do that we have to examine the delay of each configuration. The delay of the configuration is a function of both the load capacitance and the fan-in for a given power supply voltage. To explain these dependencies let us use a hypothetical case where we have only three configurations C_1, C_2 and C_3. Fig. 8.54 shows their delay as a function of the load capacitance for different fan-in conditions. The superiority of one configuration over the others is a function of the capacitance. For example in Fig. 8.54(a), C_1 is faster than C_2 and C_3 if the load is less than C_{L1}. Between C_{L1} and C_{L2} the second configuration (C_2) has the least delay and so on. However, as the fan-in changes the crossover capacitances may shift [Fig. 8.54(b)]. There is a possibility that some configurations are always slower than others as is the case for C_2 in Fig. 8.54(c). It is obvious that the load capacitance (C_L) and the fan-in (FI) are independent random variables. FI is considered to be a discrete random variable, while C_L is a continuous one. The randomness of C_L can best be described by its probability density ($f_{CL}(c_l)$) or by its distribution function. These distribution functions vary from one application to the other. Fig. 8.55 shows two distributions[36], the first one [Fig. 8.55(a)] has low probability of heavy loading and the second [Fig. 8.55(b)] has higher probability of large capacitive loads as is the case in bus intensive structures. The fan-in, which is a discrete random variable, can be similarly described by a probability histogram as shown in Fig. 8.56[36]. The delay being a function of two random variables becomes a random variable by itself. The average delay of a given strategy can be given by[36] :

$$t_{strategy} = \sum_{fi=1}^{fi=n} P_{FI}(fi) \int_0^{\infty} \min[(delay(fi,c_l)] \, f_{C_L}(c_l) \, dc_l \tag{8.8}$$

where P_{FI} is the probability that the fan-in is equal to fi, f_{C_L} is the probability density of C_L. The $\min[delay(fi,c_l)]$ is the minimum possible delay in a strategy over the full range of C_L. For example if a strategy includes C_1, C_2 and C_3 [Fig. 8.54], then the minimum delay would be the delay of C_1 for $0<C_L<C_{L1}$ and the delay of C_2 for $C_{L1}<C_L<C_{L2}$ and the delay of C_3 for $C_L>C_{L2}$.

The ratio of the different configuration in each strategy is given by[36] :

$$Ratio_{Ci} = \sum_{fi=1}^{fi=n} P_{FI}(fi) \int_{CL1}^{CL2} f_{C_L}(c_l) \, dc_l \tag{8.9}$$

where $[CL1, CL2]$ is the interval of C_L where the configuration C_i gives the shortest delay of all configurations in the considered strategy. The values of $CL1$ and $CL2$ are different for the different fan-in's [see Fig. 8.54].

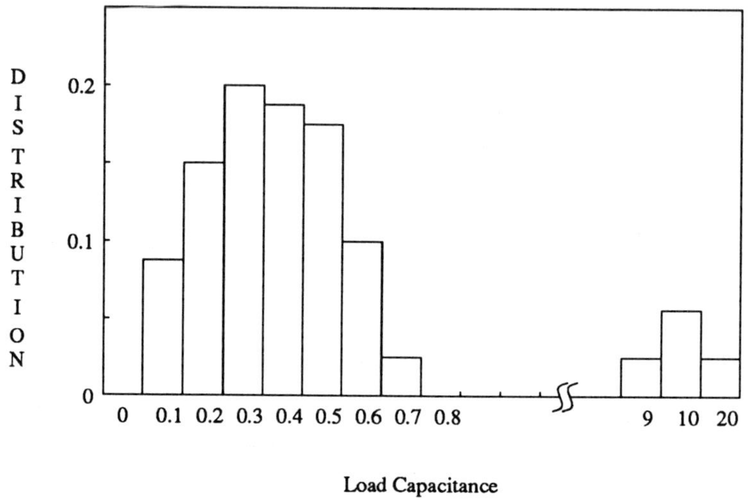

(a)

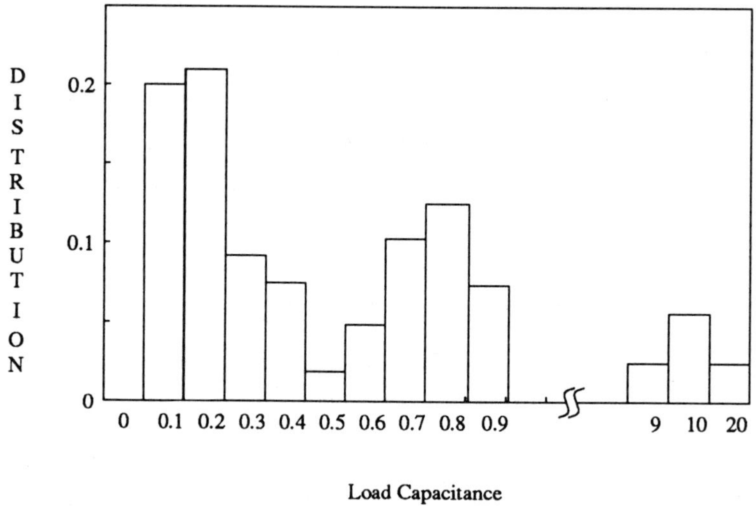

(b)

Figure 8.55 Load capacitance distributions.

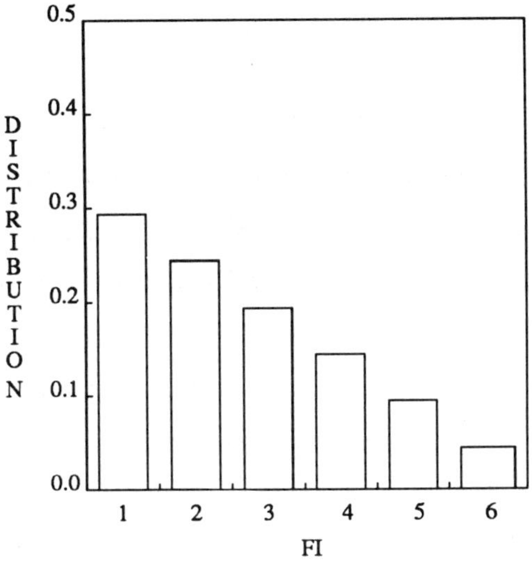

Figure 8.56 Fan-in Distribution.

8.8. Chapter Summary

In this chapter, we have demonstrated that BiCMOS can be used for a variety of applications. In most of the applications considered in this chapter we have seen that the CMOS version can be enhanced by adding the bipolar transistors either for driving or sensing purposes. The use of bipolar allows for small signal swings whose small variations can be easily sensed. This feature improves the access time of memories, register files, ALUs etc. BiCMOS also allows mixing circuits with different swings. This was demonstrated by the merged MOS/bipolar CML adders, ECL SRAM and gate arrays. This technique gives the designer more flexibility to manage power and speed of a system. Finally, we may conclude that a large portion of the systems will be implemented in CMOS, while bipolar transistors will be used whenever needed for driving or sensing purposes. The use of bipolar transistors will, in most cases, improves the speed of the system at the expense of extra power dissipation especially if it is used for sensing or in mixed swing circuits.

References

1. N. Weste and K. Eshraghian, *Principles of CMOS VLSI Design,* Addison-Wesley, Reading, MA (1985).

2. W. Heimsch, et al., "Merged CMOS/Bipolar Current Switch Logic (MCSL)," *IEEE Journal of Solid-State Circuits,* **Vol. SC-24** pp. 1307-1311 (Oct. 1989).

3. T. Hotta, et al., "A 70MHz-32b Microprocessor with 1.0um BiCMOS Macrocell Library," *International Solid-State Circuits Conference Tech. Dig.,* pp. 124-125 (Feb. 1989).

4. T. Hotta, et al., "1.3 μm CMOS/Bipolar Macrocell Library for VLSI Computers," *IEEE Journal of Solid-State Circuits,* **SC-23, No. 2** pp. 500-506 (Apr. 1988).

5. G.P. Rosseel, et al., "A Single-Ended BiCMOS Sense-Circuit for Digital Circuits," *International Solid-State Circuits Conference Tech. Dig.,* pp. 114-115 (February 1989).

6. M. Annaratone, *Digital CMOS Design,* Kluwer Academic Publishers (1986).

7. K. Hwang, *Computer Arithmetic: Principles, Architecture, and Design,* John Wiley & Sons (1979).

8. K. Kurita, et al., "Very High-Speed ROM using Bipolar/CMOS Technology," *Electronics and Communications in Japan,* **Vol. 71** pp. 18-26 (1988).

9. T. Ikeda, A. Watanabe, Y. Nishio, I. Masuda, N. Tamba, M. Okada, and K. Ogiue, "High-Speed BiCMOS Technology with a Buried Twin Well Structure," *IEEE Trans. on Electron Devices,* **Vol. ED-34, No. 6** pp. 1304-1309 (June 1987).

10. R.A. Kertis, et al., "A 12-ns ECL I/O 256Kx1-bit SRAM using a 1-μm BiCMOS Technology," *IEEE Journal of Solid-State Circuits,* **Vol. SC-23** pp. 1048-1053 (Oct. 1988).

11. H. Tran, et al., "An 8-ns 256K SRAM with CMOS Memory Array and Battery Backup Capability," *IEEE Journal of Solid-State Circuits,* **Vol. SC-23** pp. 1041-1047 (Oct. 1988).

12. H. Tran, et al., "An 8ns BiCMOS 1Mb ECL SRAM with a Configurable Memory Array size," *International Solid-State Circuits Conference Tech. Dig.*, pp. 36-37 (Feb. 1989).

13. M. Takada, et al., "A 5-ns 1-Mb ECL BiCMOS SRAM," *IEEE Journal of Solid-State Circuits*, **Vol. SC-25** pp. 1057-1062 (Oct. 1990).

14. M. Suziki, et al., "A 3.5-ns, 500-mW 16-kbit BiCMOS ECL SRAM," *IEEE Journal of Solid-State Circuits*, **Vol. SC-24** pp. 1233-1237 (Oct. 1989).

15. W. Heimsch, et al., "A 3.8-ns 16k BiCMOS SRAM," *IEEE Journal of Solid-State Circuits*, **Vol. SC-25** pp. 48-54 (Feb. 1990).

16. S. Miyaoka, et al., "A 7-ns, 350-mW 64-kbit ECL-Compatible RAM," *IEEE Journal of Solid-State Circuits*, **Vol. SC-22** pp. 847-849 (Oct. 1987).

17. M. Kubo et al., "Perspective on BiCMOS VLSI's," *IEEE Journal of Solid-State Circuits*, pp. 5-11 (Feb. 1988).

18. T. Douseki et. al., "BiCMOS Circuit Technology for a High-Speed SRAM," *IEEE Journal of Solid-State Circuits*, pp. 68-73 (Feb. 1988).

19. G. Kitsukawa, et al., "An Experimental 1-Mb BiCMOS DRAM," *IEEE Journal of Solid-State Circuits*, **Vol. SC-22** pp. 657-662 (Oct. 1987).

20. S. Watanabe, et al., "BiCMOS Circuit Technology for High Speed DRAMs," *Symposium on VLSI Circuits Dig. Tech.*, pp. 79-80 (1987).

21. G. Kitsukawa, et al., "A 1-Mb BiCMOS DRAM using Temperature Compensation Circuit Techniques," *IEEE Journal of Solid-State Circuits*, **Vol. SC-24** pp. 597-602 (June 1989).

22. T. Watanabe, et al, "Comparison of CMOS and BiCMOS 1-Mb DRAM Performance," *IEEE Journal of Solid-State Circuits*, **Vol. SC-24** pp. 771-778 (June 1989).

23. G. Kitsukawa, et al., "A 23-ns 1-Mb BiCMOS DRAM," *IEEE Journal of Solid-State Circuits*, **Vol. SC-25** pp. 1102-1111 (Oct. 1990).

24. K. Itoh, "Trends in Megabit DRAM Circuit Design," *IEEE Journal of Solid-State Circuits*, **Vol. SC-25** pp. 778-788 (June 1990).

25. H. Kadota, et al., "A 32-bit Microprocessor with On-Chip Cache and TLB," *IEEE Journal of Solid-State Circuits*, **Vol. SC-22** pp. 800-807 (Oct. 1987).

26. A.J. Smith, "Cache Memories," *Computing Surveys*, **Vol. 14** pp. 473-530 (Sept. 1982).

27. L. Tamura, et al., "A 4-ns BICMOS TLB," *IEEE Journal of Solid-State Circuits*, **Vol. SC-25** pp. 1093-1101 (Oct. 1990).

28. C.A. Mead, and L.A. Conway, *Introduction to VLSI Systems*, Addison-Wesley (1980).

29. R.W. Sherburne, et al., "Datapath Design for RISC," *Proc. Conf. Advanced Research in VLSI*, pp. 53-62 (1982).

30. R.W. Sherburne, et al., "A 32-bit NMOS Microprocessor with a Large Register File," *IEEE Journal of Solid-State Circuits*, **Vol. SC-19** pp. 682-689 (Oct. 1984).

31. F. Murabayashi et al., "A 0.5 micron BiCMOS Channelless Gate Array," *IEEE Custom Integrated Circuits Conference Tech. Dig.*, pp. 8.7.1-8.7.4 (1989).

32. J. D. Gallia et al., "High-Performance BiCMOS 100K-Gate Array," *IEEE Journal of Solid-State Circuits*, pp. 142-149 (FEb. 1990).

33. A. El Gamal et al., "BiNMOS: A Basic Cell for BiCMOS Sea-of-Gates," *IEEE Custom Integrated Circuits Conference Tech. Dig.*, pp. 8.3.1-8.3.4 (1989).

34. T. Hanibuchi et al., "A Bipolar-PMOS Merged Basic Cell for 0.8 micron BiCMOS Sea of Gates," *IEEE Journal of Solid-State Circuits*, pp. 427-431 (March 1991).

35. H. Hara et al., "A 350 ps 50K 0.8 micron BiCMOS Gate Array with Shared Bipolar Cell Structure," *IEEE Custom Integrated Circuits Conference Tech. Dig.*, pp. 8.5.1-8.5.4 (1989).

36. P.P. Duchene and M.J. Declercq, "Strategies for CMOS/BiCMOS Gate Usage on Sea-of-Gates Arrays," *IEEE Custom Integrated Circuits Conference Tech. Dig.*, pp. 14.2.1-14.2.4 (1991).

Index